I0063651

LC-MS/MS Method for Mycotoxin Analysis

Special Issue Editor
Aldo Laganà

MDPI • Basel • Beijing • Wuhan • Barcelona • Belgrade

MDPI

Special Issue Editor
Aldo Laganà
Università di Roma "La Sapienza"
Italy

Editorial Office
MDPI AG
St. Alban-Anlage 66
Basel, Switzerland

This edition is a reprint of the Special Issue published online in the open access journal *Toxins* (ISSN 2072-6651) from 2016–2017 (available at: http://www.mdpi.com/journal/toxins/special_issues/LCMS).

For citation purposes, cite each article independently as indicated on the article page online and as indicated below:

Author 1; Author 2. Article title. *Journal Name* **Year**, *Article number*, page range.

First Edition 2017

ISBN 978-3-03842-606-6 (Pbk)
ISBN 978-3-03842-607-3 (PDF)

Articles in this volume are Open Access and distributed under the Creative Commons Attribution license (CC BY), which allows users to download, copy and build upon published articles even for commercial purposes, as long as the author and publisher are properly credited, which ensures maximum dissemination and a wider impact of our publications. The book taken as a whole is © 2017 MDPI, Basel, Switzerland, distributed under the terms and conditions of the Creative Commons license CC BY-NC-ND (http://creativecommons.org/licenses/by-nc-nd/4.0/).

Table of Contents

About the Special Issue Editor

Aldo Laganà ull professor of analytical chemistry, since 2002. At present, he is the Director of the Chemistry Department of Università di Roma "La Sapienza" and President of the Executive Board of the Analytical Chemistry Division of the Italian Chemical Society. He is author of more than 200 scientific publications in international journals with high impact factor, 8 international book chapters, and 3 national books; his score on Scopus database are more than 5100 citations and h-index of 39. Since 2014, he is Editor of Toxins (section Mycotoxins) and member of the International Advisory Board of Analytical and Bionalytical Chemistry journal. His research activity is focused on the development and validation of innovative analytical methods based on mass spectrometry coupled with separating techniques for the characterization and determination of substances of natural or anthropogenic origin in environmental, food, vegetal or biological matrices. In the last years, his interest also has been focused in proteomics and metabolomics fields, employing nanoseparative techniques coupled to high resolution mass spectrometry.

Preface to "LC-MS/MS Method for Mycotoxin Analysis"

The mycelium structure of a variety of filamentous fungi are able to produce secondary metabolites. Among these secondary metabolites, those that can elicit deleterious effects on other organisms are classified as mycotoxins, the biochemical significance of which in fungal growth and development has not always been fully clarified.

The main fungal species that can produce mycotoxins belong to Fusarium, Aspergillus, and Penicillium genera; other are Claviceps and Alternaria genera. The same mycotoxin class can be produced by more species of the same mold genus, whereas other mycotoxins (e.g., ochratoxin and patulin) can be produced by different mold genera. The known mycotoxins are generally low-molecular weight compounds characterized by very heterogeneous chemical structures, since they are, for example, macrocyclic β-resorcyclic acid lactones (zearalenone and its metabolites/derivatives), trichothecenes (including several tens of toxins with a common tricyclic 12,13-epoxytrichothec-9-ene structure), coumarins (e.g., the difuranocoumarin derivatives aflatoxins and the dihydrocoumarin ochratoxin A), etc. Consequently to this chemical variety, also the toxic effects the mycotoxins can induce in humans and animals are as much various, from immunosuppressive, to estrogenic, teratogenic, and cancerogenic effects.

The most known and important mycotoxins from the toxicological point of view are aflatoxins, ochratoxin, trichothecenes A and B, fumonisins, and macrocyclic lactones.

The main route of human exposure to mycotoxins is the intake of contaminated food, directly from contaminated agricultural products or indirectly from residues and metabolites present in foods of animal origin (e.g., meat, milk and egg). Seldom, inhalation and dermal contact can represent other routes of exposure.

The mycotoxin-induced diseases are studied since the early 1960s, when aflatoxin exposure was recognized as the cause of a severe poultry disease. More recently, the carcinogenic properties of afalatoxins have been recognized by IARC. Other diseases, such as Balkan endemic nephropathy (BEN), endemic squamous cancer of the esophagus, various hemorrhagic syndromes, have been linked to various mycotoxin intake.

Although immunochemical methods are routinely used for fast screening of mycotoxin presence, however, analytical methods based on high-performance liquid chromatography (LC) are preferred for confirmation purposes, especially when coupled with tandem mass spectrometry (MS/MS), which allows the determination of multiclass mycotoxins in a single analysis. Moreover, the technical innovations available in LC-MS/MS instrumentation are prompting its application in monitoring the presence of contaminants in food and feed.

The aim of this paper collection, constituted by ten research articles and one review, is to provide to the reader an overview about the novelties and capabilities in LC-MS/MS-based multi-mycotoxin methods, including the investigation of emerging and modified mycotoxins.

Aldo Laganà
Special Issue Editor

toxins

MDPI

Article

Development of a LC-MS/MS Method for the Multi-Mycotoxin Determination in Composite Cereal-Based Samples

Barbara De Santis *, Francesca Debegnach, Emanuela Gregori, Simona Russo, Francesca Marchegiani, Gabriele Moracci and Carlo Brera

GMO and Mycotoxin Unit, Food Safety, Nutrition and Veterinary Public Health Department, Istituto Superiore di Sanità, Viale Regina Elena 299, Rome 00161, Italy; francesca.debegnach@iss.it (F.D.); emanuela.gregori@iss.it (E.G.); simon_says_go@hotmail.it (S.R.); francesca.marchegiani@libero.it (F.M.); gabriele.moracci@iss.it (G.M.); carlo.brera@iss.it (C.B.)
* Correspondence: barbara.desantis@iss.it; Tel.: +39-06-4990-2367; Fax: +39-06-4990-2363

Academic Editor: Aldo Laganà
Received: 7 April 2017; Accepted: 10 May 2017; Published: 18 May 2017

Abstract: The analytical scenario for determining contaminants in the food and feed sector is constantly prompted by the progress and improvement of knowledge and expertise of researchers and by the technical innovation of the instrumentation available. Mycotoxins are agricultural contaminants of fungal origin occurring at all latitudes worldwide and being characterized by acute and chronic effects on human health and animal wellness, depending on the species sensitivity. The major mycotoxins of food concern are aflatoxin B1 and ochratoxin A, the first for its toxicity, and the second for its recurrent occurrence. However, the European legislation sets maximum limits for mycotoxins, such as aflatoxin B1, ochratoxin A, deoxynivalenol, fumonisins, and zearalenone, and indicative limits for T-2 and HT-2 toxins. Due to the actual probability that co-occurring mycotoxins are present in a food or feed product, nowadays, the availability of reliable, sensitive, and versatile multi-mycotoxin methods is assuming a relevant importance. Due to the wide range of matrices susceptible to mycotoxin contamination and the possible co-occurrence, a multi-mycotoxin and multi-matrix method was validated in liquid chromatography-tandem mass spectrometry (LC-MS/MS) with the purpose to overcome specific matrix effects and analyze complex cereal-based samples within the Italian Total Diet Study project.

Keywords: mycotoxin; aflatoxins; ochratoxin A; deoxynivalenol; fumonisins; T-2 and HT-2 toxins; zearalenone; Total Diet Study; liquid chromatography-tandem mass spectrometry

---

1. Introduction

As is known, the change of climatic conditions of the planet will determine a warming of the eco-system leading to an unavoidable increase of the probability of the occurrence of fungal attack and mycotoxin production in the majority of crops worldwide [1–5]. The more immediate fallout is the increase of the menace of a further limitation of food availability prejudicing food security firstly and food safety secondly, as recently reported for the most dangerous hazard among mycotoxins, namely aflatoxin B1 [6]. Since the entire agri-food system is involved in this challenge, any stakeholder belonging to any position and role must deserve the highest attention in encouraging the adoption of preventive actions aimed at minimizing the phenomenon as much as possible; this, in view of guaranteeing the availability of safe feed and food products to the final consumer. In addition, the inherent health implications due to the consumption of mycotoxin-contaminated feed and food products pose a real alarm both for animals wellness, with concomitant consequences for the productive yields, the economic gain and the quality of processed foods and the consumer health with a direct impact towards the more sensitive consumer groups such as infants, children, and adolescents together with other situations regarding specific status such as the pregnancy and people affected by coexistent pathologies such as celiac disease.

These aspects are even more relevant in consideration of the probability to have the concomitant presence of more than a mycotoxin in a feed or food product [7,8] that can pose a real, and still now underestimated, increased level of risk for the end-consumer due to any potential additive, antagonistic, or synergistic toxic effects. Recently, a combined toxicity of deoxynivalenol (DON) and fumonisins (FBs), or aflatoxins (AFs) and fumonisin B1 (FB1), in the livers of piglets caused higher histopathological lesions and immune suppression [9,10]. Lee and Ryu summarized the most relevant studies reporting additive or synergistic effects due to the co-occurrence of mycotoxins and their interactive toxicity [11]. Severe reductions in growth and immune response were found in broilers by dietary combinations of AFs and ochratoxin A (OTA) [12].

For the abovementioned reasons, the need of the availability of accurate precise and sensitive analytical methods able to detect the mixture of mycotoxins in a reliable way, plays a pivotal role both for toxicological and exposure assessment issues.

To date, maximum levels (MLs) have been set for the majority of countries worldwide in different food products for different individual mycotoxins with recognized adverse effects, but a new scenario could be depicted in the near future, by the reconsideration of these levels in light of the proved increase of the toxicity due to the co-presence of mycotoxins in comparison with the one derived from the individual toxins. Furthermore, all of the Health-Based Guidance Values (HBGV) have been established for individual toxins and not for their forms of mixture.

Another new aspect to be considered is related to the co-presence of the well-known and studied mycotoxins with the so-called emerging toxins, such as enniatins and T-2 and HT-2 toxins, even if, in a recent review from EFSA, the European Authority of Food Safety, no conclusion on risk assessment for the co-presence of enniatins and beauvericin was possible due to the paucity of data available in feeds and food products.

Hence, it is important to fill the gap by carrying out validated analytical tools aimed at obtaining the accurate assessment of human and animal exposure to this group of mycotoxins by determining their levels in feeds and foodstuffs.

For the determination and quantification of mycotoxins in complex matrices, analytical methods based on liquid chromatography-tandem mass spectrometry (LC-MS/MS) have been extensively used [13–16]. With the aim to reduce the matrix effects as much as possible by reducing the interferences from the extraction step, a wide variety of sample preparations, such as liquid-liquid extraction, solid phase extraction (SPE), accelerated solvent extraction (ASE), matrix solid-phase dispersion, and dilute and shoot approaches, have been reported [17].

In a recent review [18], it was outlined that, as far as multi-mycotoxin methods, despite the advantages of these multi-mycotoxin methods with respect to conventional methods due to the superior specificity, sensitivity, and fast data acquisition features, which allow simplified sample preparation, improvements in accuracy and efficiency, as well as in the management of matrix effects are still needed [19].

As known, the matrix effects represent one of the most challenging issues, as well LOQ value, to be solved depending on the final endpoint of the analytical research. It should be underlined that multi-mycotoxin analyses must deserve high flexibility from the researchers' point of view in order to choose the more suitable approach to match the performance characteristics with the different targeted use [18].

This aspect is even more crucial if the composition and the physical nature of the sample to be analyzed corresponds to a quite high complexity of the matrix as the one investigated in a total diet study (TDS) for which the proposed analytical methodology has been set up. The cited TDS, "The Italian Total Diet Study 2012–2014" that included mycotoxins due to their high toxicity and wide frequency in the dietary foods, was organized for the first time in Italy with the aim to develop and spread the TDS methodology on the basis of harmonized principles in terms of study design, sampling and exposure assessment [20,21].

From the above, within the implementation of the Italian TDS, a versatile LC-MS/MS multi-mycotoxin and multi-matrix method was validated with the purpose to guarantee a reliable analysis of cereal-based samples characterized by high sensitivity and applicability.

2. Results and Discussion

2.1. LC-MS/MS Optimization

The optimization of the LC-MS/MS parameters was conducted by directly applying tuning solutions of the selected mycotoxins. Mycotoxin determination was performed in positive ionization mode after testing the negative ionization mode, especially for DON and ZEA evaluation. Formic acid and ammonium formate were added to facilitate the formation of the protonated precursor ion or the ammonium adduct. Only for T-2 toxin, the sodium adduct, usually considered not suitable, was selected as precursor ion. Tests using different concentrations of $HCOOH/NH_4COOH$ and adding acid and ammonium salt only to the aqueous component of the mobile phase were performed. The highest peak intensity was obtained by adding the formate buffer to both the eluents at a concentration of 0.3% of HCOOH and 5 mM of NH_4COOH. Different injection volumes were also tested and, finally, a 10 μL volume was chosen due to a remarkable improvement of the performance in terms of repeatability, 10 μL being the total loop injection mode. Both acetonitrile and methanol were tested as organic modifiers for the mobile phase composition and since no dramatic difference was evidenced, methanol was preferred because it is more environmental friendly in view of laboratory waste disposal.

For identification purpose, the two most intense transitions of the parent compound were selected, while for mycotoxin quantification only the most intense peak (quantifier) was used. In addition, retention time (RT) and ion ratio (IR) variations measured in the samples met the requested criteria (±0.1 min for RT and $\pm30\%$ for IR), when compared with the value obtained for the calibration standard. In Table 1, precursor and product ions of the tested mycotoxins and the specific MS/MS parameters are presented.

Table 1. Optimized MS/MS parameters for the analyzed mycotoxins.

Mycotoxin	Retention Time (min)	Precursor Ion (m/z)	Product Ions (m/z) [a]	Collision Energy (V) [a]	Cone Voltage (V)
AFB1	8.15	313.2 $[M + H]^+$	285.3/241	21/35	45
OTA	7.35	404.1 $[M + H]^+$	238.8/358.1	25/14	25
DON	3.89	297.3 $[M + H]^+$	203.5/249.5	15/10	22
FB1	5.55	723.3 $[M + H]^+$	335.1/353.5	40/30	50
T-2	6.24	489.6 $[M + Na]^+$	245.1/327.1	30/20	30
HT-2	5.77	447.2 $[M + NH4]^+$	285.1/345.1	15/10	30
ZEA	6.79	319.3 $[M + H]^+$	282.9/301.1	10/10	22

[a] Numerical value are given in the order quantifier/qualifier.

In Figure 1, chromatograms obtained for the investigated mycotoxins in a fortified sample belonging to the model matrix "Wheat other cereals and flours" are shown.

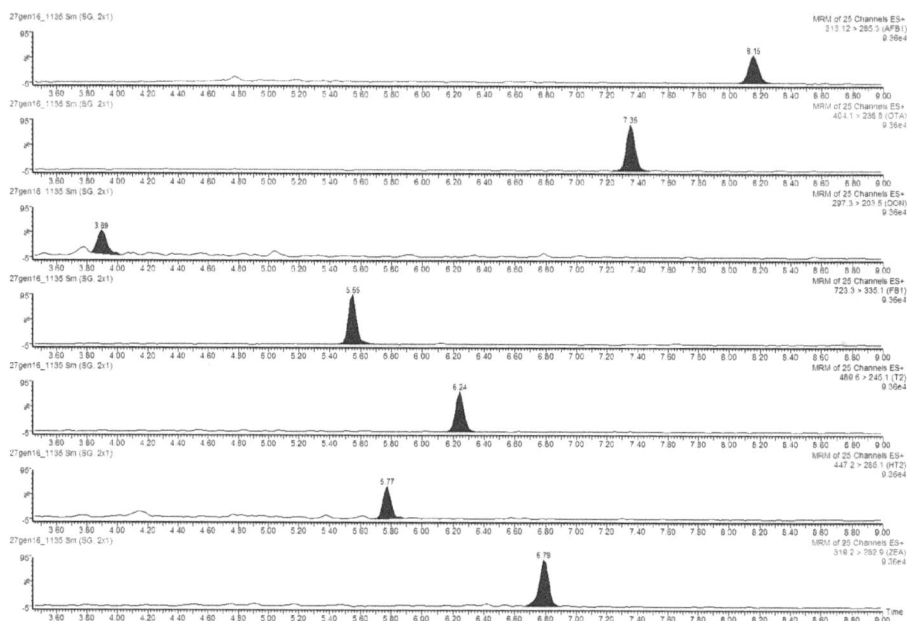

Figure 1. UPLC-MS/MS chromatograms for the considered mycotoxins in the model matrix "Wheat, other cereals and flours" fortified at 0.4 µg/kg for AFB1, 2.5 µg/kg for OTA, 125 µg/kg for DON, 50 µg/kg for FB1, 25 µg/kg for T-2, 125 µg/kg for HT-2, and 10 µg/kg for ZEA.

2.2. Optimization of the Extraction Step

LC-MS/MS methods for mycotoxin determination in food samples are often developed with the aim to reduce sample treatment prior to the injection step. Dilute and shoot is a strategy frequently applied since it allows a reduction in time and cost of the analysis and to retain groups of mycotoxins with high chemical diversity. However, the simple dilution of the sample extract may result in an increased limit of quantification (LoQ); on the other hand, the concentration of the sample can lead to strong matrix effect seriously affecting the performance of the method, if no clean-up is applied. The use of isotope dilution is also a strategy to overcome the matrix effect but the entailed costs and availability of isotopologue standards are a real inconvenience to take into account. Thus, in order to avoid sample treatment and to reduce the cost of the analysis, the matrix-matched approach was used. Different extraction mixtures were tested paying attention to the extraction efficiency of the mycotoxins from the composite cereal based food samples and to the co-extraction of interfering compounds that could reasonably result in disturbing matrix effects. In particular, three solvent mixtures, selected from the literature, were tested, namely $CH_3OH:H_2O$ 80:20 [22]; $CH_3CN:H_2O:CH_3COOH$ 79:20:1 [15], and $EtOH:H_2O$ 2:1 [23]; the evaluation of the performance was made on six replicates of spiked blank samples, comparing the recovery factor values and the associated relative standard deviation of repeatability (RSD$_r$) obtained. The matrix used for optimizing the extraction phase consisted of pooled extra material obtained from the preparation of the cereal based TDS samples (see Section 3.2). The two first extraction approaches were performed by shaking the sample (1 g) for 30 min with the solvent mixture. After extraction, the samples were treated as described in Section 3.3. The third tested extraction was conducted as reported by Breidbach et al. [23], firstly vortexing the sample with water and then extracting with EtOH in a wrist shaker for 30 min. Magnesium sulfate was added for the salting out effect and, after centrifugation, an aliquot of the supernatant was dried and redissolved with the injection solvent for LC-MS/MS analysis. The obtained results, in terms of recovery factors and RSD_r, are reported in Table 2. Accordingly, with the data presented in Table 2, the extraction with the mixture with acidified water/organic solvents, namely $CH_3CN:H_2O:CH_3COOH$ 79:20:1, was finally selected. Additionally, the ratio between the weighed sample and the amount of extraction solvent was evaluated. The small amount of sample weighted for the analysis (1 g) was not considered a critical factor in terms of representativity because, in this specific case, the TDS samples originated from thoroughly mixed and processed preparations (see Section 3.2), bypassing the issue of the heterogeneous distribution of the mycotoxins in the matrix. With respect to the amount of the extraction solvent, different quantities were tested; the best results were obtained for the weighted ratio sample:extraction solvent 1:8. Taking into account the nine subcategories to be analyzed for the TDS purpose, six model matrices were chosen as representative of the validation. On the basis of the processing and grain ingredients, the validation was performed on a selected number of model matrices, more specifically, pasta was chosen as a model for the analysis of pasta and pizza samples; bakery products were chosen as a model matrix for biscuits, savory fine bakery products, cakes, and sweet snacks; finally, breakfast cereals, bread, rice, wheat, and other cereals and flours were the remaining matrices used for validation.

Table 2. Solvent mixtures tested for extraction step optimization, recovery factors (%) and RSD of repeatability (%). Results for aflatoxin B1 (AFB1) and OTA were grouped, as well as for FB1, DON, T-2, and HT-2 toxins and ZEA.

Extraction Solvent	Recovery Factors (RSD$_r$ [a])	
	AFB1, OTA	FB1, DON, T-2 and HT-2, ZEA
$CH_3OH:H_2O$ 80:20	75 (20)	30 (19)
$CH_3CN:H_2O:CH_3COOH$ 79:20:1	65 (15)	80 (15)
H_2O 4mL + EtOH 8mL	25 (20)	80 (19)

[a] Relative standard deviation, calculated on six replicates processed under repeatability operating conditions.

2.3. Method Performance

2.3.1. Linearity

Linearity was tested by the evaluation of determination coefficients (R^2). The linear range was estimated for curves prepared in neat solvent, as well as for spiked extracts and spiked sample in Table 3. The target value for acceptability of the curve was a $R^2 > 0.99$, while the residuals were all matrix-matched calibration curves prepared for the six analyzed matrices. Results for R^2 are reported visually checked to be randomly distributed. As shown from the data reported in Table 3, satisfactory R^2 values were obtained for all the targeted mycotoxins in all the tested matrices.

Table 3. Determination coefficient (R^2) obtained for the targeted mycotoxins in the six validated cereal based matrices.

Mycotoxin	R^2						
	Neat Solvent	Bread	Pasta	Rice	Wheat	Bakery Products	Breakfast Cereal
AFB1	0.9995	0.9956	0.9997	0.9901	0.9913	0.9974	0.9995
OTA	0.9976	0.9958	0.9995	0.9994	0.9903	0.9965	0.9989
FB1	0.9975	0.9982	0.9964	0.9968	0.9982	0.9993	0.9969
DON	0.9941	0.9993	0.9997	0.9993	0.9978	0.9927	0.9914
T-2	0.9937	0.9983	0.9925	0.9929	0.9969	0.9996	0.9906
HT-2	0.9973	0.9909	0.9927	0.9952	0.9919	0.9935	0.9950
ZEA	0.9926	0.9963	0.9937	0.9939	0.9902	0.9942	0.9902

2.3.2. Apparent Recovery, Matrix Effect, and Extraction Recovery

Results obtained from the injection of calibration curves prepared in neat solvent, spiked extract and spiked samples are presented as apparent recovery (R_A), signal suppression/enhancement (SSE), and extraction recovery (R_E), in Table 4. The value of RSD_r calculated for R_A is shown in Table 4. Since calibration curves were prepared and injected in replicates ($n = 6$) for validation purposes, the LoQ for each mycotoxin, in all of the validated matrices, was assessed as the first point of the spiked extract calibration curve. The validated LoQs are reported in Table 4 for each tested model matrix.

Table 4. Apparent recovery, signal suppression/enhancement, extraction recovery, and relative standard deviation of repeatability of R_A obtained for all of the selected mycotoxins and for each validated matrix, together with the validated LoQs.

Matrix		AFB1	OTA	FB1	DON	T-2	HT-2	ZEA
	LoQ (µg/kg)	0.13	0.8	16	20	8	20	3.2
	R_A (%)	61	64	84	62	65	62	65
Bread	SSE (%)	83	98	113	61	55	63	75
	R_E (%)	73	65	74	100	119	99	87
	RSD_r (%)	14	10	15	18	17	15	20
	LoQ (µg/kg)	0.06	0.4	8	20	4	20	1.6
	R_A (%)	61	79	116	69	75	79	75
Pasta	SSE (%)	52	102	120	66	96	109	83
	R_E (%)	117	78	96	105	81	73	90
	RSD_r (%)	11	12	11	12	16	15	16
	LoQ (µg/kg)	0.06	0.4	8	20	4	20	1.6
	R_A (%)	88	79	93	70	65	63	78
Rice	SSE (%)	105	102	101	107	74	64	75
	R_E (%)	83	78	92	65	88	98	104
	RSD_r (%)	11	10	10	15	14	13	15
	LoQ (µg/kg)	0.06	0.4	8	20	4	20	1.6
	R_A (%)	69	76	122	66	61	66	65
Wheat	SSE (%)	73	92	123	71	64	73	65
	R_E (%)	94	83	99	92	95	90	101
	RSD_r (%)	12	10	12	15	15	15	15
	LoQ (µg/kg)	0.13	0.8	16	20	8	20	3.2
	R_A (%)	66	85	76	67	65	63	64
Backery products	SSE (%)	84	108	123	61	64	55	74
	R_E (%)	79	79	61	109	102	116	87
	RSD_r (%)	13	10	16	18	15	15	18
	LoQ (µg/kg)	0.13	0.8	16	20	8	20	3.2
	R_A (%)	65	68	63	66	69	69	68
Breakfast cereals	SSE (%)	59	95	123	74	72	68	71
	R_E (%)	110	71	52	89	96	101	96
	RSD_r (%)	16	6	15	14	19	11	20

In order to obtain very low LoQ values, a concentration of the sample was applied; as a consequence an increase of the matrix effect was observed, as shown from the SSE values reported in Table 4. The influence of the matrix on the sample ionization depends on the mycotoxin and on the matrix components co-extracted during the analysis. For better evaluation of the influence of the matrix effect, SSEs for different mycotoxin/matrix combinations are reported in Figure 2.

The depicted scenario of SSEs is quite satisfactory, being that the effect of the matrix on the mycotoxin signal is in the range of 60–90% or 110–120% for the majority of the analyzed mycotoxin/matrix combinations. SSE values in these ranges were considered acceptable provided that a matrix match approach is used for the correction of the response. Moreover, the 24% of the tested combinations showed a SSE value in the range 90–110%; in this range SSE may be considered as not affected by matrix effects, according to Malachová et al. [24]. On the other hand, seven out of 42 tested mycotoxin/matrix combinations were below 60% (four combinations) or above 120%

(three combinations). However, despite these unfavorable SSE cases, the good performances, in terms of R_A, R_E, and RSD_r, shown in Table 4, support the method reliability when critical values of SSE were also observed.

As for recovery and precision, Annex II of the Commission Regulation (EC) No 401/2006 [25] establishes the performance criteria to which a method shall comply. All of the recovery experiments performed at the LoQ levels for all of the model matrices gave results in the range of acceptability for the tested mycotoxins (namely: >50–120% for OTA and AFB1; 60–110% for DON; 60–120% for ZEA and FBs; and 60–130% for T-2 and HT-2). The calculated relative standard deviations of repeatability were all below 22%, thus confirming a satisfactory performance of method precision even though occasionally a suppression (e.g., AFB1 in pasta or breakfast cereals, T-2 in bread, and HT-2 in bakery products) or enhancement (e.g., FB1 in pasta, wheat, bakery products, and breakfast cereals) was registered.

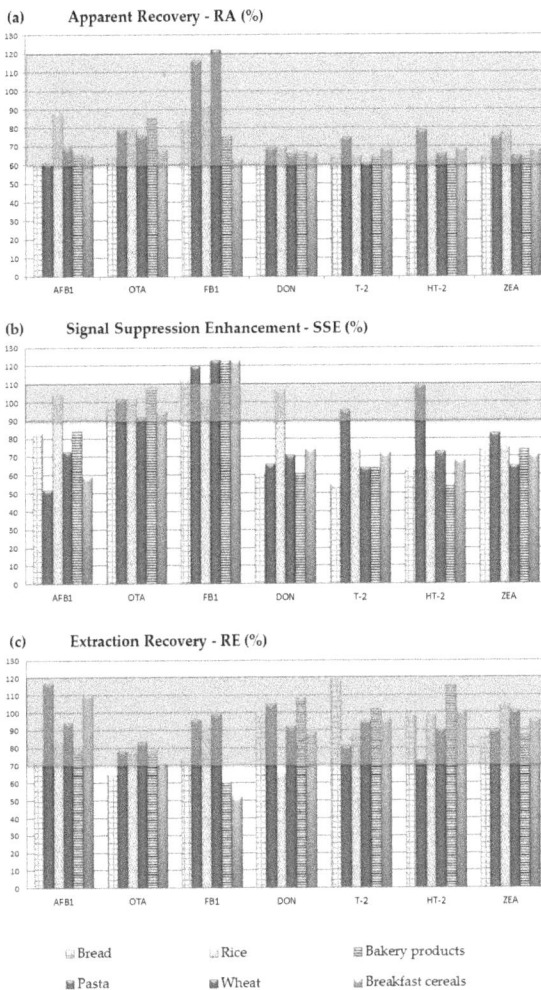

Figure 2. Histograms of the apparent recovery, signal suppression/enhancement, and extraction recovery obtained for all the selected mycotoxins and for each validated matrix. The grey stripe highlights the range 60–120% in (**a**); 90–120% in (**b**); and the range 70–120% in (**c**).

2.3.3. Application to TDS Samples

The validated method was applied to the analysis of 36 pooled samples obtained within the 2012–2014 Italian Total Diet Study [20]. Examining the levels of contamination of the pooled samples, the 25% resulted lower than LoQ, for all the investigated mycotoxins, the 47% of the samples were contaminated with only one mycotoxin and the 28% of the samples were positive to two or more mycotoxins. In relation to the presence of one or more mycotoxins, DON and ZEA were the most frequently found mycotoxins. Estimated concentrations of DON ranged from LoQ to a maximum value of 200 µg/kg in pooled samples of the "Wheat, other cereals, and flours" subcategory. ZEA was determined at concentrations between LoQ and 40 µg/kg, but no toxin was found in "pasta", "rice", "biscuits", and "savory fine bakery products".

Moreover, the co-occurring DON + ZEA + FB1 was found in two samples out of 36 of the "Wheat, other cereals, and flours" subcategory.

It is noteworthy to underline that only one pooled sample exceeded the EU maximum levels, namely a sample of subcategory "bread" where the OTA contamination was 18.7 µg/kg.

Finally, considering the high toxicity of aflatoxin B1, it is remarkable that only one sample was positive at a concentration level close to the LoQ value.

3. Materials and Methods

3.1. Chemicals and Reagents

Methanol (MeOH, Chromasolv® for HPLC, ≥99.9%), acetonitrile (AcCN Chromasolv® for HPLC, ≥99.9%), ammonium formate, and formic acid were purchased from Sigma-Aldrich (St. Louis, MO, USA). Ultra-pure water was produced by a Milli-Q system (Millipore, Bedford, MA, USA).

The certified standard solutions were purchased from Romer Labs Diagnostic GmbH (Tulln, Austria). A composite standard working solution of all of the mycotoxins was prepared by dissolving appropriate volumes of each compound in a mobile phase mixture, A:B, 50:50 *v:v* (H_2O (A) and MeOH (B), both containing 5 mmol·L^{-1} ammonium formate and 0.3% (*v/v*) formic acid). Stock solutions were then diluted with the mobile phase mixture, in order to obtain the appropriate working solutions for the calibration. All solutions were stored at –20 °C in amber glass vials and darkness before use.

3.2. Samples

The investigated food samples were obtained from the stock supplied by the "2012–2014 Italian Total Diet Study" [20]. The optimization and validation was conducted on those samples of the food category "Cereals, cereal products, and substitutes" which, in turn, was composed by nine subcategories as follows: "bread"; "pasta"; "pizza"; "rice"; "wheat, other cereals, and flours"; "breakfast cereals"; "biscuits"; "savory fine bakery products"; and "cakes and sweet snacks". Each subcategory represented a composite core food, which was obtained by pooling a number (up to eight) different "individual samples", selected according to market share and processing (packed food), origin, and species (fresh food). Each individual sample was, in turn, formed by the combination of a fixed number of "elementary samples" (from 16 to 32) that belonged to the established sampling program conducted within the Italian territory. In some cases the individual samples were prepared and cooked according to normal consumer practices (e.g., pasta and rice were boiled) and then were freeze-dried to enable long-term storage for the study purpose.

The sampling program considered the collection of elementary samples from the four main geographical areas in Italy (northeast, northwest, center, and south), thus, after pooling the samples of nine subcategories representative of four geographical areas, the "Cereals, cereal products, and substitutes" food category summed up a grand total of 36 samples.

As for the method optimization, a suitable cereal based sample was prepared by combining extra material obtained from the preparation of the TDS samples. The validation was performed on six model matrices (namely bread, pasta, rice, wheat, breakfast cereal, and bakery products) considered as representative of the nine subcategories.

The analytical method was finally applied to all the 36 samples of the "Cereals, cereal products, and substitutes" category, as obtained by the "2012–2014 Italian Total Diet Study" [20], for the multi-mycotoxin determination. Water loss as a consequence of freeze-drying was measured and fresh/dry weight ratios calculated.

3.3. Sample Preparation

For all of the tested samples one gram (1.0 g ± 0.1 g) of freeze-dried sample was accurately weighed into a 15 mL centrifuge tube. Samples were extracted by shaking with 8 mL of AcCN:H_2O (80:20) 1% HCOOH on a mechanical wrist shaker for 30 min. Extracted samples were centrifuged at 10,000 rpm for 10 min. Four milliliters of supernatant were dried under a stream of nitrogen, redissolved with 400 µL of injection solution (mobile phase A:B, 50:50), and then centrifuged at 10,000 rpm for 10 min. Prior to injection on the LC-MS/MS system, the samples were filtered through a 4 mm, 0.2 µm polytetrafluoroethylene (PTFE) syringe filter (Sartorius, Göttingen, Germany.

3.4. LC-MS/MS Analysis

A Waters UPLC system (Waters, Milford, MA, USA) was used to perform a reverse phase chromatography separation of the selected mycotoxins. The separation was achieved by a Kinetex Biphenyl column (50 mm × 3 mm i.d., 2.6 µm particle size) preceded by a SecurityGuard™ ULTRA Holder pre-column, both supplied by Phenomenex (Phenomenex, Torrance, CA, USA). The mobile phase was a time-programmed gradient using H_2O (eluent A) and MeOH (eluent B), both containing 5 mmol L^{-1} ammonium formate and 0.3% (*v/v*) formic acid. Gradient elution was started isocratically with 95% A for 1 min. Then, B was linearly increased to 100% within 7.5 min and kept constant for 2 min. Finally, B was decreased linearly to 5% in 1.0 min and equilibrated for 5 min. The flow rate was set at 300 µL min^{-1}.

The LC system was coupled with a Waters Quattro Premier XE TQ mass spectrometer (Waters, Manchester, UK) equipped with an ESI source operating in positive ionization mode (ESI+). ESI-MS/MS was performed in multiple reaction monitoring (MRM). The MassLynx v4.1 (Waters, Milford, MA, USA) was used in order to control the UPLC-MS/MS system. Capillary voltage, source temperature, desolvation gas flow rate, and its temperature were set at 3 kV, 120 °C, 600 L h^{-1}, and 350 °C, respectively. Collision-induced dissociation was performed using argon as collision gas at a pressure of 3.5 × 10^{-3} mbar in the collision cell. Cone voltage (CV) and collision energy (CE) values were optimized for each precursor ion and different product ions. For each compound, at least one precursor and two product ions for both identification and quantification purposes were identified, selecting the most abundant product ion for quantification and the second one for confirmation purposes. The precursor ion and the optimized MS/MS parameters (cone voltage and collision energy) for each analyte are summarized in Table 1.

3.5. Method Performance

Linearity, apparent recovery, matrix effect, and recovery of extraction were evaluated by preparing a set of three calibration curves prepared (i) in neat solvent; (ii) by spiking the extract of a blank sample (spiked extract curve); and (iii) by spiking a blank sample before the extraction step (spiked sample curve).

The matrix-matched calibration curves (spiked extract and spiked samples curves) were prepared for each validated matrix. The described set of calibration curves was processed in replicates (*n* = 6) for repeatability evaluation. The spiked samples used for the spiked calibration curve were also used as spiking experiments for the assessment of trueness and precision.

Limits of detection (LoDs) based on a signal-to-noise (S/N) ratio of 3:1, and LoQs on a S/N ratio of 6:1, were calculated by injecting neat solvent standard solutions at different concentration levels.

The calculated LoQs were than verified and validated on the different analyzed matrices, being one of the selected concentration levels of the replicated (n = 6) spiking experiments.

3.5.1. Linearity

Linearity was evaluated by preparing a six-concentration levels calibration curve. The concentration range for each mycotoxin is reported in Table 5. The spiked sample curves were used for quantification, each working day the calibration curve was constructed in duplicate and the average values were considered. The matrix-matched calibration curves were prepared for all of the six validated matrices.

Table 5. Concentration levels prepared for the selected mycotoxins.

Mycotoxin	Concentration Ranges	
	ng/mL	ng/g
AFB1	0.075–2.000	0.06–1.60
OTA	0.5–12.5	0.4–10.0
FB1	10–250	8–200
DON	25–625	20–500
T-2	5–125	4–100
HT-2	25–625	20–500
ZEA	2–50	1.6–40

3.5.2. Apparent Recovery, Matrix Effect, Recovery of Extraction

The evaluation of apparent recovery, signal suppression/enhancement due to matrix effects and extraction recovery were calculated from the six points calibration curves as described in Section 3.5.1, as follows [15]:

$$R_A \ (\%) = 100 \times \frac{slope_{spiked\ sample}}{slope_{neat\ solvent}} \tag{1}$$

$$SE \ (\%) = 100 \times \frac{slope_{spiked\ exrtact}}{slope_{neat\ solvent}} \tag{2}$$

$$R_E \ (\%) = 100 \times \frac{R_A}{SSE} \tag{3}$$

where $slope_{spiked\ sample}$, $slope_{neat\ solvent}$, and $slope_{spiked\ extract}$ represent the values of the gradient of the three calibration curves obtained by plotting peak areas: (1) of the spiked samples; (2) of the neat standards; and (3) of the spiked extracts versus the analyte concentration. Each curve was run in duplicate and each spiking experiment was replicated six times to assess repeatability.

Trueness was assessed following EURACHEM criteria [26] by performing replicated spiking experiments (n = 6) at the LoQ levels for each of the model matrix for all of the tested mycotoxins. The average values of recovery were calculated and the standard deviations of repeatability from the six replicates were taken as a measure of the precision of the method.

With respect to the identification requirements for mass spectrometric detection of mycotoxins, the retention times of the analytes in the sample extract were checked to correspond to that of the average of the calibration standards measured in the same sequence with a tolerance of ±0.1 min; the ion ratio (defined as the response of the peak with the lower area divided by the response of the peak with the higher area) was checked to be within ±30% (relative) to that obtained from the average of the calibration standards from the same sequence.

4. Conclusions

The proposed method represents an example of multi-mycotoxin determination in composite cereal-based samples fit for the purpose of a TDS study. The "Cereals, cereal products, and substitutes" was the category of reference in which nine kinds of subcategories (i.e., "bread"; "pasta"; "pizza";

"rice"; "wheat, other cereals, and flours"; "breakfast cereals"; "biscuits"; "savory fine bakery products"; and "cakes and sweet snacks") were included. The number of publications available for LC-MS/MS techniques around food and feed subject matter are extensive [17,18,23,24], and all kind of references criteria are defined on the basis of "real life" scenarios very much depending on either the specific group (or number) of toxins or combination (and kinds) of matrices. The need to reach low levels of LoQ and satisfactory performances for all the considered mycotoxins was successfully achieved by means of a robust matrix matched validation experiments performed on a selected model matrix samples. The lack of validation criteria/acceptability criteria specifically addressed to mycotoxins in the context of LC-MS method, pushed us to assess the outputs on the basis of the general criteria of precision and recovery [27] and the obtained results are encouraging in terms of acceptable and satisfactory performances to be considered in the context of a multi-toxin analytical method for complex matrices. However, the dynamism in terms of extension to emerging toxins, MS method development, and the setting of new maximum limits compels the researchers to constantly improve method performances, especially in those remarkable cases, such as baby foods, for which the requested LoQs represent a real challenge.

Acknowledgments: This work has been implemented within the "The Italian Total Diet Study 2012-2014" funded by the Italian Ministry of Health (grants CUP I85J12002030005 and I85I14000680005; ISS files 3M23, 4M04, and 6M02), and headed by Francesco Cubadda belonging to the ISS team. Maria Cristina Barea Toscan (ISS), Giuliana Verrone (ISS), and Marilena D'Amato (ISS) are acknowledged for their technical assistance.

Author Contributions: Barbara De Santis, Francesca Debegnach, Emanuela Gregori, and Carlo Brera conceived and designed the experiments; Simona Russo, Gabriele Moracci, and Francesca Marchegiani performed the experiments; Barbara De Santis, Francesca Debegnach, and Emanuela Gregori analyzed the data; Gabriele Moracci, Francesca Marchegiani, and Simona Russo contributed reagents/materials/analysis tools; Barbara De Santis, Francesca Debegnach, Emanuela Gregori, and Carlo Brera wrote the paper.

Conflicts of Interest: The authors declare no conflict of interest.

References

1. Miraglia, M.; De Santis, B.; Brera, C. Climate change: Implications for mycotoxin contamination of foods. *J. Biotechnol.* **2008**, *136*, S715. [CrossRef]
2. Miraglia, M.; Marvin, H.; Kleter, G.; Battilani, P.; Brera, C.; Coni, E.; Cubadda, F.; Croci, L.; De Santis, B.; Dekkers, S. Climate change and food safety: An emerging issue with special focus on Europe. *Food Chem. Toxic.* **2009**, *47*, 1009–1021. [CrossRef] [PubMed]
3. Tirado, M.; Clarke, R.; Jaykus, L.A.; McQuatters-Gollop, A.; Frank, J. Climate change and food safety: A review. *Food Res. Intern.* **2010**, *43*, 1745–1765. [CrossRef]
4. Van der Fels-Klerx, H.J.; Liu, C.; Battilani, P. Modelling climate change impacts on mycotoxin contamination. *World Mycotoxin J.* **2009**, *9*, 717–726. [CrossRef]
5. Paterson, R.R.M.; Lima, N. How will climate change affect mycotoxins in food? *Food Res. Int.* **2010**, *43*, 1902–1914. [CrossRef]
6. Battilani, P.; Toscano, P.; Van der Fels-Klerx, H.J.; Moretti, A.; Camardo Leggieri, M.; Brera, C.; Rortais, A.; Goumperis, T.; Robinson, T. Aflatoxin B1 contamination in maize in Europe increases due to climate change. *Sci. Rep.* **2016**, *6*, 24328. [CrossRef] [PubMed]
7. Bennett, J.W.; Klich, M. Mycotoxins. *Clin. Microbiol. Rev.* **2003**, *16*, 497–516. [CrossRef] [PubMed]
8. Hajslova, J.; Zachariasova, M.; Cajka, T. Analysis of multiple mycotoxins in food. *Methods Mol. Biol.* **2011**, *747*, 233–258. [PubMed]
9. Grenier, B.; Loureiro-Bracarense, A.P.; Lucioli, J.; Pacheco, G.D.; Cossalter, A.M.; Moll, W.D.; Schatzmayr, G.; Oswald, I.P. Individual and combined effects of subclinical doses of deoxynivalenol and fumonisins in piglets. *Mol. Nutr. Food Res.* **2011**, *55*, 761–771. [CrossRef] [PubMed]
10. Harvey, R.B.; Edrington, T.S.; Kubena, L.F.; Elissalde, M.H.; Rottinghaus, G.E. Influence of aflatoxin and fumonisin B1-containing culture material on growing barrows. *Am. J. Vet. Res.* **1995**, *56*, 1668–1672. [PubMed]
11. Lee, H.J.; Ryu, D. Worldwide Occurrence of Mycotoxins in Cereals and Cereal-Derived Food Products: Public Health Perspectives of Their Co-occurrence. *J. Agric. Food Chem.* **2017**. [CrossRef] [PubMed]

12. Boeira, L.S.; Bryce, J.H.; Stewart, G.G.; Flannigan, B. The effect of combinations of *Fusarium* mycotoxins (deoxynivalenol, zearalenone and fumonisin B1) on growth of brewing yeasts. *J. Appl. Microbiol.* **2000**, *88*, 388–403. [CrossRef] [PubMed]

13. Verma, J.; Johri, T.S.; Swain, B.K.; Ameena, S. Effect of graded levels of aflatoxin, ochratoxin and their combinations on the performance and immune response of broilers. *Br. Poult. Sci.* **2004**, *45*, 512–518. [CrossRef] [PubMed]

14. Serrano, A.B.; Capriotti, A.L.; Cavaliere, C.; Piovesana, S.; Samperi, R.; Ventura, S.; Laganà, A. Development of a Rapid LC-MS/MS method for the determination of emerging *Fusarium* mycotoxins enniatins and beauvericin in human biological fluids. *Toxins* **2015**, *7*, 3554–3571. [CrossRef] [PubMed]

15. Capriotti, A.L.; Caruso, G.; Cavaliere, C.; Foglia, P.; Samperi, R.; Laganà, A. Multiclass mycotoxin analysis in food, environmental and biological matrices with chromatography/mass spectrometry. *Mass Spectrom. Rev.* **2012**, *31*, 466–503. [CrossRef] [PubMed]

16. Sulyok, M.; Berthiller, F.; Krska, R.; Schuhmacher, R. Development and validation of a liquid chromatography/tandem mass spectrometric method for the determination of 39 mycotoxins in wheat and maize. *Rapid Commun. Mass Spectrom.* **2006**, *20*, 2649–2659. [CrossRef] [PubMed]

17. Serrano, A.B.; Font, G.; Mañes, J.; Ferrer, E. Comparative assessment of three extraction procedures for determination of emerging *Fusarium mycotoxins* in pasta by LC-MS/MS. *Food Control.* **2013**, *32*, 105–114. [CrossRef]

18. Krska, R.; Schubert-Ullrich, P.; Molinelli, A.; Sulyok, M.; Macdonald, S.; Crews, C. Mycotoxin analysis: An update. *Food Addit. Contam. Part A* **2008**, *25*, 152–163. [CrossRef] [PubMed]

19. Berthiller, F.; Brera, C.; Iha, M.H.; Krska, R.; Lattanzio, V.M.T.; MacDonald, S.; Malone, R.J.; Maragos, C.; Solfrizzo, M.; Stranska-Zachariasova, M.; et al. Developments in mycotoxin analysis: An update for 2015–2016. *World Mycotoxin J.* **2016**, *9*, 5–30. [CrossRef]

20. Zhang, K.; Wong, J.W.; Krynitsky, A.J.; Trucksess, M.W. Perspective on advancing FDA regulatory monitoring for mycotoxins in foods using liquid chromatography and mass spectrometry. *J. AOAC Int.* **2016**, *99*, 890–894. [CrossRef] [PubMed]

21. D'Amato, M.; Turrini, A.; Aureli, F.; Moracci, G.; Raggi, A.; Chiaravalle, E.; Mangiacotti, M.; Cenci, T.; Orletti, R.; Candela, L.; et al. Dietary exposure to trace elements and radionuclides: the methodology of the Italian Total Diet Study 2012–2014. *Annali dell'Istituto Super. di Sanità* **2013**, *49*, 272–280.

22. Cubadda, F.; D'Amato, M.; Aureli, F.; Raggi, A.; Mantovani, A. Dietary exposure of the Italian population to inorganic arsenic: The 2012–2014 Total Diet Study. *Food Chem. Toxicol.* **2016**, *98*, 148–158. [CrossRef] [PubMed]

23. *EN 14123:2007-Foodstuffs-Determination of Aflatoxin B1 and the Sum of Aflatoxins B1, B2, G1 and G2 in Hazelnuts, Peanuts, Pistachios, Figs and Paprika Powder—High Performance Liquid Chromatographic Method With Post-Column Derivatisation and Immunoaffinity Column Cleanup*; BSI: London, UK, 2008; ISBN 978-0-580-58514-2.

24. Breidbach, A.; Bouten, K.; Kroeger-Negoita, K.; Stroka, J.; Ulberth, F. LC-MS Based Method of Analysis for the Simultaneous Determination of Four Mycotoxins in Cereals and Feed: Results of a Collaborative Study. 2013. Available online: http://publications.jrc.ec.europa.eu/repository/handle/JRC80176 (accessed on 15 May 2017).

25. Malachová, A.; Sulyok, M.; Beltrán, E.; Berthiller, F.; Krska, R. Optimization and validation of a quantitative liquid chromatography-tandem mass spectrometric method covering 295 bacterial and fungal metabolites including all regulated mycotoxins in four model food matrices. *J. Chrom. A* **2014**, *1362*, 145–156. [CrossRef] [PubMed]

26. Commission Regulation (EC) No 401/2006. Laying down the methods of sampling and analysis for the official control of the levels of mycotoxins in foodstuffs. *Off. J. Eur. Union* **2006**, *L70*, 12–34.

27. Magnusson, B.; Örnemark, U. (Eds.) *Eurachem Guide: The Fitness for Purpose of Analytical Methods—A Laboratory Guide to Method Validation and Related Topics*, 2nd ed.; 2014; ISBN 978-91-87461-59-0. Available online: www.eurachem.org (accessed on 15 May 2017).

© 2017 by the authors. Licensee MDPI, Basel, Switzerland. This article is an open access article distributed under the terms and conditions of the Creative Commons Attribution (CC BY) license (http://creativecommons.org/licenses/by/4.0/).

toxins

MDPI

Article

Simultaneous Determination of Multi-Mycotoxins in Cereal Grains Collected from South Korea by LC/MS/MS

Dong-Ho Kim [1,†], Sung-Yong Hong [2,†], Jea Woo Kang [2], Sung Min Cho [2], Kyu Ri Lee [2], Tae Kyung An [3], Chan Lee [3,*] and Soo Hyun Chung [2,*]

[1] National Agricultural Products Quality Management Service, 5-3 Gimcheon innocity, Nam-myeon, Gimcheon City Gyeongsangbuk-do Province 39660, Korea; anoldmu@korea.kr

[2] Department of Integrated Biomedical and Life Science, Korea University, Seoul 02841, Korea; lunohong@korea.ac.kr (S.-Y.H.); wodn5000@gmail.com (J.W.K.); gangnangie@korea.ac.kr (S.M.C.); leekr0511@naver.com (K.R.L.)

[3] Department of Food Science and Technology, Chung-Ang University, Anseong-si, Gyeonggi-do 17546, Korea; cobyatk@naver.com

* Correspondence: chanlee@cau.ac.kr (C.L.); chungs59@korea.ac.kr (S.H.C.); Tel.: +82-31-670-3035 (C.L.); +82-2-3290-5641 (S.H.C.); Fax: +82-31-676-8865 (C.L.); +82-927-7825 (S.H.C.)

† These authors contributed equally to this work.

Academic Editor: Aldo Laganà
Received: 4 January 2017; Accepted: 8 March 2017; Published: 16 March 2017

Abstract: An improved analytical method compared with conventional ones was developed for simultaneous determination of 13 mycotoxins (deoxynivalenol, nivalenol, 3-acetylnivalenol, aflatoxin B_1, aflatoxin B_2, aflatoxin G_1, aflatoxin G_2, fumonisin B_1, fumonisin B_2, T-2, HT-2, zearalenone, and ochratoxin A) in cereal grains by liquid chromatography-tandem mass spectrometry (LC/MS/MS) after a single immunoaffinity column clean-up. The method showed a good linearity, sensitivity, specificity, and accuracy in mycotoxin determination by LC/MS/MS. The levels of 13 mycotoxins in 5 types of commercial grains (brown rice, maize, millet, sorghum, and mixed cereal) from South Korea were determined in a total of 507 cereal grains. Mycotoxins produced from *Fusarium* sp. (fumonisins, deoxynivalenol, nivalenol, and zearalenone) were more frequently (more than 5%) and concurrently detected in all cereal grains along with higher mean levels (4.3–161.0 ng/g) in positive samples than other toxins such as aflatoxins and ochratoxin A (less than 9% and below 5.2 ng/g in positive samples) from other fungal species.

Keywords: aflatoxins; deoxynivalenol; fumonisins; HT-2; nivalenol; ochratoxin A; T-2; zearalenone; grains; LC/MS/MS

1. Introduction

Mycotoxins are biologically active secondary metabolites produced by a variety of fungi such as *Aspergillus*, *Penicillium*, and *Fusarium* sp. To date, approximately 400 compounds have been identified as mycotoxins such as aflatoxins (AFs), ochratoxin A (OTA), fumonisins (Fs), nivalenol (NIV), deoxynivalenol (DON), zearalenone (ZEN), T-2 toxin (T-2), and HT-2 toxin (HT-2) [1]. The International Agency for Research on Cancer [2] assigned major mycotoxins into one of 5 groups based on their carcinogenicity. Aflatoxin B_1 (AFB_1) is categorized as a Group 1 human carcinogen due to its potent carcinogenic properties in liver [2]. OTA, fumonisin B_1 (FB_1), and fumonisin B_2 (FB_2) are classified as possible carcinogens (Group 2B) to human since OTA causes nephrotoxicity, immune suppression, carcinogenicity, and teratogenicity in laboratory animals, and it has been associated with a fatal human kidney disease known as Balkan Endemic Nephropathy, and FB_1 and FB_2 cause equine leukoencephalomalacia (ELEM) in horses and porcine pulmonary edema (PPE) in pigs [3]. However, DON, ZEN, and T-2 are categorized into Group 3 (not classifiable as to its carcinogenicity to humans) because there is no evidence in their mutagenicity and carcinogenicity [4].

These mycotoxins occur in agricultural crops during pre-harvest and storage. As the mycotoxins are chemically very stable, they are not degraded during food processing, causing a variety of adverse and toxic health effects in target organs such as liver, kidney, and nerve systems in human. Thus, most of the countries started to reinforce management of mycotoxins in foods and feeds, and the European Union and the Codex have made efforts to set common regulatory limits for mycotoxins. South Korea has also set and regulated the maximum allowable limits for important mycotoxins in foods and feeds based on Food Sanitation Act and Control of Livestock and Fish Feed Act. Currently, risks of mycotoxins found in agricultural crops are assessed by analyses of only one toxin among several mycotoxins, which can contaminate the same agricultural crops, and the maximum allowable limits and analytical methods for major mycotoxins are established only for individual toxin in South Korea as well as other countries. Moreover, these mycotoxins found in a contaminated agricultural commodity can cause serious synergistic effects in human and animals when consumed simultaneously by them. Boeira and co-workers reported a synergistic toxicity between *Fusarium* mycotoxins (DON and ZEA) on the growth of brewing yeasts [5]. Other researchers have demonstrated that a combined toxicity of either DON and Fs or AFs and FB_1 in liver in piglets or barrows caused higher histopathological lesion and immune suppression [6,7]. Severe reductions in growth and immune response were found in broilers by dietary combinations of AFs and OTA [8]. Also, synergistic cytotoxic effects of mycotoxin combinations were shown in mammalian cell [9–12]. Because co-occurrence of mycotoxins is very common in agricultural commodities, a reliable and sensitive analytical method is needed for simultaneous determination of multi-mycotoxins. In addition, a prerequisite for the analyses of multi-toxins is good recoveries of the toxins at toxin extraction and clean-up steps. Most of recently developed multi-mycotoxin analytical methods employ acetonitrile-water mixtures for co-extraction of mycotoxins at toxin extraction steps [13]. In order to purify toxin extracts at toxin clean-up steps, commercial immunoaffinity columns (IAC) have been successfully applied for simultaneous determination of mycotoxins by liquid chromatography-tandem mass spectrometry (LC/MS/MS). There are two toxin elution methods at IAC steps; single elution [13] and double elution [14]. In the current study, in order to develop analytical methods for rapid and efficient determination of major mycotoxins in food, we developed a new rapid method to co-elute all 13 major mycotoxins by one step using 5 mL of 80% methanol (MeOH) containing 0.5% acetic acid at the same time and established an analytical method for simultaneous determination of the 13 mycotoxins by LC/MS/MS. The method was successfully applied for rapid and simultaneous determination of the 13 mycotoxins in grains collected from retail markets in South Korea. To the best of our knowledge, this is the first report on improved co-extraction of 13 mycotoxins in a variety of grains and simultaneous analyses of the multi-mycotoxins in the cereal grains collected from markets in South Korea by LC/MS/MS.

2. Results

2.1. Linearity of Calibration Curves for 13 Mycotoxins

The linearity of 13 mycotoxins in the analytical method was assessed by each standard curve using 5 levels of standard solutions for each toxin. An extract ion chromatogram (EIC) of 13 mycotoxins analyzed by the LC/MS/MS is shown in Figure 1. The linearity of the calibration curves was determined by linear regression analysis. The curves for all 13 mycotoxins showed r^2 values of 0.9932–1.0000 (Figure S1). Therefore, we concluded that the standard curves for all 13 mycotoxins were linear in the range of 1.3–53 ng/mL.

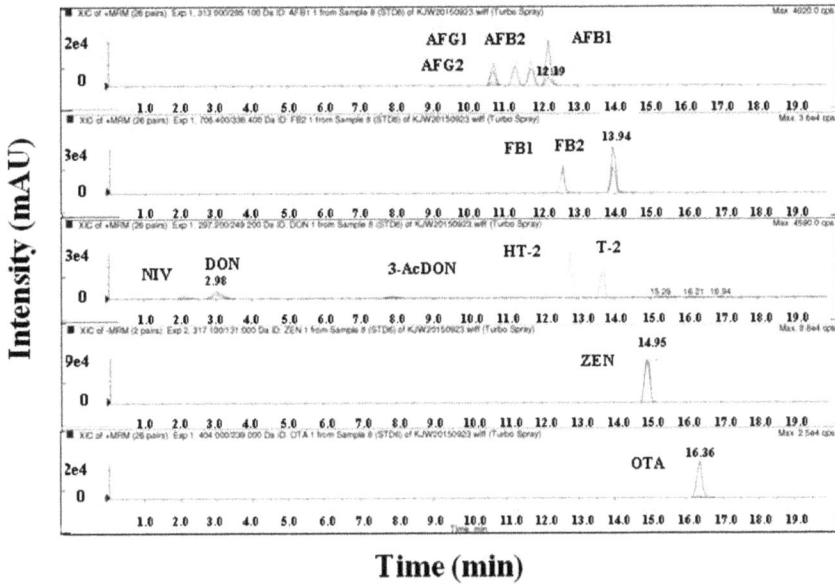

Figure 1. Chromatograms of 13 mycotoxins using LC/MS/MS. Extract ion chromatogram of 13 mycotoxins. Ten microliters of the toxin mixture standard was injected into the LC/MS/MS system. The retention times of peaks corresponding to each toxin are as follow: NIV, 2.08 min; DON, 2.98 min; 3-AcDON, 7.95 min; AFG$_2$, 10.67 min; AFG$_1$, 11.25 min; AFB$_2$, 11.71 min; AFB$_1$, 12.19 min; FB$_1$, 12.54 min; HT-2, 12.77 min; T-2, 13.68 min; FB$_2$, 13.94 min; ZEN, 14.95 min; OTA, 16.36 min.

2.2. Extraction of 13 Mycotoxins from 5 Different Matrices and the Effects of the Matrices on Toxin Extraction

Previously, toxins have been extracted by two steps using water and MeOH, which require time-consuming and laborious shaking and clean-up [14,15]. In this study our method describes a rapid and efficient co-extraction and co-elution of 13 mycotoxins in 5 different matrices including mixed cereal with organic solvents containing acetic acid. Our one step method for extraction of multi-toxins shortened the toxin extraction procedure.

Extraction of toxins from grain samples with organic solvents entails the possibility of analytical problems (matrix effects) due to the co-extraction of matrix components in the samples. The matrix effects can affect the ionization efficiency of toxins, leading to suppression or enhancement of the signal in LC/MS/MS depending on combinations of types of toxins and matrices. Thus, the effects of matrices on the determination of 13 mycotoxins in 5 types of grains were evaluated.

The signal suppression/enhancement was calculated by the following equation:

$$SSE = \frac{Slope\ of\ a\ standard\ curve\ using\ a\ sample\ spiked\ with\ a\ toxin\ x \times 100}{Slope\ of\ a\ standard\ curve\ using\ a\ toxin\ standard\ solution}$$

Five types of grain samples showed minor matrix effects on the determination of all 13 toxins in the samples (74.5%–112.2%) (Figure 2). Of these toxins, signals for levels of 3-AcDON in all 5 types of samples were slightly enhanced by the matrices (102.9%–112.2%), whereas those for all other toxins in the samples were a little suppressed by the matrices (74.5%–90.0%).

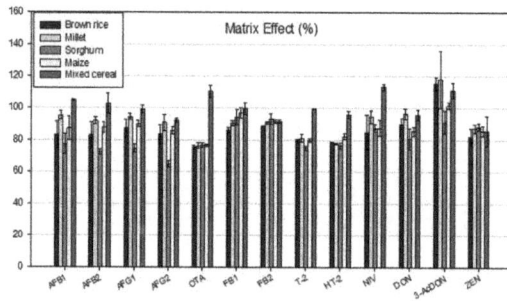

Figure 2. Signal suppression/ enhancement (SSE) effect of 13 mycotoxins in 5 types of matrices. Five levels (1.325, 2.65, 13.25, 26.5, and 53 ng/mL) of standard solutions for each toxin were prepared by mixing the extract from each type of grain sample. The standard solutions for each toxin were injected into LC/MS/MS in triplicate.

2.3. Recovery of 13 Mycotoxins from 5 Different Matrices

Recoveries for 13 mycotoxins extracted from each matrix spiked with each toxin standard solution were evaluated by using our newly developed co-extraction and co-elution method.

The recoveries were calculated by the following equation:

$$\text{Recovery} = \frac{\text{Each toxin concentration equivalent to the peak area measured from the spiked sample} \times 100}{\text{Each toxin concentration used for spiking the sample}}$$

The recoveries were measured by injecting toxins extracted from each matrix, which was naturally uncontaminated with 13 mycotoxins and spiked with 1.2–326.1 ng/mL of each toxin standard solution as described in materials and methods, into LC/MS/MS. The recovery rates for 13 mycotoxins in 5 types of matrices were 73.9%–133.0% along with relative standard deviation (RSD_r) of 0.1%–14.3% (Figure 3). These recovery rates of 13 mycotoxins satisfied allowable limits of the recovery and RSD recommended by Codex or Association of Official Analytical Chemists (AOAC) [16,17]. The Codex recommends 60%–120% of recovery rates in food samples contaminated with 1–10 μg/kg of mycotoxins and the guideline for the recoveries by AOAC is 70%–125% in food samples contaminated with 10 μg/kg of mycotoxins. In addition, RSD_r values of the 13 mycotoxins (0.1%–14.3%) were below 15%, which is recommended for food samples contaminated with 10 μg/kg of mycotoxins by AOAC. Thus, we concluded that the analytical method by co-extraction and co-elution using MeOH containing acetic acid had good recoveries from 5 types of matrices.

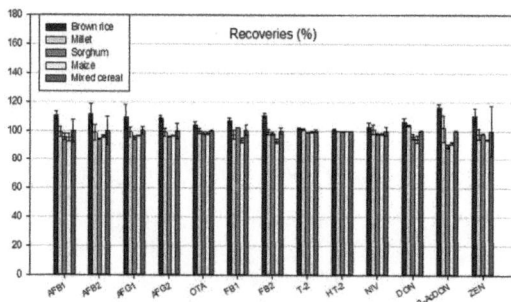

Figure 3. Recovery of 13 mycotoxins from 5 types of matrices. Each matrix, which was naturally uncontaminated with 13 mycotoxins, was spiked with 1.2–326.1 ng/mL of each toxin standard solution as described in materials and methods. Toxins extracted from each matrix were injected into LC/MS/MS.

2.4. LOD and LOQ of an Analytical Method for Determination of Levels of 13 Mycotoxins

The sensitivity of the analytical method using LC/MS/MS was determined by a limit of detection (LOD) and limit of quantification (LOQ). They were calculated as signal-to-noise (S/N) ratios of 3 and 10, respectively, which were measured by using Analyst 1.6 software. The LODs of the analytical method for 13 mycotoxins in 5 types of cereal grains were between 0.1 and 18.1 ng/g, whereas the LOQs of the method for the mycotoxins in cereal grains were between 0.4 and 54.8 ng/g (Table S1). They were as low as those for detection of trace amounts of the toxins. It indicates that the method using LC/MS/MS is highly sensitive for determination of all 13 mycotoxins in cereal grains.

2.5. Monitoring Levels of 13 Mycotoxins in Commercial Cereal Grains

The analytical method validated above was used for the determination of levels of 13 mycotoxins in 5 types of 507 cereal grains (brown rice, maize, millet, sorghum, and mixed cereal) collected from local markets in South Korea. The occurrence and levels of all 13 toxins in the commercial products are summarized in Table 1. Mycotoxins produced from *Fusarium* sp. (FB_1, FB_2, DON, NIV, and ZEN) were more frequently detected in all grain samples than other toxins such as AFs and OTA from other fungal species.

Table 1. Mean and range of levels and incidence of 13 mycotoxins in a total of 507 brown rice, millet, sorghum, maize, and mixed cereal collected from retail market.

Mycotoxin[1]		Grain Sample				
	Item	Brown Rice	Millet	Sorghum	Maize	Mixed Cereal
AFB_1	Incidence (%)	1	9	4	1	4
	Mean[2] (ng/g)	1.1	1.3	1.0	5.2	4.3
	Range (ng/g)	1.1	0.4–5.6	0.7–1.7	5.2	0.7–12.4
AFB_2	Incidence (%)	0	1	0	0	0
	Mean[2] (ng/g)	0	0.5	0	0	0
	Range (ng/g)	-	0.5	-	-	-
FB_1	Incidence (%)	42	52	95	47	74
	Mean[2] (ng/g)	13.6	12.4	160.8	136.5	17.3
	Range (ng/g)	2.1–22.8	2.0–32.6	5.8–890.0	3.8–2990.0	3.1–80.1
FB_2	Incidence (%)	44	50	89	59	58
	Mean[2] (ng/g)	9.6	12.0	42.8	45.2	14.6
	Range (ng/g)	1.6–18.8	1.6–31.1	4.0–223.5	1.9–620.0	1.8–22.1
T-2	Incidence (%)	0	0	0	2	0
	Mean[2] (ng/g)	0	0	0	10.0	0
	Range (ng/g)	-	-	-	6.4–13.7	-
HT-2	Incidence (%)	0	0	0	0	1
	Mean[2] (ng/g)	0	0	0	0	4.3
	Range (ng/g)	-	-	-	-	4.3
OTA	Incidence (%)	0	0	0	0	1
	Mean[2] (ng/g)	0	0	0	0	0.5
	Range (ng/g)	-	-	-	-	0.5
DON	Incidence (%)	7	25	70	13	19
	Mean[2] (ng/g)	5.6	46.5	64.0	180.4	41.5
	Range (ng/g)	6.0–12.3	12.1–212.0	18.1–257.0	17.0–1405.0	14.5–162.0
NIV	Incidence (%)	5	16	53	18	40
	Mean[2] (ng/g)	26.3	45.3	48.2	116.1	50.2
	Range (ng/g)	16.3–36.8	15.7–102.0	18.1–211.5	12.7–570.0	13.8–175.0
3-AcDON	Incidence (%)	0	0	0	2	0
	Mean[2] (ng/g)	0	0	0	17.8	0
	Range (ng/g)	-	-	-	7.9–27.7	-
ZEN	Incidence (%)	32	14	62	7	47
	Mean[2] (ng/g)	5.2	7.4	37.5	4.3	6.1
	Range (ng/g)	0.4–37.6	0.7–61.5	0.9–313.0	0.9–14.7	0.2–36.0

[1] AFG_1 and AFG_2 were not detected in all grain samples; [2] Mean indicates an average in positive samples.

The AFB_1 showed 1%–9% of incidence in all cereal grains (Table 1). The highest levels of AFB_1 in maize and millet were 5.2 and 5.6 ng/g, respectively, which were below the maximum allowable limit (10 ng/g) set by the Korean Food and Drug Administration (KFDA) [18]. However, levels of AFB_1 in mixed cereal (12.4 ng/g) exceeded the allowable limit. The occurrence of AFB_2 was much lower than that of AFB_1; the occurrence of AFB_2 was below 1% in millet. AFG_1 and AFG_2 were not detected in any types of grain samples.

The incidence (42%–95%) and levels of FB_1 and FB_2 were similar to each other in the same grain group, and the levels of Fs were in the range of 1.6–2990 ng/g in positive samples (Table 1). In particular, Fs were detected in relatively high concentrations (1.9–2990 ng/g of the range and 42.8–160.8 ng/g of the mean levels in positive samples) in maize and sorghum compared to those (1.6–80.1 ng/g of the range and 9.6–17.3 ng/g of the mean levels in positive samples) in other types of cereal grains. Nevertheless, the levels of FB_1 and FB_2 detected in all types of grains were below the maximum allowable limit (4 µg/g for FB_1 and FB_2 in maize) set by the KFDA.

Of trichothecenes (TCs), DON and NIV were more frequently detected in all types of cereal grains than T-2, HT-2, and 3-AcDON; the incidence of DON and NIV was between 16%–70% in all grains except 7% for DON and 5% for NIV in brown rice, and the mean levels of DON and NIV were in the range of 6.0-1405 ng/g in positive samples (Table 1). Except the levels of DON in maize, the levels of DON and NIV detected in all types of cereal grains were below the maximum allowable limit (1 µg/g for DON in grains and their products) set by the KFDA when the limit of DON is used for levels of NIV since the legal limits of NIV are not set yet in South Korea [18]. However, the occurrence of T-2, HT-2, and 3-AcDON was less than 2% in the cereal grains, and the ranges of those in positive samples were lower than 27.7 ng/g. The ranges of T-2, HT-2, and 3-AcDON were far below than 1 µg/g, which was set for DON as the legal limit by KFDA [18]. Thus, the levels of DON and NIV as well as those of 3AcDON, T-2, and HT-2 in all types of cereal grains except for DON in maize could not pose a health risk in South Korea.

The incidence of ZEN, one of mycotoxins from *Fusarium* sp., was between 7%-62% in all types of cereal grains, and its range was between 0.2–313.0 ng/g (Table 1). In particular, the levels of ZEN detected in sorghum were above the maximum allowable limit (200 ng/g) set by the KFDA [18]. Therefore, the level of ZEN in sorghum could pose a risk to public health in South Korea.

Finally, OTA was detected rarely (0%–1% of incidence) in the cereal grains, and its range was 0–0.5 ng/g. The levels of OTA detected in all grains were below the maximum allowable limit (5 ng/g) set by the KFDA [18].

The co-occurrence of the mycotoxins in 507 cereal grains was as follows; 30% of 2 groups of micotoxins, 22% of 3 groups of mycotoxins, and 3% of 4 groups of mycotoxins as shown in Figure 4A. In samples which showed co-occurrence of 2 groups of mycotoxins, Fs and TCs were frequently detected. Fs, TCs, and ZEN were major mycotoxins in the category of 3 groups of mycotoxins. These data on the co-occurrence of the mycotoxins in the same grain sample indicate that mycotoxins produced from *Fusarium* sp. (FB_1, FB_2, HT-2, T-2, DON, NIV, 3-AcDON, and ZEN) were concurrently detected in the same samples. In addition, they were detected more frequently along with AFB_1 or AFB_2 in the same samples than along with OTA, and any samples were not co-contaminated with AFs and OTA; AFs with *Fusarium* mycotoxins in 17 samples and OTA with *Fusarium* mycotoxins in only 1 mixed cereal sample (Figure 4B).

Overall, although the levels of AFB_1 (except for mixed cereal), total AFs, FB_1, FB_2, and OTA detected in all 5 types of grain samples were below the maximum allowable limit set by the KFDA, they should have been much lower than observed levels because even low exposure to dietary toxins could pose a carcinogenic risk to human. In addition, the levels of DON and NIV in all 5 types of grains (except the levels of DON in maize) as well as those of 3AcDON, T-2, and HT-2 in the cereal grains were below the maximum allowable limit set for DON by KFDA. Also, the level of ZEN detected in sorghum was above the maximum allowable limit set by KFDA. Therefore, monitoring the levels of 13 mycotoxins in cereal grains marketed in South Korea should be continued.

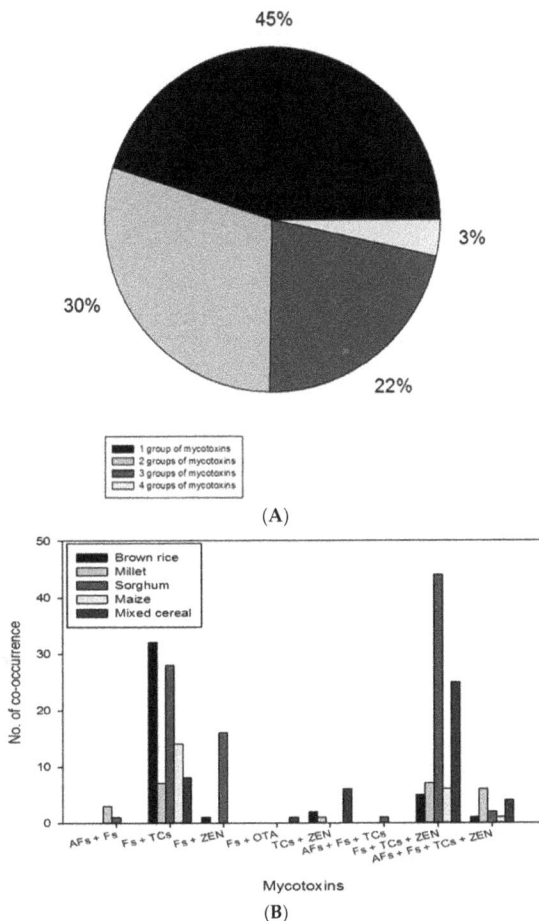

Figure 4. Co-occurrence of mycotoxins in the same sample of 5 types of cereal grains collected from retail market. (**A**) Percentage of co-occurrence of mycotoxins in a total of 507 cereal grains; (**B**) Number of co-occurrence of mycotoxins in 5 types of cereal grains.

3. Discussion

In this study we developed an improved rapid one-step method to co-elute all 13 major mycotoxins from Myco6in1 with IAC using 5 mL of 80% MeOH containing 0.5% acetic acid and established an analytical method for simultaneous determination of the mycotoxins by LC/MS/MS. The method was then used for rapid and simultaneous determination of levels of the 13 mycotoxins in grains collected from retail markets in South Korea. For 100% elution of 13 mycotoxins from IACs, all bonds between analytes and antibodies should be broken. In our one-step elution, 5 mL of 80% MeOH can provide enough time to break the bonds, and the acidic condition produced by 0.5% acetic acid makes the bond breakage occur easily, which improves the recoveries of the mycotoxins in this study. The recovery rates for 13 mycotoxins in 5 types of matrices were 73.9%–133.0% using the improved toxin extraction method (Figure 3). In contrast, a previous study described by Lattanzio and collaborators showed that recovery rates were relatively low (63%–93%) in simultaneous determination of multi-mycotoxins in corn and wheat samples [14].

Our data showed that the incidence of AFB_1 in all cereal grains was from 1% to 9% (Table 1). Of these grains, the occurrence of AFB_1 in maize was 1% in our study, while AFB_1 was not detected in corn collected in South Korea in one study described by Park and co-workers [19]. One of the reasons for this discrepancy may have come from the small sample size ($n = 18$) in the previous study.

Also, the occurrences of FB_1 and FB_2 in brown rice and millet were about 50% (42% in brown rice and 52% in millet for FB_1, and 44% in brown rice and 50% in millet for FB_2) in the current study (Table 1), while both mycotoxins were not detected in brown rice and millet collected in South Korea in one study described by Seo and collaborators [20]. Again, the small sample size ($n = 12$) in the previous study may have been one of the reasons for this discrepancy.

Table 1 showed that the incidence and the mean levels of DON and NIV in maize were similar to each other (13% and 180 ng/g for DON, and 18% and 116 ng/g for NIV). These results show lower occurrences and levels of DON and NIV compared to those detected in corn collected in South Korea in 1993, in which authors reported occurrences of 65.2% of DON (310 ng/g of the mean level) and 34.8% of NIV (77 ng/g of the mean level) [21]. One of possible reasons for this discrepancy may have been due to different climate when the maize was harvested in the field or differences between the regions where the maize was harvested.

The incidence and the mean level of ZEN, another mycotoxin from *Fusarium* sp., in maize were 7% and 4.3 ng/g in the present study (Table 1). The level of ZEN in maize was lower than that (151 ng/g of the mean level) detected in corn collected in South Korea in 1993 as described by Kim and co-workers, while the occurrence of ZEN in maize in our study was similar to that (8%) detected in the previous study [21]. The same reason (different climate when maize was harvested or different regions where maize was harvested) as that for DON, and NIV may be able to explain this discrepancy.

Finally, OTA was detected rarely (1% of incidence) in all cereal grains, and its mean level was 0–0.5 ng/g in our study (Table 1). Of these grains, OTA was not detected in maize in this study, which was the same as that (0%) in corn collected in South Korea in 2002 [19].

On the other hand, Figure 4B showed that 17 samples were co-contaminated with AFs and mycotoxins from *Fusarium* sp. (FB_1, FB_2, HT-2, T-2, DON, NIV, 3-AcDON, and ZEN), whereas only one sample (mixed cereal) was concurrently contaminated with OTA and *Fusarium* mycotoxins. These data are in agreement with Ok and co-workers' study in which they detected DON and ZEN concurrently in 12 out of 70 corn and barley samples collected from South Korea in 2005 and 2006 [22]. In addition, another study from South Korea reported the co-occurrence of AFB_1, FB_1, and ZEN in corn and barley [19]. Moreover, studies from other countries also showed the co-occurrence of the mycotoxins from *Fusarium* sp. One study from Finland reported that the same samples of cereal grains such as barley, oats, and wheat were contaminated simultaneously with DON, 3-AcDON, HT-2, T-2, and ZEN [23]. Another study by Ali and collaborators also reported that Fs, DON, NIV, and ZEN were co-contaminated with AFs in corn from Indonesia, and they isolated *Aspergillus flavus* and *Aspergillus parasiticus* along with *Fusarium moniliforme* in the same samples [24]. These data suggest that our cereal samples may also have been co-infected with the *Aspergillus* sp. and *Fusarium* sp.

Because the co-occurrence of these mycotoxins is common in cereal grains and they can cause synergistic effects to human and animals, it is necessary that efficient control methods are developed to prevent and monitor contamination of multi-mycotoxins and fungi in grains.

4. Conclusions

In our current study we developed a highly sensitive and reliable analytical method for simultaneous determination of levels of 13 mycotoxins (AFB_1, AFB_2, AFG_1, AFG_2, DON, NIV, 3-AcDON, FB_1, FB_2, T-2, HT-2, ZEN, and OTA) in cereal grains by LC/MS/MS after IAC clean-up. We were able to minimize any interfering materials against determination of levels of the mycotoxins in grains using the Myco6in1 with IAC columns and were able to elute all of the mycotoxins simultaneously from the IAC using 5 mL of 80% MeOH containing 0.5% acetic acid. The analytical method established in this study showed a good linearity, sensitivity, specificity, and accuracy in determination of levels of the mycotoxins by LC/MS/MS. The recovery rates of the mycotoxins in rice were 73.9%–133.0% along with RSD_r of 0.1%–14.3%, which satisfied the legal limits of the recovery and RSD recommended by Codex or AOAC. The LODs of the analytical method for all of the mycotoxins were in the range of 0.1–18.5 μg/kg at a signal-to-noise (S/N) ratio of 3, and the LOQs of the method for the mycotoxins were in the range of 0.4–56.1 μg/kg at an S/N ratio of 10.

Finally, we investigated levels of 13 mycotoxins in 5 types of commercial cereal grains (brown rice, maize, millet, sorghum, and mixed cereal) collected from local markets in South Korea using the analytical method established in this study. The levels of DON and NIV in all types of cereal grains (except levels of DON in maize) and those of 3-AcDON, T-2, and HT-2 in cereal grains were below the maximum allowable legal limit (1 μg/g) set for DON or NIV by the KFDA. The levels of AFB_1 (except for mixed cereal), total AFs, FB_1, FB_2, ZEN (except for sorghum) and OTA in the cereal grains were also below the maximum allowable limit set by the KFDA in South Korea. Because levels of DON in maize, those of AFB_1 in mixed cereal, and those of ZEN in sorghum were higher than the maximum legal limits set by KFDA, extensive and active research should be continued for monitoring all 13 mycotoxins in cereal grains. Furthermore, establishment of legal limits of trichothecenes including NIV, 3-AcDON, T-2, and HT-2 for grains marketed in South Korea is required for monitoring them.

5. Experimental Sections

5.1. Samples

Brown rice, millet, sorghum, maize, and mixed cereal were used for the development of the analytical method and the determination of levels of 13 mycotoxins in these grains. The 5 types of 507 cereal grains were purchased from retail markets in South Korea. The grain samples were stored at 4 °C until use.

5.2. Standard Solutions and Reagents

The standards of Aflatoxins Mix 1 (2 μg/mL for AFB_1 and AFG_1, and 0.5 μg/mL for AFB_2 and AFG_2), OTA (10 μg/mL), Fumonisin Mix 3 (50 μg/mL for FB_1 and FB_2), DON (100 μg/mL), NIV (100 μg/mL), and 3-AcDON (100 μg/mL) were obtained from Biopure (Cambridge, MA, USA), while those of ZEN (100 μg/mL), T-2 (100 μg/mL), and HT-2 (100 μg/mL) were obtained from Sigma-Aldrich (St. Louis, MO, USA). The standard of 3 TCs mixture (DON, NIV, and 3-AcDON) was prepared by mixing 100 μL of each TC (101.9 μg/mL of DON, 100.9 μg/mL of NIV, and 100.6 μg/mL of 3-AcDON, which were diluted with acetonitrile [ACN]) with 490 μL of ACN, followed by dilution of the mixture with 410 μL of MeOH. The phosphate buffered saline (PBS, pH 7.4) for pH adjustment of toxin extracts was purchased from Sigma-Aldrich (St. Louis, MO, USA). HPLC grade ACN and MeOH were obtained from Merck (Darmstart, Germany). Stock solutions for each toxin were prepared by dilution of the standard solutions with 100% ACN (except for FB_1 and FB_2 for which 50% ACN was used), and working solutions with a series of toxin concentrations were made by dilution with 50% MeOH containing 1% acetic acid.

5.3. Assessment of the Precision, Linearity, and Sensitivity of the Analytical Method forDetermination ofLevels of 13 Mycotoxins

The linearity of a series of concentrations of 13 toxins in the analytical method was assessed by each standard curve using five levels (1.325, 2.65, 13.25, 26.5, and 53 ng/mL) of standard solutions for each toxin. The calibration curve for each toxin was constructed by plotting the peak areas (*y* axis) versus concentrations of each toxin (*x* axis) in LC/MS/MS analyses (Figure S1).

Five types of cereal grains that were not naturally contaminated with toxins were spiked with a mixed standard solution including 13 mycotoxins to determine the precision of the analytical method at levels of the toxins as follows: NIV, 321.0 ng/g; DON, 326.1 ng/g; 3-AcDON, 321.9 ng/g; AFB_1, 8.5 ng/g; AFB_2, 8.5 ng/g; AFG_1, 8.2 ng/g; AFG_2, 8.0 ng/g; FB_1, 163.2 ng/g; FB_2, 160.3 ng/g; HT-2,

161.6 ng/g, T-2, 64.6 ng/g; ZEN, 64.3 ng/g; OTA, 16.1 ng/g. Extraction and clean-up of analytes from the spiked samples were performed in triplicate by the procedures described below.

The sensitivity of the analytical method using LC/MS/MS was determined by LOD and LOQ. They were calculated as signal-to-noise (S/N) ratios of 3 and 10, respectively, which were determined by using HPLC software (Analyst 1.6 software program, Sciex, Framingham, MA, USA).

5.4. Toxin Extraction Procedure

Each sample (12.5 g) was weighed and placed in 100 mL of Erlenmeyer flasks after being ground into powders by a food grinder (Hallde, KISTA, Sweden). Fifty milliliters of ACN containing acetic acid (ACN: water: acetic acid = 79.5:20.0:0.5; v:v:v) as a selected solvent was added to it, and the 13 toxins were extracted by shaking at 320 rpm for 1 h with a wrist action shaker (EYELA, Tokyo, Japan). After the extracts were centrifuged at $3000 \times g$ for 5 min under 4 °C, supernatants were filtered through Whatman No. 4 filter paper (Whatman^{TM}, Maidstone, UK). Five milliliters of each filtrate was diluted with 75 mL of PBS (pH 7.4) and then filtered through a GF/A filter paper (Whatman^{TM}, Maidstone, UK). Sixty-five milliliters of the filtrate was loaded onto an IAC (Myco6in1+ column, VICAM, Milford, MA, USA) and passed through at a flow rate of one—two drops/sec. The column was washed with 10 ml of PBS and distilled water until 2–3 mL of air passed through it, and toxins were finally eluted from the column with 5 mL of 80% MeOH containing 0.5% acetic acid. The eluates were evaporated to dryness under a stream of N_2 at 50 °C using a vacuum manifold (Agilent, Santa Clara, CA, USA), and the residues were re-dissolved in 1 mL of 50% MeOH containing 0.5% acetic acid. The solutions were vortexed for 1 min and filtered through a 0.22 μm syringe filter.

5.5. Matrix Effects on Toxin Extraction

After extraction of matrix components from the 5 types of cereal grains by the same procedure as the toxin extraction method described above, 265 ng/mL of each toxin standard solution was mixed with the extract from each type of grain sample at a ratio (*v:v*) of 4 to 1 to make 53 ng/mL of each toxin standard solution containing the extracted matrix components. Then, a series of five levels (1.325, 2.65, 13.25, 26.5, and 53 ng/mL) of standard solutions for each toxin were prepared by serial dilution of 53 ng/mL of the standard solution with the extract containing matrix components. After LC/MS/MS analyses, each calibration curve for each toxin was constructed as described above. The matrix-matched calibration curves were used for calculation of recovery rates of 13 mycotoxins from 5 types of matrices.

5.6. LC/MS/MS Conditions

HPLC (1260 series, Agilent, Santa Clara, CA, USA) equipped with a AB SCIEX QTRAP mass spectrometer (AB 3200, Applied Biosystems, Foster city, CA, USA) was used to detect the 13 toxins. Separation of the toxins was carried out on a Scherzo SM-C18 column (3 mm × 150 mm, 3 μm particle size; Imtakt, Kyoto, Japan). The two elution solutions used were (A) 0.1% formic acid in water containing 2 mM ammonium acetate and (B) 0.1% formic acid in MeOH containing 2 mM ammonium acetate. The solutions were pumped at a flow rate of 0.5 mL/min and a gradient elution program was applied as shown in Table 2. The injection volume of the samples was 10 μL. Analysis software (version 1.6, Sciex, Framingham, MA, USA) was used to control the LC/MS/MS system and to acquire and process data. The mass spectrometer was operated in the positive ESI (electrospray ionization) mode with MRM (multiple reaction monitoring) at unit resolution. The main MS parameters were optimized and finally set as follows: curtain gas, 20 psi; collision gas (CAD), medium; capillary temperature (TEM), 500 °C; ion spray voltage, ± 4500 V; ion source gas 1 (GS1), 50 psi; ion source gas 2 (GS2), 50 psi; interface heater (ihe), on. Nitrogen was used as the nebulizer, heater, curtain, and collision gas.

Table 2. A gradient condition of a mobile phase composed of two solutions in analyses of multi-mycotoxins by HPLC.

Total Time (min)	Flow Rate (mL/min)	A Solution (%)	B Solution (%)
0	0.5	70	30
3	0.5	70	30
13	0.5	10	90
16	0.5	10	90
18	0.5	70	30
20	0.5	70	30

MRM parameters for detection of 13 mycotoxins by the mass spectrometer are shown in Table 3.

Table 3. Multiple reaction monitoring (MRM) parameters for detection of 13 mycotoxins by the mass spectrometer.

Mycotoxin	Precursor ion	Q1 (*m/z*)	Q3 (*m/z*)	Time (msec)	DP (volts)	EP (volts)	CE (volts)	CXP (volts)
NIV	$[M + CH_3COO^-]^-$	313	175	100	41	5	19	6
		313	295	100	41	5	15	6
DON	$[M + CH_3COO^-]^-$	297.2	249.2	100	40	7.5	12	4
		297.2	231.2	100	40	7.5	12	4
3-AcDON	$[M + CH_3COO^-]^-$	339	231	100	15	3.5	15	2
		339	213	100	15	3.5	15	2
AFB$_1$	$[M + H]^+$	313	285.1	100	70	11	51	4
		313	115	100	70	11	89	4
AFB$_2$	$[M + H]^+$	313	287.1	100	70	10	50	3
		313	259.1	100	70	10	50	3
AFG$_1$	$[M + H]^+$	329	243.0	100	70	11	25	4
		329	215.1	100	70	10	50	4
AFG$_2$	$[M + H]^+$	331	189	100	70	10.0	50	3
		331	257	100	70	10.5	47	4
FB$_1$	$[M + H]^+$	722.4	334.3	100	91	8	53	6
		722.4	352.4	100	91	8	43	6
FB$_2$	$[M + H]^+$	706.4	336.4	100	70	9	45	6
		706.4	318.2	100	70	9	47	6
HT-2	$[M + NH_4]^+$	442	263	100	26	3	19	4
		442	215	100	26	3	19	4
T-2	$[M + NH_4]^+$	484.1	215.2	100	21	6	29	4
		484.1	185.2	100	21	6	31	4
ZEN	$[M - H]^-$	317.1	131	100	−50	−4.5	−40	−2
		317.1	175	100	−50	−4.5	−34	−2
OTA	$[M + H]^+$	404	239	100	41	5	31	4
		404	220.9	100	41	5	19	4

Supplementary Materials: The following are available online at www.mdpi.com/2072-6651/9/3/106/s1, Figure S1: Calibration curves of 13 mycotoxins, Table S1: LODs and LOQs of 13 mycotoxins in 5 types of cereal grains.

Acknowledgments: This work was supported by the National Agricultural Products Quality Management Service and the Chung-Ang University Research Scholarship Grants in 2015.

Author Contributions: D.-H.K. and S.H.C. conceived and designed the experiments; D.-H.K., S.-Y.H., and J.W.K. performed the experiments; D.-H.K., S.-Y.H., J.W.K., S.M.C., K.R.L., T.K.A., and S.H.C. analyzed the data; S.-Y.H., C.L., and S.H.C. wrote the paper.

Conflicts of Interest: The authors declared no conflict of interests.

Abbreviations

The following abbreviations are used in this manuscript:

LC/MS/MS	Liquid Chromatography-Tandem Mass Spectrometry
AFs	aflatoxins
DON	deoxynivalenol
3-AcDON	3-Acetyldeoxynivalenol
OTA	ochratoxin A
Fs	fumonisins
NIV	nivalenol
ZEN	zearalenone
T-2	T-2 toxin
HT-2	HT-2 toxin
TCs	Trichothecenes
TIC	Total Ion Chromatogram
EIC	Extract Ion Chromatogram
RSD$_r$	relative standard deviation
LOD	Limit of Detection
LOQ	Limit of Quantification
ACN	acetonitrile
MeOH	methanol
MRM	multiple reaction monitoring

References

1. Binder, E.M.; Tan, L.M.; Chin, L.J.; Handl, J.; Richard, J. Worldwide occurrence of mycotoxins in commodities, feeds and feed ingredients. *Anim. Feed Sci. Technol.* **2007**, *137*, 265–282. [CrossRef]
2. IARC. *Monograph on the Evaluation of Carcinogenic Risk to Humans*; World Health Organization: Lyon, France, 2002; pp. 171–175.
3. Park, J.W.; Chung, S.H.; Kim, Y.B. Ochratoxin a in Korean food commodities: Occurrence and safety evaluation. *J. Agric. Food Chem.* **2005**, *53*, 4637–4642. [CrossRef] [PubMed]
4. Ok, H.E.; Kim, H.J.; Cho, T.Y.; Oh, K.S.; Chun, H.S. Determination of deoxynivalenol in cereal-based foods and estimation of dietary exposure. *J. Toxicol. Environ. Health A* **2009**, *72*, 1424–1430. [CrossRef] [PubMed]
5. Boeira, L.S.; Bryce, J.H.; Stewart, G.G.; Flannigan, B. The effect of combinations of *Fusarium* mycotoxins (deoxynivalenol, zearalenone and fumonisin b1) on growth of brewing yeasts. *J. Appl. Microbiol.* **2000**, *88*, 388–403. [CrossRef] [PubMed]
6. Grenier, B.; Loureiro-Bracarense, A.P.; Lucioli, J.; Pacheco, G.D.; Cossalter, A.M.; Moll, W.D.; Schatzmayr, G.; Oswald, I.P. Individual and combined effects of subclinical doses of deoxynivalenol and fumonisins in piglets. *Mol. Nutr. Food Res.* **2011**, *55*, 761–771. [CrossRef] [PubMed]
7. Harvey, R.B.; Edrington, T.S.; Kubena, L.F.; Elissalde, M.H.; Rottinghaus, G.E. Influence of aflatoxin and fumonisin b1-containing culture material on growing barrows. *Am. J. Vet. Res.* **1995**, *56*, 1668–1672. [PubMed]
8. Verma, J.; Johri, T.S.; Swain, B.K.; Ameena, S. Effect of graded levels of aflatoxin, ochratoxin and their combinations on the performance and immune response of broilers. *Br. Poult. Sci.* **2004**, *45*, 512–518. [CrossRef] [PubMed]
9. Ruiz, M.J.; Macakova, P.; Juan-Garcia, A.; Font, G. Cytotoxic effects of mycotoxin combinations in mammalian kidney cells. *Food Chem. Toxicol.* **2011**, *49*, 2718–2724. [CrossRef] [PubMed]
10. Tammer, B.; Lehmann, I.; Nieber, K.; Altenburger, R. Combined effects of mycotoxin mixtures on human t cell function. *Toxicol. Lett.* **2007**, *170*, 124–133. [CrossRef] [PubMed]
11. Creppy, E.E.; Chiarappa, P.; Baudrimont, I.; Borracci, P.; Moukha, S.; Carratu, M.R. Synergistic effects of fumonisin B1 and ochratoxin A: Are in vitro cytotoxicity data predictive of in vivo acute toxicity? *Toxicology* **2004**, *201*, 115–123. [CrossRef] [PubMed]
12. Klaric, M.S.; Rumora, L.; Ljubanovic, D.; Pepeljnjak, S. Cytotoxicity and apoptosis induced by fumonisin B1, beauvericin and ochratoxin a in porcine kidney pk15 cells: Effects of individual and combined treatment. *Arch. Toxicol.* **2008**, *82*, 247–255. [CrossRef] [PubMed]

13. Vaclavikova, M.; MacMahon, S.; Zhang, K.; Begley, T.H. Application of single immunoaffinity clean up for simultaneous determination of regulated mycotoxins in cereals and nuts. *Talanta* **2013**, *117*, 345–351. [CrossRef] [PubMed]

14. Lattanzio, V.M.; Ciasca, B.; Powers, S.; Visconti, A. Improved method for the simultaneous determination of aflatoxins, ochratoxin A and *Fusarium* toxins in cereals and derived products by liquid chromatography-tandem mass spectrometry after multi-toxin immunoaffinity clean up. *J. Chromatogr. A* **2014**, *1354*, 139–143. [CrossRef] [PubMed]

15. Lattanzio, V.M.; Solfrizzo, M.; Powers, S.; Visconti, A. Simultaneous determination of aflatoxins, ochratoxin A and *Fusarium* toxins in maize by liquid chromatography/tandem mass spectrometry after multitoxin immunoaffinity cleanup. *Rapid Commun. Mass Spectrom.* **2007**, *21*, 3253–3261. [CrossRef] [PubMed]

16. AOAC. *Sub-Committee on Feed Additives and Contaminants*; AOAC: Rockville, MD, USA, 2008; pp. 41–42.

17. Food and Agriculture Organization. *Codex General Standard for Contaminants and Toxins in Food And Feed (Codex Stan 193–1995)*; Food and Agriculture Organization: Geneva, Switzerland, 1995.

18. KFDA. *Food Standards Codex*; Korea Foods Industry Association: Seoul, Korea, 2010; pp. 10–16.

19. Park, J.W.; Kim, E.K.; Shon, D.H.; Kim, Y.B. Natural co-occurrence of aflatoxin B1, fumonisin B1 and ochratoxin A in barley and corn foods from Korea. *Food Addit. Contam.* **2002**, *19*, 1073–1080. [CrossRef] [PubMed]

20. Seo, E.; Yoon, Y.; Kim, K.; Shim, W.B.; Kuzmina, N.; Oh, K.S.; Lee, J.O.; Kim, D.S.; Suh, J.; Lee, S.H.; et al. Fumonisins B1 and B2 in agricultural products consumed in South Korea: An exposure assessment. *J. Food Prot.* **2009**, *72*, 436–440. [CrossRef] [PubMed]

21. Kim, J.C.; Kang, H.J.; Lee, D.H.; Lee, Y.W.; Yoshizawa, T. Natural occurrence of *Fusarium* mycotoxins (trichothecenes and zearalenone) in barley and corn in Korea. *Appl. Environ. Microbiol.* **1993**, *59*, 3798–3802. [PubMed]

22. Ok, H.E.; Chang, H.J.; Choi, S.W.; Lee, N.; Kim, H.J.; Koo, M.S.; Chun, H.S. Co-occurrence of deoxynivalenol and zearalenone in cereals and their products. *J. Food Hyg. Saf.* **2007**, *22*, 375–381.

23. Eskola, M.; Rizzo, A.; Soupas, L. Occurrence and amounts of some *Fusarium* toxins in finnish cereal samples in 1998. *Acta Agric. Scand.* **2001**, *50*, 183–186.

24. Ali, N.; Sardjono; Yamashita, A.; Yoshizawa, T. Natural co-occurrence of aflatoxins and *Fusarium* mycotoxins (fumonisins, deoxynivalenol, nivalenol and zearalenone) in corn from Indonesia. *Food Addit. Contam.* **1998**, *15*, 377–384. [CrossRef] [PubMed]

© 2017 by the authors. Licensee MDPI, Basel, Switzerland. This article is an open access article distributed under the terms and conditions of the Creative Commons Attribution (CC BY) license (http://creativecommons.org/licenses/by/4.0/).

toxins

MDPI

Article

A Greener, Quick and Comprehensive Extraction Approach for LC-MS of Multiple Mycotoxins

Andreas Breidbach

Andreas Breidbach, European Commission—Joint Research Centre—Directorate F—Health, Consumers, and Reference Materials, Retieseweg 111, B-2440 Geel, Belgium; Andreas.breidbach@ec.europa.eu; Tel.: +32-14-571-205

Academic Editor: Aldo Laganà
Received: 12 January 2017; Accepted: 1 March 2017; Published: 7 March 2017

Abstract: In food/feed control, mycotoxin analysis is often still performed "one analyte at a time". Here a method is presented which aims at making mycotoxin analysis environmentally friendlier through replacing acetonitrile by ethyl acetate and reducing chemical waste production by analyzing four mycotoxins together, forgoing sample extract clean-up, and minimizing solvent consumption. For this, 2 g of test material were suspended in 8 mL water and 16 mL ethyl acetate were added. Extraction was accelerated through sonication for 30 min and subsequent addition of 8 g sodium sulfate. After centrifugation, 500 μL supernatant were spiked with isotopologues, dried down, reconstituted in mobile phase, and measured with LC-MS. The method was validated in-house and through a collaborative study and the performance was fit-for-purpose. Repeatability relative standard deviation (RSDs) between 16% at low and 4% at higher contaminations were obtained. The reproducibility RSDs were mostly between 12% and 32%. The trueness of results for T-2 toxin and Zearalenone were not different from 100%, for Deoxynivalenol and HT-2 toxin they were larger than 89%. The extraction was also adapted to a quick screening of Aflatoxin B1 in maize by flow-injection–mass spectrometry. Semi-quantitative results were obtained through standard addition and scan-based ion ratio calculations. The method proved to be a viable greener and quicker alternative to existing methods.

Keywords: mycotoxins; LC-MS; green analytical chemistry

---

1. Introduction

Mycotoxins are toxic secondary metabolites of certain moulds whose occurrence in food and feed is difficult to avoid. They present a potential risk to the health of consumers. Therefore, many countries have regulated the occurrence of mycotoxins [1,2]. A wealth of methods of analysis exists to enforce these regulations. Many of these methods rely on mixtures of acetonitrile (ACN)/water for extraction of the mycotoxins from the solid food/feed matrix [3]. To minimize sub-sampling uncertainties, test portion sizes of 25–50 g are not uncommon, which are then extracted with tens or hundreds of mL of extraction solvent [4]. In the end, only an aliquot of the extract is cleaned up and actually used for the determination. The rest is chemical waste needing proper disposal.

A compounding factor for the waste problem is the need to prepare the test material several times to determine several mycotoxins in the "classical" way. Classical approaches in mycotoxin analysis make use of immuno-affinity clean-up to obtain the necessary selectivity prior to separation/detection with HPLC with ultraviolet (UV) or fluorescence detector. These methods are mostly used for single or, at most, very few, very similar analytes.

"Green Analytical Chemistry" (GAC) is a rapidly developing trend that is rooted in the desire to make chemical analysis environmentally friendlier. Key principles are, amongst others, chemical waste reduction and the use of "safer" solvents [5]. One way of reducing the amount of chemical waste produced during food/feed analysis is certainly decreasing test portion size. With a smaller test portion, the volume of extraction solvent can be reduced while maintaining a high solvent/sample ratio beneficial for good extraction yields. This requires a higher effort in homogenizing the sample which is delivered to the laboratory to avoid sub-sampling uncertainties which might exceed the measurement uncertainty. Sample materials milled to particle sizes smaller than 500 μm and thoroughly mixed usually fulfil this requirement. Another approach would be to prepare aqueous slurries. For this, the test material and a certain amount of water are high-speed blended to create a well-mixed dispersion of very small particles. Such slurries display very little heterogeneity and thus minimize the sub-sampling uncertainty as shown by Spanjer et al. for 10 kg samples [6]. Preparing slurries of 10–50 g of test material and then using a test portion which is the slurry equivalent of 1 or 2 g of test material would be a compromise. The limited amount of unused slurry consisting of water and test material is easy to dispose of and the small aliquot of slurry which is extracted only requires a relatively small volume of organic solvent. Yet the sub-sample size is that of older, proven methods.

Increasing the number of analytes determined per analysis from the same test material preparation also reduces the generation of chemical waste. In mycotoxin analysis, modern generation LC-MS with high sensitivity and selectivity enable this [7] and the selectivity of LC-MS also allows the analyst to forgo sample extract clean-up, which eliminates another waste-producing step [8,9]. Another advantage of LC-MS is its preference for lower flow rates. While in the classical HPLC analysis, analytical columns of 4.6 mm i.d. at flow rates of 1 mL/min are used; LC-MS, with the most commonly used electrospray ionization (ESI), works better at lower flow rates. Therefore, analytical columns of 2.1 mm i.d. are being used which can reduce the mobile phase consumption by a factor of five or more.

The other principle of GAC listed above was the use of "safer" solvents. Mixtures of ACN/water are still the preferred extraction solvent in mycotoxin analysis. In a proficiency test about the determination of multiple mycotoxins in cereals executed in 2013 by the European Reference Laboratory (EURL) for Mycotoxins, close to 70 laboratories participated [10]. Of those, only 22 laboratories used a multi-mycotoxin method of analysis and 17 of the 22 used an ACN/water mixture for extraction. A "safer" alternative is ethyl acetate (EtOAc). It is less toxic, readily available and less expensive. Of course, to be useful as extraction solvent it needs to provide sufficient solubility for the analytes. That it is not miscible with water is not a disadvantage for extraction.

In this paper, a method of analysis is presented for the simultaneous determination of the mycotoxins Deoxynivalenol (DON), HT-2 toxin, T-2 toxin, and Zearalenone (ZON) in unprocessed cereals. Extraction is accomplished with an EtOAc/water system. To limit EtOAc consumption, a small test portion size is used. The method has been thoroughly validated in-house and through a collaborative trial and shown to perform well. The logarithm of the partition coefficient (octanol-water; log P_{OW}) of the mycotoxins listed above ranges from −0.7 (DON) to 3.6 (ZON). It is safe to assume that for other mycotoxins with log P_{OW} within the stated range, the extraction will work as well. To display the flexibility of the extraction approach the results of a quick screening method for Aflatoxin B1 (AFB1) in maize based on slight modifications of this extraction approach and flow injection-mass spectrometry (FI-MS) are also presented.

2. Results & Discussion

2.1. LC-MS Multi-Mycotoxin Method

The aim of this method of analysis is to be a tool for enforcement of existing (DON, ZON) [2] and tentative (HT-2, T-2) [11] legal limits in unprocessed cereals. At the same time, it should be quick, easy to apply and be a "greener" alternative to existing methods. The results of the in-house validation and the collaborative study will be presented and discussed together below. The collaborative study data represents results from 21 laboratories which have analyzed ten coded samples (five different materials as blind duplicates). Full details of the in-house [12] and the collaborative study [13] validation are available online.

2.1.1. Isotopologues as Internal Standard (ISTD)/Matrix Effects

The use of isotopologues in MS analyses is called isotope dilution mass spectrometry (IDMS) and its utility for accurate determinations of organic substances has been recognized in the late '80s of the last century [14]. IDMS is a method of analysis of very high metrological order and enables exceptional accuracies. Yet true IDMS requires addition of the isotopologue to the test material and thorough equilibration with the analyte before extraction. This is difficult to achieve and control. Moreover, it can be prohibitively expensive. Those issues preclude this approach from being used in a routine analysis context.

We used a different approach in that we added the isotopologues to only an aliquot of the sample extract. Used in such a way, the isotopologues can only be employed for correction of matrix effects which are a major concern in LC-MS. Upstream effects like extraction efficiency and/or losses during transfer or dilution steps are not covered. Yet if the extraction conditions are under good control then this mode of usage offers the benefit of cancellation of matrix effects at very reasonable expenses [9].

For determination of the matrix effects, analyte-free materials were extracted and the extracts were spiked at four different levels with the analytes and at constant levels with the isotopologue ISTD mix [15]. After determination of the ion ratios, the matrix effects (ME) were calculated with the following formula:

$$ME = B/C, \tag{1}$$

with B = slope of regression fit of ion ratios of analyte added to extracts of analyte-free material, C = slope of regression fit of ion ratios of analyte in neat solvent. A value of 1 indicates no matrix effect, values > 1 indicate ion enhancement, values < 1, ion suppression. For the four analytes in the six tested materials, MEs between 0.81 and 1.04 were calculated. Considering the uncertainties of the modelling and the pipetting steps, none of these values were significant with the exception of the values in oat. Also, those are still within an acceptable range. This is evidence of the validity of the approach to add the isotopologues after extraction.

2.1.2. Recovery/Trueness

The recovery (Rec), the ratio of observed content to expected, of this method of analysis was determined following a similar scheme as above. The six tested analyte-free materials were spiked with the four analytes at four different levels before extraction. All of this was done in duplicate, resulting in 48 samples. After sufficient equilibration (data not shown), these samples were extracted and ion ratios were determined. The method recovery was then calculated as:

$$Rec = A/C, \tag{2}$$

with A = slope of regression fit of ion ratios of analyte added to analyte-free material, C = slope of regression fit of ion ratios of analyte in neat solvent. Recoveries between 0.79 and 1.07 were calculated. This is well within accepted ranges [16].

Of more importance is the trueness of a method of analysis. While recovery can be estimated from spiked materials with the spiked amount being the expected amount, the determination of trueness necessitates the knowledge of a "true" value. This true value can be the certified value of a Certified Reference Material (CRM) or a value determined with a reference method [17].

To this end, for the collaborative study two test materials for which reference values were determined by exact-matching double IDMS (EMD-IDMS) were included. The process of EMD-IDMS is described in detail by Breidbach et al. [18] and in the study report [13]. Table 1 displays the results of the determination of the two reference materials (RM). The bias between the study results and the assigned values is only significant (confidence interval does not include zero) for DON and HT-2 and amounts to −11% in RM I and −8% in RM II for DON, and −11% in RM II for HT-2. This demonstrates better than recovery experiments the small to negligible systematic error achievable by the combination of EtOAc as extraction solvent, isotopologues as surrogate ISTD and LC-MS for detection.

Table 1. The assigned reference values from EMD-IDMS and the collaborative study results.

Analyte	Assigned Value		Study Result					
	xa (µg/kg)	$u(xa)$ [1] (µg/kg)	Overall Mean (µg/kg)	sR [2] (µg/kg)	(µg/kg) [3]	A [4]	$-A\ sR$ [5] (µg/kg)	$+A\ sR$ [6] (µg/kg)
			Reference Material I (EFL2)					
DON	282	13	250	33	−32	0.47	−47	−16
HT-2	51	3	49	12	−2	0.48	−8	4
T-2	18	1	18	4	0	0.47	−2	2
ZON	28	2	30	6	2	0.46	−1	5
			Reference Material II (EFL3)					
DON	605	24	559	67	−46	0.46	−77	−15
HT-2	201	7	178	23	−23	0.45	−34	−13
T-2	52	2	50	6	−2	0.46	−5	1
ZON	445	8	430	49	−15	0.46	−38	7

[1] combined standard uncertainty of the assigned value; [2] reproducibility standard deviation; [3] bias (overall mean—assigned value); [4] factor for ca. 95% confidence interval; [5] lower limit 95% confidence interval around bias; [6] upper limit 95% confidence interval around bias.

2.1.3. Precision

The variability in the results of a method of analysis, its precision, can be determined under two extreme conditions: repeatability and reproducibility. Repeatability conditions describe the minimum variability inherent in such a method determined through repeated measures in a short span of time keeping other contributing factors, such as operator, instrumentation, calibration, etc., constant. Maximum variability will be recorded under reproducibility conditions when all those factors are varied during a collaborative study. The between-laboratory variability adds then onto the repeatability variability.

The repeatability standard deviations were determined in-house through analysis of three different naturally-contaminated test materials (designated as EFL1, EFL2, and EFL3 in Table 2) 20 times each. For DON, the relative repeatability standard deviations (RSD_r) ranged from 15% at 73 µg/kg to 6% at 427 µg/kg. The ranges for HT-2 were from 11% at 27 µg/kg to 6% at 174 µg/kg; for T-2, from 16% at 7 µg/kg to 6% at 45 µg/kg; and for ZON, from 18% at 6 µg/kg to 4% at 483 µg/kg.

The results for the collaborative study are listed in Table 2. It is apparent that the RSD_r values are in line with the values found in-house with three exceptions: T-2 with values of 27% and 35%, and ZON with 32%. All three exceptions are connected to very low contamination levels. A similar picture unfolds for the relative reproducibility standard deviations (RSD_R). For most of the analyte/matrix combinations, the values are at acceptable levels except for T-2 with 44% and 88%, and ZON with 98% and 65%. The two T-2 values and the ZON value of 98% are related to the same low contamination levels as mentioned above.

The second ZON value of 65% in material IRMMFEED is apparently not related to the contamination level but to the complexity of the matrix. There is a second material (EFL1) with a very similar contamination level for which the RSD_R is less than half at 31%. That for both materials the RSD_r is very similar points to a large between-laboratory variability for IRMMFEED. The higher matrix complexity of IRMMFEED was a challenge for the separation capabilities. Based on prescribed resolution requirements, the results of the laboratories could be separated into two groups: those for which the requirements were met and those for which they were not. For test material EFL1, there was no significant difference between the overall means of the two groups; for test material IRMMFEED there was. The existence of two distinct groups explains the much higher between-laboratory variability and underlines the need for proper chromatographic separation.

Table 2. Results of the collaborative study.

Material [1]	Overall Mean [2] (µg/kg)	s_r [3] (µg/kg)	RSD$_r$ (%)	s_R [4] (µg/kg)	RSD$_R$ (%)
		DON			
EFL1	88.5	9.5	11	17.0	19
EFL2	250.0	13.6	6	33.3	13
EFL3	558.6	30.1	5	66.9	12
IRMMCER	135.8	8.2	6	23.0	17
IRMMFEED	281.8	19.9	7	33.1	12
		HT-2			
EFL1	38.0	3.4	9	6.2	16
EFL2	49.1	3.4	7	12.0	25
EFL3	177.6	13.5	8	23.2	13
IRMMCER	53.1	8.1	15	12.4	24
IRMMFEED	22.0	3.3	15	6.3	29
		T-2			
EFL1	12.1	1.7	14	3.9	32
EFL2	17.7	1.6	9	4.4	25
EFL3	50.3	3.1	6	6.5	13
IRMMCER	7.0	1.8	27	3.1	44
IRMMFEED	3.5	1.2	35	3.1	88
		ZON			
EFL1	13.9	2.0	15	4.3	31
EFL2	30.5	2.9	10	6.0	20
EFL3	430.0	25.0	6	49.3	12
IRMMCER	3.4	1.1	32	3.3	98
IRMMFEED	15.9	1.7	11	10.4	65

[1] more information about the materials in [13]; [2] average of all retained lab results; [3] repeatability standard deviation; [4] reproducibility standard deviation.

2.1.4. Other Method Validation Parameters

In-house, the following other validation parameters were determined: for selectivity, the six materials, which were free of all analytes, were extracted as is and after spiking with the analytes at the lowest calibration level. Analyte peaks were detected in none of the chromatograms of the unspiked and in all of the spiked materials. During calibration experiments at six equidistant levels within the working ranges (in µg/kg: DON 200–2560, HT-2 25–400, T-2 15–240, ZON 50–240), performed on six different days, no deviations from linearity were found.

Limits of detection (LOD) were determined from the data of the recovery experiments based on ISO 11843 Part 2 [19]. Since the estimated LODs (in µg/kg: DON < 47, HT-2 < 9, T-2 < 3, ZON < 5) were much smaller than the targeted working ranges, no more resources were invested in determining more precise LOD values. The same was the case for the determination of the limit of quantification (LOQ). These values were also estimated from the recovery data. Since almost all LOQs, with the exception of the value for DON in maize, were below the targeted working range, no more resources were invested. The LOQ for DON in maize was 340 µg/kg and almost twice as large as the next lower value (190 µg/kg in rice). The reason was an inconsistency in the recovery data for maize. Since the legislative limit for DON in unprocessed maize is 1750 µg/kg, even a "true" LOQ of 340 µg/kg would not pose a problem. Figure 1 shows a total ion current (TIC) chromatogram of the separation of a cereal mix with a fractional content of 23% maize. The peaks of the four analytes are well shaped and resolved with a total run time of 8.7 min. Figure 2 depicts the extracted ion current (EIC) chromatograms of the four analytes and their respective isotopologues of the same run as in Figure 1. The top chromatogram in panel (a) of Figure 2 shows the EIC of DON at a contamination level of about 90 µg/kg. Obviously, the signal-to-noise ratio is already very favorable at this level which indicates that the LOQ of 340 µg/kg is a very conservative estimate.

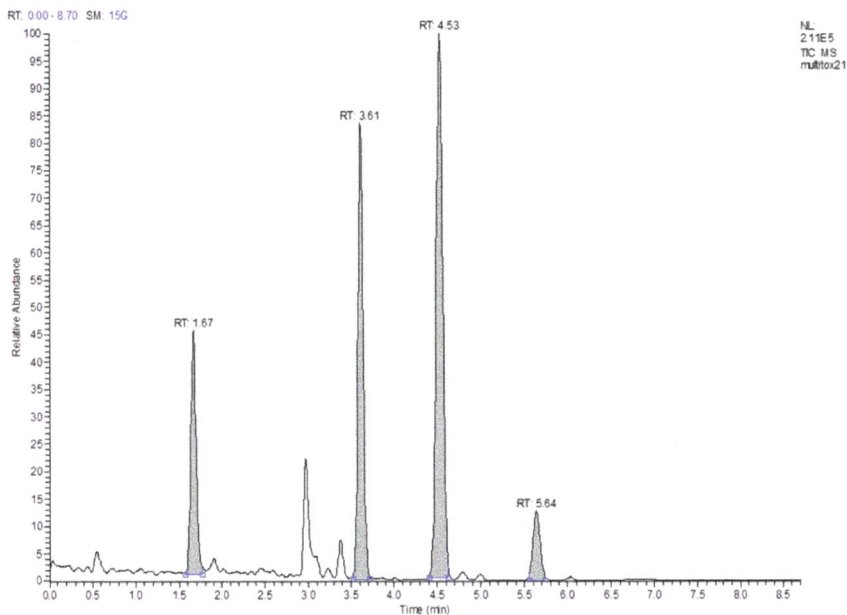

Figure 1. Total ion current chromatogram of a cereal mix with low contamination of DON, HT-2, T-2, and ZON; the majority of the peak signals stems from the isotopologues.

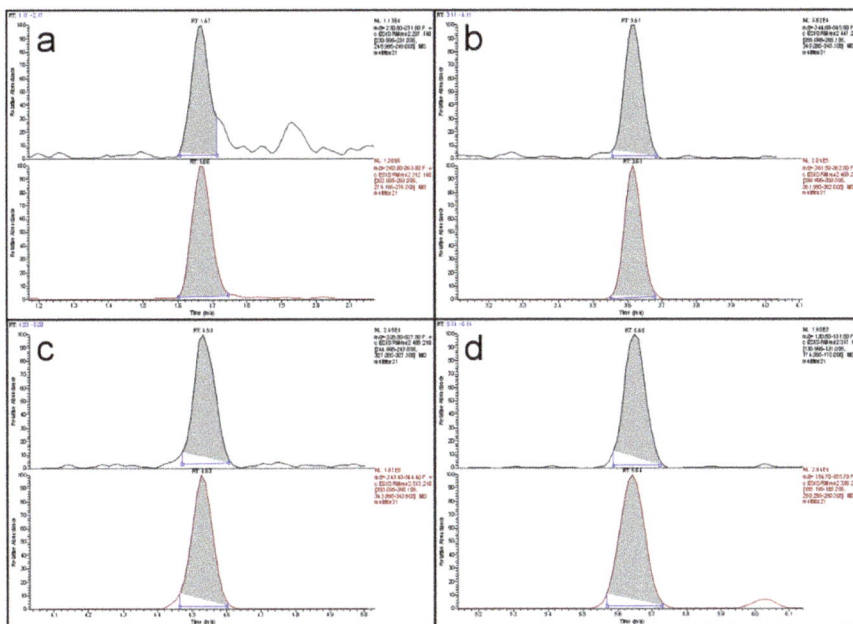

Figure 2. Extracted ion current chromatograms of: (**a**) DON; (**b**) HT-2; (**c**) T-2; and (**d**) ZON of a cereal mix with low contamination; in each panel the analyte (**top**) is depicted above its respective isotopologue (**bottom**).

2.2. Quick Screening Method

To investigate the distribution of AFB1 in about 200 kg of contaminated whole grain maize, a quick screening method for 10 g samples was needed. Since the EtOAc extraction approach worked so well, it was adapted and tested in this context. To keep overall solvent consumption low, the method should work with small test portions without generating a high sub-sampling variability. Therefore, the preparation of aqueous slurries of the 10 g samples was tested.

Nine test portions equivalent to 1 g of maize each out of a slurry prepared from 10 g contaminated whole grain maize were measured with a routine LC-MS method to assess the sub-sampling variability. The relative standard deviation (RSD) of results of the nine determinations was 3% based on the peak area of AFB1. This was seen as sufficiently small to continue with slurries and test portions of 1 g maize equivalent.

2.2.1. Extraction Efficiency

To speed up the extraction process, the sodium sulfate (Na_2SO_4), which was previously added after sonication in the LC-MS procedure, adding 10–20 min of crystallization time to the extraction, was now added before avoiding the additional wait. Whether this change influenced the extraction efficiency (EE) was determined following a similar scheme as described above. Out of a slurry of contaminated whole grain maize, four aliquots were spiked with AFB1 at two levels in duplicate before extraction and four were spiked at the same levels after extraction. The spiked and two unspiked extracts were then measured with LC-MS to obtain the peak areas of AFB1 and to calculate EE with the following formula:

$$EE = A/B, \tag{3}$$

with A = slope of regression fit of peak areas of AFB1 added before extraction and B = slope of regression fit of peak areas of AFB1 added after extraction. An extraction efficiency of 89% $\pm$ 5% (95% confidence) was calculated, which is in line with findings of Mol et al. [20] and shows that the early addition of the Na_2SO_4 has no significant negative effect.

2.2.2. Flow Injection-MS

While chromatographic separation before MS detection has many advantages and is strongly recommended for quantitative assessments, it is time consuming. One of those advantages is certainly the reduction of matrix effects because fewer compounds will coelute with the analyte. However, for a qualitative screening method, this can be sacrificed to speed up analysis even more.

Therefore, the test solutions were directly delivered to the MS by injecting them into the spray solvent flow. Since ionization efficiencies of electrospray improve at lower flow rates, and to save solvent, a flow rate of 20 µL/min was chosen. This led to broad, irregular elution profiles (see Figure 3) which could not be integrated automatically. To obtain a concentration-dependent response, a different approach was chosen: Scan-based ion ratios, which had shown to lead to exceptional accuracies in IDMS measurements [21], were utilized to automatically evaluate the flow injection measurements. For this, the ion ratio of the AFB1 signal over an ISTD signal (Caffeine was chosen as ISTD) was calculated as follows: for each scan event measuring the transition m/z 313->241 (AFB1) the signals of the preceding and the successive scan event measuring the transition m/z 195->138 (Caffeine) were averaged. The signal of m/z 313->241 was then divided by the average of m/z 195->138 to obtain the ion ratio for the respective scan. When plotting the signal of the ISTD, a broad peak with flat top could be observed (Figure 3). Averaging all ion ratios belonging to that flat-top region then provides a mean ion ratio which is a representation of the amount-dependent analyte response.

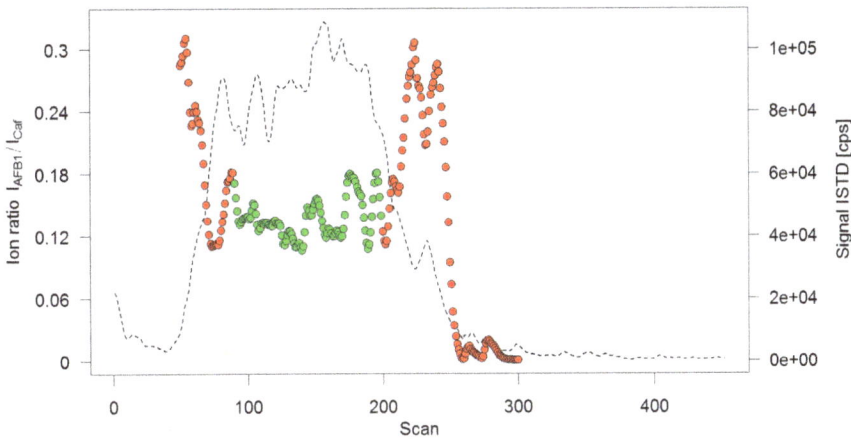

Figure 3. Ion ratio profile of a flow injection analysis. Solid circles depict the calculated ion ratios per scan, only the green ones have been retained for averaging. Superimposed is the chronogram of the internal standard (ISTD) caffeine (broken line).

2.2.3. AFB1 Distribution and Standard Addition

Panel (a) of Figure 4 shows the distribution of AFB1 over ten 10 g samples drawn at random from 10 kg of crushed maize and submitted to the flow injection analysis. The 10 kg were sampled from the 200 kg of contaminated whole grain maize according to the incremental sampling scheme (100 × 100 g) described in the corresponding European Union (EU) regulation [16] for large lots. Next to samples containing very little AFB1, there is one sample containing very much. This is evidence of the known heterogeneity of AFB1 contamination. To obtain an idea of the contamination levels, the slurries of samples 2, 6, and 10 were submitted to standard addition (STDADD) analysis.

The use of STDADD became necessary because of the uncontrolled and severe matrix effects that come along with the flow injection measurement. The additional effort was limited because several aliquots were available for extraction from each slurry. For STDADD experiments, six aliquots, representing 1 g of maize each, were used. Two were extracted in their native state, the other four were spiked with AFB1 at two levels in duplicate. For sample 2 (Panel (b), Figure 4), one medium and three high-level spikes were produced due to a pipetting error.

The calculated mass fractions of AFB1 in the three samples were: 4 (0–76) µg/kg in sample 2; 28 (16–40) µg/kg in sample 6; and 100 (59–165) µg/kg in sample 10. The values in parentheses represent the ~95% confidence interval around the calculated value. These intervals have been determined from the computed prediction intervals of the STDADD experiments. The confidence intervals around the predicted value are unsymmetrical by definition. The value zero in the interval of sample 2 shows that the calculated value is below the detection capabilities of the analytical method. The width of the intervals indicates the semi-quantitative nature of this approach. Yet it appears to be valuable as a tool to screen large numbers of samples. In this study, the focus was on AFB1 which is notoriously heterogeneously distributed and has low legal limits. This approach could easily be adapted to other mycotoxins, for instance fumonisins, and certainly DON, HT-2, T-2, and ZON. The use of 1% trifluoroacetic acid in the slurries boosts extraction efficiencies of fumonisins and prevents, due to the low pH, enzymatic activities which appear to convert T-2 toxin through deacetylation to HT-2 toxin.

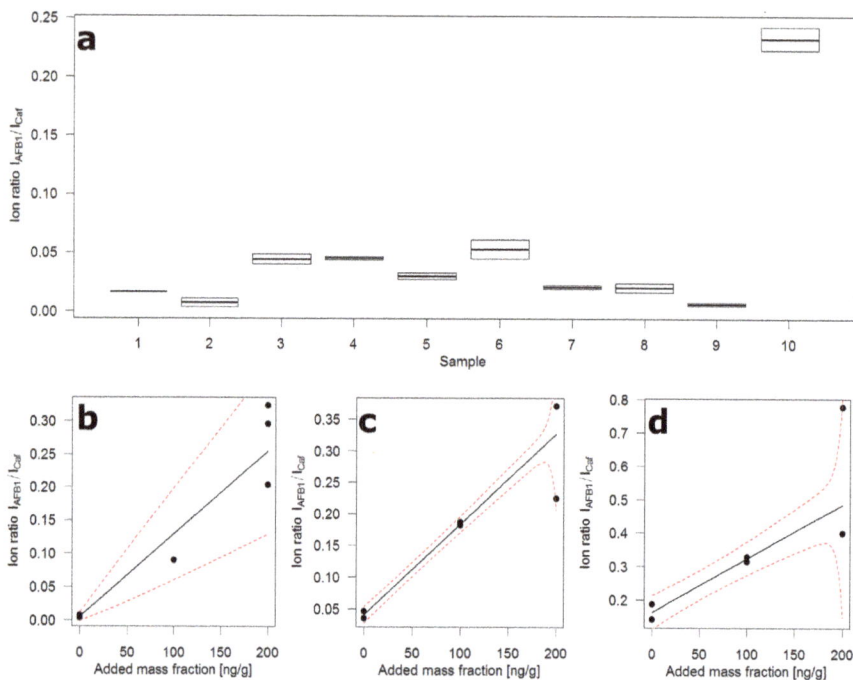

Figure 4. Ion ratio distribution over ten 10 g samples of contaminated crushed maize and standard addition plots. (**a**) Box plot of duplicate flow injections of ten random sub-samples, boxes extend to the two measured ion ratios, horizontal line depicts the mean of the two ion ratios; (**b**) Standard addition plot of sample 2 in (**a**); (**c**) Standard addition plot of sample 6 in (**a**); (**d**) Standard addition plot of sample 10 in (**a**); black circles—ion ratio (mean of two injections), black line—regression fit, red broken line—prediction interval.

3. Conclusions

In conclusion, it can be said that the presented approach is:

- greener than many existing approaches because it replaces ACN through EtOAc, minimizes the consumption of organic solvent, and produces less chemical waste;
- simple and quick because a time-consuming clean-up step is avoided and sample processing is easily scaled up or down;
- adaptable to other applications without too much effort;
- showing favorable performance characteristics and is fit-for-purpose as a tool to enforce EU mycotoxin legislation.

The fitness-for-purpose is also evidenced through the fact that the presented LC-MS method of analysis has been published by the European Committee for Standardization (CEN) as European standard method of analysis [22].

4. Materials and Methods

4.1. Chemicals and Materials

All chemicals were purchased from either Sigma-Aldrich (Overijse, Belgium) or VWR (Leuven, Belgium) and were of at least analytical grade. For mobile phases LC-MS CHROMASOLV grade water and methanol (MeOH), formic acid (FA), ammonium formate (NH$_4$FA) (all LC-MS grade, Fluka, Sigma-Aldrich), and LC-MS grade acetonitrile (ACN; VWR) were used. Deionised water was generated by a MilliQ system (Millipore, Overijse, Belgium). All tested materials came from the material pool of the EURL for Mycotoxins of the Joint Research Centre (JRC) of the European Commission (EC).

The mycotoxins Deoxynivalenol (DON), HT-2 toxin, T-2 toxin, Zearalenone (ZON), Aflatoxin B1 (AFB1), and the isotopologues $^{13}C_{15}$-DON, $^{13}C_{22}$-HT-2, $^{13}C_{24}$-T-2, and $^{13}C_{18}$-ZON were purchased from Biopure (Romer Labs, Getzersdorf, Austria) as ready-to-use solutions and were stored at the recommended temperatures.

From those solutions, a mixed stock solution of 3.2 µg/mL DON, 0.5 µg/mL HT-2 toxin, 0.3 µg/mL T-2 toxin, and 0.3 µg/mL ZON in acetonitrile was prepared and stored. This stock solution was freshly diluted for every calibration task. A mixed ISTD solution with the same concentrations of the respective ^{13}C-isotopologues in acetonitrile was also prepared and used undiluted. These two solutions were stable for at least three months in the dark between 2 °C and 8 °C.

For buffered mobile phases, an equimolar mix of NH$_4$FA and FA (pH 3.7), adjusted to 6 mol/L in water, was added to the solvents.

4.2. Sample Preparation for LC-MS

In a 50 mL conical screw-cap polyethylene centrifuge tube (VWR), 2 g of test material (comminuted to <500 µm particle size) were fully suspended in 8 mL of water. Then 16 mL ethyl acetate (EtOAc) was added and after a brief, hard shake, the mixture was sonicated for 30 min. After sonication, 8 g of Na$_2$SO$_4$ were added. The mixture was again shaken hard and then left for 10 to 20 min to allow the Na$_2$SO$_4$ to crystallize. To settle particulate matter and aid phase separation, the tube was centrifuged at a relative centrifugal force (RCF) of 3200 g for at least 1 min. 500 µL of the clear supernatant were transferred to a silylated auto sampler vial (2 mL, Supelco, Sigma-Aldrich), 25 µL of ISTD mix were added and the content of the vial was evaporated to dryness with a stream of dry nitrogen (boil-off) at 60 °C. The dry residue was reconstituted with 250 µL MeOH/FA (999/1, *v/v*) and 250 µL water/FA (999/1, *v/v*), in that order. Initial reconstitution with the pure organic mobile phase significantly improved the dissolution of the more hydrophobic analytes. Turbidity of the injection solutions, often seen in these reconstituted extracts, did not negatively affect column lifetime in our experience.

4.3. Sample Preparation for Quick Screening

Ten g of maize (whole grain or comminuted) were high-speed blended (Ultra Turrax, IKA-Werke, Germany) with 20 g water/trifluoroacetic acid (TFA) (99/1, *v/v*) to obtain a smooth, semi-viscous slurry; 3 g of that slurry (representing 1 g of test material) were transferred to a 50 mL conical screw-cap polyethylene centrifuge tube, and 8 mL ethyl acetate (EtOAc) and 4 g Na$_2$SO$_4$ were added. The tube was capped, shaken hard, and then sonicated for 30 min. Right after, the tube was centrifuged at RCF of 3200 g for 2 min. Five hundred µL of supernatant, together with 25 µL of a 400 µg/mL solution of caffeine, used as ISTD, were dried down in a silylated auto sampler vial with nitrogen as described above. The dry extract was reconstituted with 250 µL ACN/water/NH$_4$FA pH 3.7 (90/9/1, *v/v/v*) and 250 µL water/NH$_4$FA pH 3.7 (99/1, *v/v/v*).

To determine contamination levels, known amounts of AFB1 were added to aliquots of the slurry before extraction. After allowing for 60 min of equilibration between slurry and added analyte, these spiked standard addition samples were extracted as described above.

4.4. LC-MS Measurements

The LC consisted of two LC-20 AD pumps (Shimadzu, Belgium) for a high-pressure binary gradient and an Accela Auto Liquid Sampler (ALS) (Thermo Scientific, Belgium). The gradient delay volume was minimized through the use of an ASI binary static micro-mixer with a 25 µL cartridge (Supelco, Sigma-Aldrich) and 0.13 mm ID tubing for all fluid connections. The ALS was equipped with a 5 µL sampling loop. As analytical column, an Ascentis Express C18 (75 × 2.1 mm, 2.7 µm particle size; Supelco, Sigma-Aldrich) was employed. Mobile phase A was water/FA (999/1, *v/v*) and mobile phase B MeOH/FA (999/1, *v/v*). The gradient conditions were as follows: 0 min 8% B, 2 min 57% B, 6 min 61% B, 6.1 min 95% B, 7.6 min 95% B, 7.7 min 8% B, 8.7 min 8% B at a flow rate of 0.3 mL/min. The column was maintained at 40 °C during analysis. This gradient was designed with optimal resolution and shortest analysis time for just the four mycotoxins in mind and was obtained using computer-aided design [23].

A TSQ Quantum Ultra triple-quadrupole mass spectrometer with IonMax HESI2 interface (Thermo Scientific, Erembodegem-Aalst, Belgium) was used in SRM mode with the settings listed in Table 3. The MS was calibrated according to manufacturer's recommendations and source/analyser settings have been optimized using "Design of Experiments" with the full LC-MS setup. Other source settings were: vaporizer temp. 350 °C, capillary temp. 320 °C, sheath gas press. 30 arbitrary units (a.u.), auxiliary gas press. 10 a.u., sweep gas press. 10 a.u. The collision gas was argon at 1.5 mTorr. The data acquisition was segmented to limit the number of acquired transitions and enable longer dwell times per analyte while maintaining a fast-enough scan rate to obtain 15–20 scans per peak.

Table 3. Mass spectrometry (MS) source and analyser settings.

Item	Segment 1	Segment 2	Segment 3	Segment 4
Run time (min)	0–2.6	2.6–4.1	4.1–4.9	4.9–8.7
Analyte	DON + $^{13}C_{15}$-DON	HT-2 + $^{13}C_{22}$-HT-2	T-2 + $^{13}C_{24}$-T-2	ZON + $^{13}C_{18}$-ZON
Adduct	Protonated	Sodium	Sodium	Deprotonated
Transitions (m/z)(Collision Energy [eV])	297->231 (16), 297->249 (13), 312->263 (9), 312->276 (9)	447->285 (22), 447->345 (20), 469->300 (19), 469->362 (18)	489->245 (30), 489->327 (25), 513->260 (26), 513->344 (23)	317->131 (25), 317->175 (22), 335->185 (26), 335->290 (21)
Dwell time [ms]	130	130	180	180
Tube Lens [V]	80	110	140	80
Polarity	Pos	Pos	Pos	Neg
Spray Voltage [V]	2800	2800	2800	2200

4.5. Flow Injection Measurements

To perform the flow injection measurements, a LC-20 AD pump equipped with a low-pressure gradient unit and the standard mixer (Shimadzu Benelux, 's-Hertogenbosch, The Netherlands) delivered the mobile phase at a flow rate of 20 µL/min. The mobile phase consisted of ACN/water/NH$_4$FA pH 3.7 (45/54/1, *v/v/v*). A HTC PAL (CTC Analytics, Switzerland) with a 10 µL loop was used to inject test solutions into the solvent flow. The loop was overfilled three times. The same TSQ Quantum Ultra triple-quadrupole mass spectrometer with IonMax HESI2 interface (Thermo Scientific, Erembodegem-Aalst, Belgium) was used in SRM mode. The source settings were as follows: vaporizer temp. off, capillary temp. 300 °C, sheath gas press. 10 a.u., auxiliary gas press. 1 a.u., sweep gas press. 1 a.u. The collision gas was argon at 1.5 mTorr. The spray voltage was set to 3 kV and the tube lens to 110 V. The following transitions (respective collision energies in parentheses) were continuously measured for the run time of 2 min: for Caffeine m/z 195->83 (30 eV), 195->110 (30), 195->138 (30); for AFB1 m/z 313->241 (37 eV), 331->270 (29).

4.6. Computations

All statistical computations were performed using the free software "R" [24]. For all calculations involving responses, either ion ratios of peak area of analyte over peak area of respective isotopologue (LC-MS) or ion ratios of ion intensity of analyte over ion intensity of ISTD (flow injection) were used. Due to the heteroscedasticity, regression fits of those ion ratios over amount of analyte were performed using generalized least-squares modelling with variance function class "varConstPower" [25]. From the modelled variance function, prediction intervals were computed which allowed assigning proper confidence intervals around predicted amounts of analyte. The computations for LOD/LOQ and prediction intervals were based on ideas in an S-PLUS script published by O'Connell [26]. For the evaluation of the flow injection measurements, the MS data was converted to mzML [27] format, which was then further processed with a "R" script. All "R" packages and scripts used for this study are available from either [24] or on request from the author.

4.7. Method Validation

The LC-MS method underwent first a thorough single-laboratory validation. In accordance to Regulation (EC) No. 882/2004 [28], selectivity, linearity, matrix effects, working range, LOD, LOQ, repeatability, intermediate precision, recovery, trueness, and robustness were investigated. This was done in the following materials: maize, wheat, oat, rice, soy and a cereal-based compound feed. For matrix effect and method recovery determination, different amounts of the analytes and equal amounts of ISTD were either spiked into materials free of the analytes before extraction (Set A) or after extraction of the analyte-free materials (Set B). After regression analysis, the slopes of the signals of the sets A and B were then compared with the slopes of a calibration done in neat solvent (Set C). Comparing slope A and C indicated method recovery, while comparison of B and C determined the extent of matrix effects [15].

The data of these spiking experiments were also used to compute LOD based on ISO 11843 Part 2 [19]. LOQs were computed based on an extension of the concept of ISO 11843 towards variance function estimation [25]. For repeatability and intermediate precision, naturally contaminated cereal mixes were prepared and measured 20 times on the same day (repeatability) and once each on a total of eight days by three different operators (intermediate precision). A detailed validation report is available online [12]. The method was then further validated through a collaborative trial [13]. This method and the results of the collaborative trial have recently been published as a European Standard by the European Committee for Standardization (CEN).

The sole purpose of the flow injection method was to screen maize samples for the presence of AFB1. Its validation was therefore very limited and included only the determination of the extraction efficiency in maize.

Acknowledgments: The author extends his thanks to the staff of the EURL for Mycotoxins: Katrien Bouten, Katy Kröger-Negoita, and Carsten Mischke for their technical support and Jörg Stroka for helpful discussions.

Conflicts of Interest: The author declares no conflict of interest.

References

1. Food Quality and Standards Service (ESNS). *Worldwide Regulations for Mycotoxins in Food and Feed in 2003*; FAO: Rome, Italy, 2004.
2. European Commission. Commission Regulation (EC) No 1881/2006 of 19 December 2006 setting maximum levels for certain contaminants in foodstuffs (Text with EEA relevance). *Off. J. Eur. Union* **2006**, *364*, 5–24.
3. Berthiller, F.; Brera, C.; Crews, C.; Iha, M.H.; Krsha, R.; Lattanzio, V.M.T.; MacDonald, S.; Malone, R.J.; Maragos, C.; Solfrizzo, M.; et al. Developments in mycotoxin analysis: An update for 2013–2014. *World Mycotoxin J.* **2015**, *8*, 5–35. [CrossRef]
4. CEN/TC 275. *Foodstuffs—Determination of Aflatoxin B1 and the Sum of Aflatoxin B1, B2, G1 and G2 in Hazelnuts, Peanuts, Pistachios, Figs, and Paprika Powder—High Performance Liquid Chromatographic Method with Post-Column Derivatisation and Immunoaffinity Column Cleanup*; European Committee for Standardization: Brussels, Belgium, 2007.
5. Gałuszka, A.; Migaszewski, Z.; Namieśnik, J. The 12 principles of green analytical chemistry and the SIGNIFICANCE mnemonic of green analytical practices. *TrAC Trends Anal. Chem.* **2013**, *50*, 78–84. [CrossRef]
6. Spanjer, M.C.; Scholten, J.M.; Kastrup, S.; Jörissen, U.; Schatzki, T.F.; Toyofuku, N. Sample comminution for mycotoxin analysis: Dry milling or slurry mixing? *Food Addit. Contam.* **2006**, *23*, 73–83. [CrossRef] [PubMed]

7. Tamura, M.; Mochizuki, N.; Nagatomi, Y.; Harayama, K.; Toriba, A.; Hayakawa, K. A method for simultaneous determination of 20 fusarium toxins in cereals by high-resolution liquid chromatography-orbitrap mass spectrometry with a Pentafluorophenyl column. *Toxins* **2015**, *7*, 1664–1682. [CrossRef] [PubMed]
8. Sulyok, M.; Franz, B.; Rudolf, K.; Rainer, S. Development and validation of a liquid chromatography/tandem mass spectrometric method for the determination of 39 mycotoxins in wheat and maize. *Rapid Commun. Mass Spectrom.* **2006**, *20*, 2649–2659. [CrossRef] [PubMed]
9. Varga, E.; Glauner, T.; Köppen, R.; Mayer, K.; Sulyok, M.; Schuhmacher, R.; Krska, R.; Berthiller, F. Stable isotope dilution assay for the accurate determination of mycotoxins in maize by UHPLC-MS/MS. *Anal. Bioanal. Chem.* **2011**, *402*, 2675–2686. [CrossRef] [PubMed]
10. Kujawski, M.; Mischke, C.; Bratinova, S.; Stroka, J. *Report on the 2013 Proficiency Test of the European Union Reference Laboratory for Mycotoxins, for the Network of National Reference Laboratories—Determination of Fumonisin B1, Deoxynivalenol and Aflatoxin B1 in Cereals*; Publications Office of the European Union: Luxembourg, Luxembourg, 2014.
11. European Commission. 2013/165/EU: Commission Recommendation of 27 March 2013 on the presence of T-2 and HT-2 toxin in cereals and cereal products Text with EEA relevance. *Off. J. Eur. Union* **2013**, *91*, 12–15.
12. Breidbach, A. *Validation of an Analytical Method for the Simultaneous Determination of Deoxynivalenol, Zearalenone, T-2 and HT-2 Toxins in Unprocessed Cereals—Validation Report*; EC, JRC, IRMM: Geel, Belgium, 2011; Available online: http://skp.jrc.cec.eu.int/skp/showPub?id=JRC66507 (accessed on 18 February 2017).
13. Breidbach, A.; Bouten, K.; Kröger, K.; Stroka, J.; Ulberth, F. *LC-MS Based Method of Analysis for the Simultaneous Determination of four Mycotoxins in Cereals and Feed: Results of a Collaborative Study*; Publications Office of the European Union: Luxembourg (Luxembourg), 2013; Available online: http://publications.jrc.ec.europa.eu/repository/handle/JRC80176 (accessed on 18 February 2017).
14. Heumann, K.G. Isotope dilution mass spectrometry of inorganic and organic substances. *Fresenius J. Anal. Chem.* **1986**, *325*, 661–666. [CrossRef]
15. Matuszewski, B.K.; Constanzer, M.L.; Chavez-Eng, C.M. Strategies for the Assessment of Matrix Effect in Quantitative Bioanalytical Methods Based on HPLC-MS/MS. *Anal. Chem.* **2003**, *75*, 3019–3030. [CrossRef] [PubMed]
16. European Commission. Commission Regulation (EC) No 401/2006 of 23 February 2006 laying down the methods of sampling and analysis for the official control of the levels of mycotoxins in foodstuffs (Text with EEA relevance). *Off. J. Eur. Union* **2006**, *70*, 12–34.
17. International Organization for Standardization. *ISO 5725–1 Accuracy (Trueness and Precision) of Measurement Methods and Results*; ISO Copyright Office: Geneva, Switzerland, 1994.
18. Breidbach, A.; Ulberth, F. Two-dimensional heart-cut LC-LC improves accuracy of exact-matching double isotope dilution mass spectrometry measurements of aflatoxin B1 in cereal-based baby food, maize, and maize-based feed. *Anal. Bioanal. Chem.* **2015**, *407*, 3159–3167. [CrossRef] [PubMed]
19. International Organization for Standardization. *ISO 11843–2 Capability of Detection*; ISO Copyright office: Geneva, Switzerland, 2000.
20. Mol, H.G.J.; Plaza-Bolanifos, P.; Zomer, P.; de Rijk, T.C.; Stolker, A.A.M.; Mulder, P.P.J. Toward a generic extraction method for simultaneous determination of pesticides, mycotoxins, plant toxins, and veterinary drugs in feed and food matrixes. *Anal. Chem.* **2008**, *80*, 9450–9459. [CrossRef] [PubMed]
21. Breidbach, A. Improved precision of measured isotope ratio through peak parking and scan-based statistics in IDMS of small organic molecules. In Proceedings of the 20th International Mass Spectrometry Conference, Geneva, Switzerland, 24–29 August 2014.
22. CEN/TC 327 WG5. *EN 16877:2016 Animal Feeding Stuffs: Methods of Sampling and Analysis—Determination of T-2 and HT-2 toxins, Deoxynivalenol and Zearalenone, in Feed Materials and Compound Feed by LC-MS*; European Committee for Standardization: Brussels, Belgium, 2016.
23. Hendriks, G.; Franke, J.P.; Uges, D.R.A. New practical algorithm for modelling retention times in gradient reversed-phase high-performance liquid chromatography. *J. Chromatogr. A* **2005**, *1089*, 193–202. [CrossRef] [PubMed]
24. RCoreTeam R—A Language and Environment for Statistical Computing and Graphics, 2.4.0. 2007. Available online: https://www.r-project.org/ (accessed on 18 February 2017).
25. Davidian, M.; Haaland, P.D. Regression and calibration with nonconstant error variance. *Chemom. Intell. Lab.* **1990**, *9*, 231–248. [CrossRef]

26. O'Connell, M.A.; Belanger, B.A.; Haaland, P.D. Calibration and assay development using the four-parameter logistic model. *Chemom. Intell. Lab.* **1993**, *20*, 97–114. [CrossRef]

27. Chambers, M.C.; Maclean, B.; Burke, R.; Amodei, D.; Ruderman, D.L.; Neumann, S.; Gatto, L.; Fischer, B.; Pratt, B.; Egertson, J.; et al. A cross-platform toolkit for mass spectrometry and proteomics. *Nat. Biotechnol.* **2012**, *30*, 918–920. [CrossRef] [PubMed]

28. European Parliament; Council. Regulation (EC) No 882/2004 of the European Parliament and of the Council of 29 April 2004 on official controls performed to ensure the verification of compliance with feed and food law, animal health and animal welfare rules. *Off. J. Eur. Union* **2004**, *165*, 1–141.

© 2017 by the author. Licensee MDPI, Basel, Switzerland. This article is an open access article distributed under the terms and conditions of the Creative Commons Attribution (CC BY) license (http://creativecommons.org/licenses/by/4.0/).

toxins

MDPI

Article

QuEChERS Purification Combined with Ultrahigh-Performance Liquid Chromatography Tandem Mass Spectrometry for Simultaneous Quantification of 25 Mycotoxins in Cereals

Juan Sun [1,2], Weixi Li [1,2], Yan Zhang [1,2], Xuexu Hu [1,2], Li Wu [1,2] and Bujun Wang [1,2,*]

1   Institute of Crop Sciences, Chinese Academy of Agricultural Sciences, Beijing 100081, China;
     gwzx848@163.com (J.S.); liweixi@caas.cn (W.L.); zhangyan06@caas.cn (Y.Z.); huxuexu@caas.cn (X.H.);
     wuli5151@126.com (L.W.)
2   Laboratory of Quality and Safety Risk Assessment for Cereal Products (Beijing), Ministry of Agriculture,
     Beijing 100081, China
*   Correspondence: wangbujun@caas.cn; Tel.: +86-10-82105798

Academic Editor: Aldo Laganà
Received: 4 May 2016; Accepted: 9 December 2016; Published: 15 December 2016

Abstract: A method based on the QuEChERS (quick, easy, cheap, effective, rugged, and safe) purification combined with ultrahigh performance liquid chromatography tandem mass spectrometry (UPLC–MS/MS), was optimized for the simultaneous quantification of 25 mycotoxins in cereals. Samples were extracted with a solution containing 80% acetonitrile and 0.1% formic acid, and purified with QuEChERS before being separated by a C18 column. The mass spectrometry was conducted by using positive electrospray ionization (ESI+) and multiple reaction monitoring (MRM) models. The method gave good linear relations with regression coefficients ranging from 0.9950 to 0.9999. The detection limits ranged from 0.03 to 15.0 $\mu g \cdot kg^{-1}$, and the average recovery at three different concentrations ranged from 60.2% to 115.8%, with relative standard deviations (RSD%) varying from 0.7% to 19.6% for the 25 mycotoxins. The method is simple, rapid, accurate, and an improvement compared with the existing methods published so far.

Keywords: ultrahigh performance liquid chromatography tandem mass spectrometry (UPLC–MS/MS); QuEChERS; mycotoxin; cereals

1. Introduction

Mycotoxins can be analyzed by various methods, including thin layer chromatography (TLC) [1], enzyme-linked immunosorbent assay [2], gas chromatography [3,4], and immunoaffinity column/high-performance liquid chromatography with fluorescence and diode array detection [5,6]. The specific determination of multiclass mycotoxins in cereals requires highly selective and sensitive techniques such as ultrahigh-performance liquid chromatography tandem mass spectrometry (UPLC–MS/MS) [7–11] and direct analysis in real time (DART) ionization coupled to an (ultra)high-resolution mass spectrometer based on orbitrap technology (orbitrap MS) [12].

Several methods have been developed to purify mycotoxins from crude samples such as solid-phase extraction (SPE) [13], immunoaffinity columns (IACs) [14], and MycoSep columns [15]. A direct and simple method for the purification of multiple mycotoxins is challenging because of their diverse chemical structures and properties. The SPE cartridge is the most commonly used purification column, but the purification is tedious and time consuming. The most prominent feature of IACs is their low matrix effects, high selectivity, and increased rate of recovery. However, they present some drawbacks such as being too expensive and unsuitable for the determination of a large number of molecules. The MycoSep columns are efficient but can only deal with the mixtures of mycotoxins exhibiting similar structures or properties.

Recently, the method based on QuEChERS (quick, easy, cheap, effective, rugged, and safe) has attracted increasing attention in the research field of mycotoxins due to its simplicity and effectiveness for isolating mycotoxins from complex matrices. The successful application of this method has been reported in many products, including cereals and their products [16], spices [17], nuts and seeds [18], Madeira wine [19], human breast milk [20], sesame butter [21], noodles [22], eggs [23], human biological fluids [24], and baby food [25]. In order to reduce the matrix effects (MEs), which can result in unsatisfactory recoveries, additional purification with sorbents such as octadecyl silica (C18), primary secondary amine (PSA), and graphitized carbon black (GCB) are commonly used. The sorbents react differently depending on the physicochemical properties of the compounds constituting the samples [26,27].

The published analyses of multiple mycotoxins in cereals only report the determination of 14 mycotoxins, including deoxynivalenol (DON), aflatoxins (AFs), and fumonisins (FBs) [15,28–31], but do not take into account enniatins (ENNs) and beauvericin (BEA), which are commonly found in harvested grains [32]. To the best of our knowledge, few studies have been published on the evaluation of the simultaneous presence of 25 mycotoxins in cereals. Although Liu et al. [21] detected 26 mycotoxins in sesame butter, they used a two-ion mode and mobile phase system. For quality and safety risk assessment of cereals and their products, the aim of this study was to develop and validate an efficient and reliable method for the simultaneous quantitative analysis of multiple mycotoxins in cereals. The strategy exploits the efficiency of QuEChERS extractions and the sensitive and selective UPLC–MS/MS technique. We have optimized the mobile phase types, gradients of elution, and the removal of the matrix effects.

2. Results and Discussion

2.1. Optimization of the Type of the Mobile Phase

The mobile phase plays an important role in the ionization efficiency when the analytes enter the MS/MS system. Acetonitrile, methanol, water, formic acid, and buffers such as ammonium acetate are commonly used as the mobile phase in UPLC. We have selected formic acid and ammonium acetate because of their high ionization efficiency and solubility in the presence of methanol. Therefore, we investigated mixtures of methanol/water and acetonitrile/water, separately, as the mobile phase. The results showed that the response of fumonisins was 15 times higher when using methanol/water than when using acetonitrile/water, with an acceptable peak shape. Moreover, the use of methanol/water produced better separation of the mycotoxins compared with acetonitrile/water. Therefore, we selected methanol as the elution mobile phase.

Adding ammonium acetate (10 mM) and formic acid (0.1%, v/v) to the mobile phase of methanol improved the responses of all mycotoxins. Adding formic acid greatly enhanced the response especially of positively charged ENNs, BEA, and FBs. Thereby, the formic acid in water was selected to improve the mobile phase. The possible reason for the superior performance is that ionic strength can increase after adding a suitable amount of formic acid solution to the mobile phase; the change in ionic strength can affect the behavior of the separated material, which is advantageous to the positive ion protons.

Varying concentrations of formic acid solution (0.05%, 0.1%, 0.5%, 1.0%, and 5.0%, v/v) showed minimal effect, but the responses of zearalenone (ZEN), ENN A1, ENN B1, BEA, fusarenon-X (FUS-X), gliotoxin (GLT), and neosolaniol (NEO) were relatively higher with 0.5% formic acid compared to the other proportions (Figure 1). Considering the efficiency of all mycotoxins, 0.5% (v/v) formic acid in water was selected for the mobile phase.

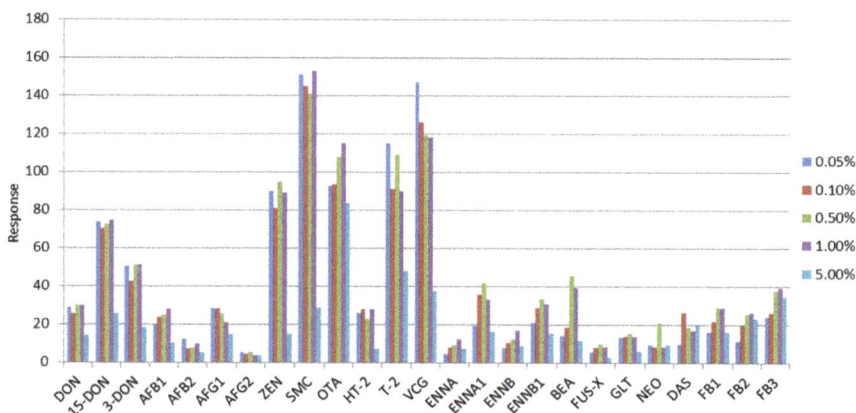

Figure 1. Response obtained for a fixed concentration of the 25 mycotoxins at different concentrations of the formic acid in water. DON: deoxynivalenol; AF: aflatoxin; ZEN: zearalenone; SMC: sterigmatocystin; OTA: ochratoxin; T-2: T-2 toxin; HT-2: HT-2 toxin; VCG: verruculogen; ENN: enniatin; BEA: beauvericin; FUS-X: fusarenon-X; GLT: gliotoxin; NEO: neosolaniol; DAS: 4,5-diacetoxyscirpenol; FB: fumonisin

2.2. Optimization of the Gradient Elution

The gradient elution program of mobile phase had effects on the retention time and peak shape of target compounds, and also influenced the ionization efficiency and sensitivity of the target compounds. The gradient elution supports an efficient separation of the multiple mycotoxins, shortens the retention time, and enhances the peak shape and sensitivity.

The initial ratio of 0.5% (v/v) formic acid water solution/methanol was set to 80:20 (v/v), resulting in a suitable response and peak shape for the ENNs, while there was no response for FBs, ochratoxin A (OTA), DON, 3-AcDON, and 15-AcDON. In this condition, the results showed that when increasing the proportion of methanol, the ENNs and BEA mycotoxins presented two peaks with varying retention times (Figure 2), which can seriously affect the accuracy of qualitative and quantitative results of the compounds. When the initial ratio of 0.5% (v/v) formic acid water solution/methanol was set to 95:5 (v/v), we could observe better-resolved peaks. The optimized gradient elution of eluent A (0.5% formic acid water solution) was as follows: starting with 5% and rapidly increasing the proportion to 85% in 5.5 min, slowly increasing to 100% within 5.8 min, then linearly reducing to 5% in 10.0 min. Under this condition, the concentration of DON, 3-AcDON, 15-AcDON, FUS-X, 4,5-diacetoxyscirpenol (DAS), GLT, FB1, FB2, and FB3 is 200 µg·L^{-1}; the concentration of ZEN, OTA, verruculogen (VCG), sterigmatocystin (SMC), T-2 toxin (T-2), HT-2 toxin (HT-2), NEO, ENNA, ENNA1, ENNB, ENNB1, and BEA is 50 µg·L^{-1}; and the concentrations of AF B1, AF B2, AF G1, and AF G2 are 10, 3, 10, and 3 µg·L^{-1}, respectively. The 25 mycotoxins could be monitored as shown in Figure 3. A good peak shape and resolution were achieved for all the target mycotoxins within 10 min except for the conjugations 3-AcDON and 15-AcDON, which were selected as different MS/MS daughter ions for quantification.

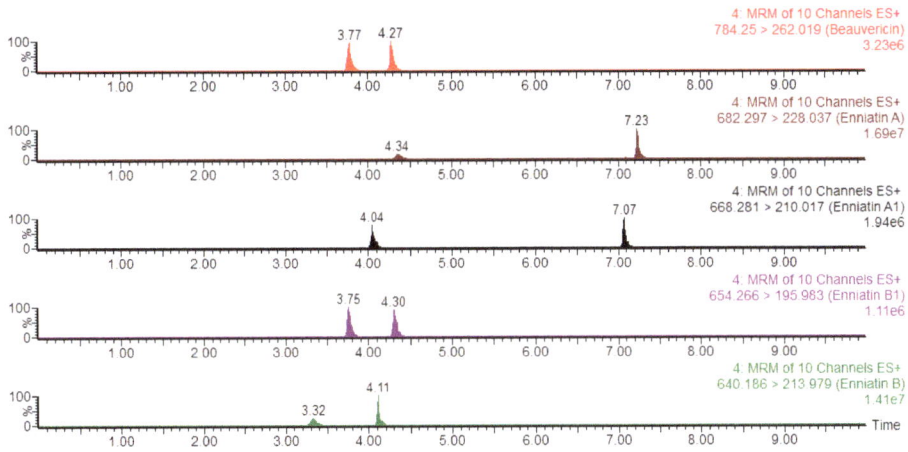

Figure 2. The change of peak shape between ENNs and BEA when using different ratios of formic acid water solution/methanol. MRM: multiple reaction monitoring.

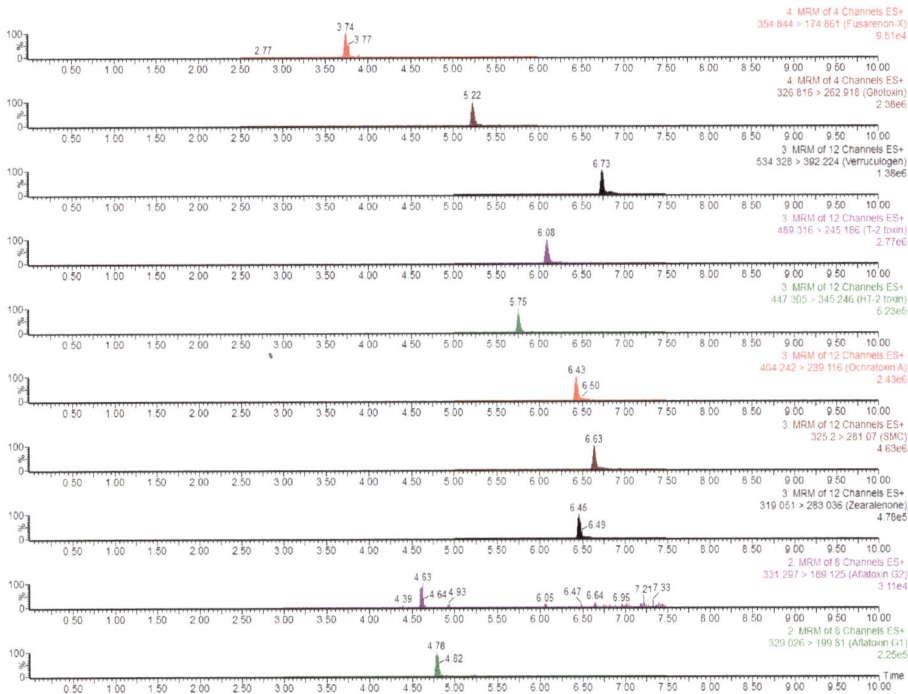

Figure 3. *Cont.*

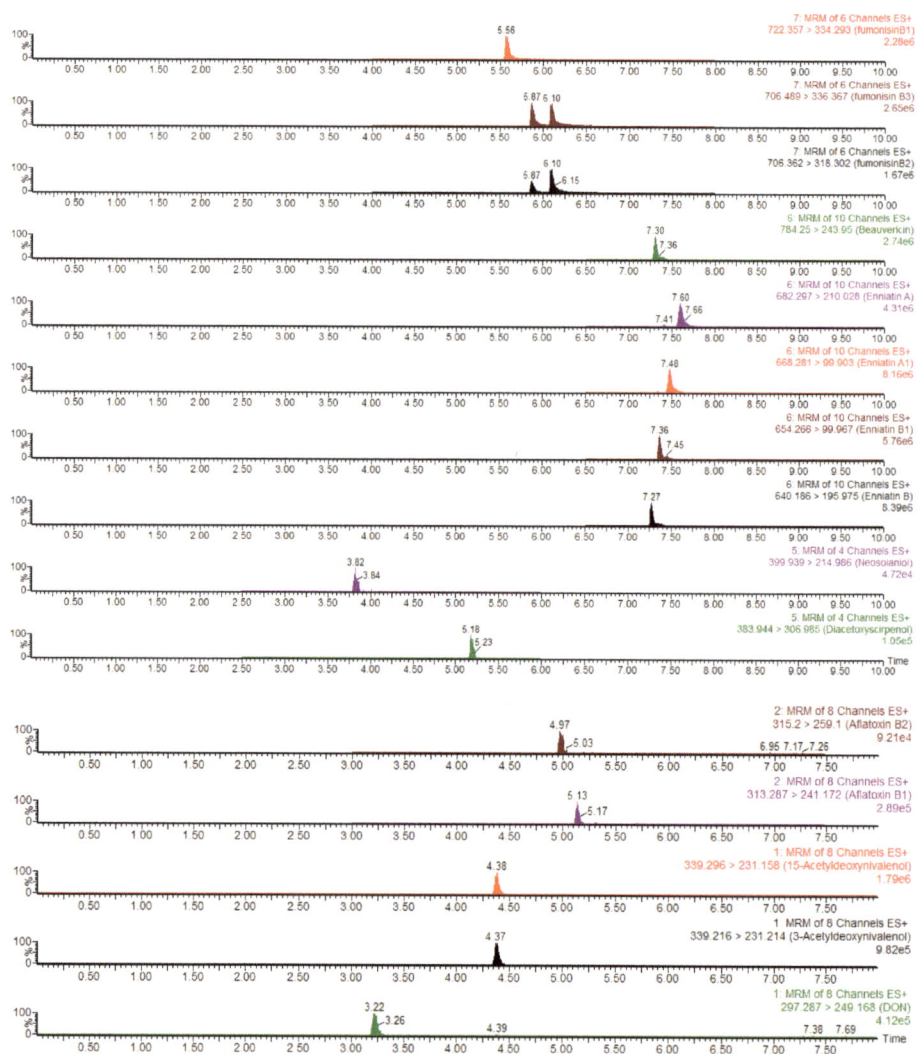

Figure 3. Multiple reaction monitoring (MRM) chromatograms of mixed solutions of 25 mycotoxins standards by positive electrospray ionization (ESI+).

2.3. Selection of the Extraction Solvent and Purification Evaluation

The extraction solvent was selected according to the method described in a previous work [15]. The extraction yield greatly improved when the acetonitrile/water content gradually increased until reaching the maximum yield at the proportions of 80% and 84%, respectively [33]. In this study, we have made a slight modification. When using pure acetonitrile as the extraction solvent, the yields of ENNs, BEA, and FBs were less than 60% in all matrices. However, when extracting with a solution containing a ratio of 99:1 (v/v) of 0.1% formic acid water solution/acetonitrile, we observed more prominent intensities of the peaks, especially for the FBs.

Sorbents C18, PSA, GCB, SPE C18, and the amino column (NH$_2$) were selected for the purification of mycotoxins. The sorbent C18 can be used to remove the fatty and nonpolar components of the matrix, whereas PSA is mainly used for the removal of organic acids, pigments, and phenol, among others, whereas GCB mainly removes pigments of the crude sample.

We have optimized the purification by separately testing 100 mg of C18, 150 mg of PSA, and 150 mg GCB. The results are summarized in Table 1. The mycotoxins used with PSA and GCB were recovered with less than 50% yield, and the extraction yield with GCB was less than 20%. The low yield can be explained by the high affinity of the mycotoxins to the sorbents, except for DON mycotoxin. Published results show that GCB is suitable for application with biscuits when using an amount smaller than 500 mg with dichloromethane/methanol (80:20, *v/v*) [26]. Nevertheless, dichloromethane is toxic and is not recommended. The sorbent GCB was not suitable for the detection of mycotoxins in pomegranate juices [17] and beer-based drinks [2] due to the sorption of the mycotoxins, so we may not choose GCB as an alternative cleaning sorbent.

Using 0.4 g of PSA produced significant improvements in the recovery, reaching yields superior to 80% for DON, AF B1, AF B2, AF G1, AF G2, T-2 toxin, HT-2 toxin, OTA, ZEN, FB1, and FB2 in rice [28], but the yields decreased when using PSA in sesame, with recoveries less than 10% for T-2, HT-2, OTA, ZEN, FB1, FB2, and FB3, and lower than 60% for AF B1. When the amount of PSA increased, the recovery of AF B2 decreased from 70.4% to 36.4%, and did not affect the recoveries of AF G1 and AF G2 [21].

Our results showed that C18 obtained satisfactory yields of about 60.0%~101.7% for most of the mycotoxins, excepted for T-2, HT-2, SMC, NEO, and DAS, which is consistent with the results obtained for sesame butter [21]. However, using a C18 sorbent concentration of 50 mg/mL did not improve the purification of human breast milk [20].

Using a combination of C18 with PSA in the range of 10, 20, 50, and 100 mg showed that when the amount of PSA increased, the recovery of most mycotoxins decreased. However, the mycotoxins VCG, FUS-X, and GLT were not affected, with extraction yields inferior to 30%. Our results demonstrated that the type and quantity of sorbents greatly influence the extraction yield.

Evaluation of the commonly used purification column MycoSep 226, 227, and 400 (Romer Labs Inc., Union, MO, USA) showed that the three columns were all suitable for DON, 3-AcDON, 15-AcDON, T-2, HT-2, VCG, FUS-X, and DAS, with recoveries from 61.3% to 125.1%, with the MycoSep 227 superior to the 226 and 400, although the three MycoSep columns were not suitable for OTA, ENNs, BEA, NEO, and FBs. Our results are consistent with the observations of Sun et al., who reported on the extraction of mycotoxins from wheat, rice, and corn with MycoSep 226 [15], and the observations reported by other researchers extracting other feedstuffs with MycoSep 400 [34].

The comparison of column C18 with amino column NH_2 showed that FBs and OTA were not properly recovered from the column NH_2, as reported in the case of multiple mycotoxins determination of sorghum [35].

Satisfactory recoveries were obtained when using column C18 for FBs, compared with a MultiSep 211 FUM purification cartridge [36], with recoveries of 80.9%–97.0%. In addition, the cost of a C18 column is much lower than an immunoaffinity column. The method thus provides a simple and cheap purification. The results showed that the amino column NH_2 is also suitable for the extraction of ENNs and BEA, although the yields ranged from 61.3% to 83.6%. This work presents the first simple and cost-effective purification of ENNs and BEA.

It can be clearly observed that by optimization of the extraction parameters, satisfactory results could be obtained for most of the mycotoxins, although a few mycotoxins presented low yields. No cleanup was applied for the following experiments of 25 mycotoxins.

Table 1. Recoveries (%) of the 25 mycotoxins using different sorbents for the purification.

Mycotoxin	100 mg C18	100 mg C18 + 10 mg PSA	100 mg C18 + 20 mg PSA	100 mg C18 + 50 mg PSA	100 mg C18 + 100 mg PSA	150 mg PSA	150 mg GCB	MycoSep 226	MycoSep 227	MycoSep 400	SPE C18	SPE NH2
DON	82.6	96.5	76.8	85.1	89.0	61.3	58.8	110.0	86.9	87.5	88.2	94.1
15-DON	79.2	81.3	77.1	84.0	78.4	34.9	18.7	125.1	113.8	74.8	78.7	90.3
3-DON	85.3	88.4	68.7	86.9	85.4	30.3	15.8	106.3	100.7	77.4	112.1	112.1
AF B1	67.7	67.9	44.7	17.1	24.2	ND	ND	73.5	ND	68.8	64.6	81.6
AF B2	65.3	62.3	61.0	13.0	15.0	ND	ND	68.7	ND	63.2	71.9	45.0
AF G1	64.7	61.2	64.5	32.7	14.7	3.2	ND	86.1	ND	63.9	67.3	36.4
AF G2	101.7	72.0	103.3	ND	ND	ND	ND	71.6	ND	60.4	61.3	72.1
ZEN	67.0	50.0	18.9	43.8	23.5	22.6	0.7	74.8	0.9	53.9	30.7	49.4
SMC	52.1	38.4	12.7	40.8	39.6	19.6	ND	62.7	ND	36.6	34.0	51.4
OTA	71.8	8.7	23.2	2.6	0.7	21.5	0.5	15.4	4.1	40.1	71.9	5.1
HT-2	36.3	50.1	8.8	42.1	41.5	13.6	8.6	70.6	72.9	67.2	29.2	33.2
T-2	51.9	49.6	14.6	54.0	53.1	22.0	11.2	86.1	94.1	61.7	34.6	37.1
VCG	63.6	67.6	60.4	68.7	85.2	28.1	1.0	69.7	65.2	61.3	26.3	40.2
ENA	82.2	39.1	16.4	34.1	29.3	16.8	4.1	6.1	ND	25.2	6.4	65.0
ENA1	60.0	30.8	15.9	23.3	24.3	26.8	8.1	8.2	ND	28.6	14.4	66.0
ENB	64.4	49.5	47.1	38.0	41.8	35.2	15.2	30.6	ND	39.1	32.1	83.6
ENB1	60.4	33.9	24.8	28.6	30.7	35.4	12.1	17.6	ND	36.1	22.2	74.5
BEA	66.2	33.2	34.6	30.4	47.5	16.2	3.4	35.5	ND	27.2	13.9	61.3
FUS-X	62.2	79.4	83.3	67.6	71.3	19.2	10.3	100.2	94.8	71.1	91.3	77.0
GLT	98.6	82.3	56.3	75.2	66.6	26.7	11.5	21.5	5.0	24.7	69.1	5.5
NEO	28.9	26.0	17.4	27.2	25.2	5.3	13.7	58.4	37.3	42.8	14.0	31.0
DAS	45.9	38.0	49.1	21.8	34.9	31.6	9.8	115.5	108.1	85.1	65.9	51.1
FB1	63.7	3.9	2.5	1.3	0.5	0.2	7.6	2.4	0.8	1.5	117.0	3.5
FB2	89.5	8.2	5.1	3.0	1.4	0.6	4.9	6.4	0.6	4.2	113.3	1.1
FB3	84.5	5.6	4.3	3.0	1.8	0.9	8.5	2.8	0.9	11.7	114.4	3.6

ND: not detected; C18: octadecyl silica; PSA: primary secondary amine; GCB: graphitized carbon black; SPE: solid-phase extraction.

2.4. Evaluation of the Matrix Effect

Matrix effects (MEs) are unavoidable, and the removal of the matrix interference is challenging. At present, reports from scientific literature offer solutions such as the internal standard method using the internal standards zearalenone (ZEN) and deepoxy-deoxynivalenol (DOM) [35] or isotopically labeled standards [37]. Although using a standard curve with an acceptable linear relation presents the advantage of high precision, the choice of an appropriate internal standard for a multicomponent analysis is often difficult and expensive. The response of the target mycotoxins can be suppressed or enhanced on account of the interfering matrix components. The ME was calculated by the Equation (1) for different blanks of wheat, corn, and rice samples as shown in Figure 4.

$$\text{ME} = 100 \times \left(1 - \frac{\text{area of mycotoxin seandard in blank matrix}}{\text{area of mycotoxin standard in solent}}\right) \tag{1}$$

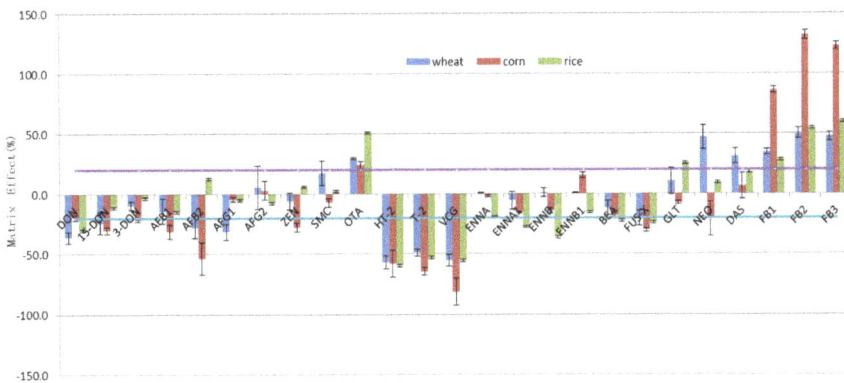

Figure 4. Matrix effects of blank cereals (wheat, corn, and rice) on the response of 25 mycotoxins. The two dashed lines show the tolerance level of the matrix effect.

It can be observed that the signal enhancement was prominent for the FBs, with an ME ranging from 28.4% to 132.1%. The effect is emphasized in the case of corn, which exhibited an ME of about twice that of wheat or rice. The signal enhancement of the FBs was also reported in spices (enhancement of about 20%) [38] and sesame butter [21], but the was suppressed in white rice (ME between 13.2% and 17.5%) [9]. The MEs of OTA and DAS were slightly enhanced, presenting values ranging from 24.3% to 50.9% and 6.8% to 31.4%, respectively.

Most of the mycotoxins had a signal suppression effect and were close to the tolerable range between +20% and −20% [23]. However, the signals of HT-2, T-2, and VCG were significantly suppressed by 48.8%–81.3%, as reported for cereal syrups [38].

It is worth mentioning that AF B1 and B2 in corn presented a strong signal suppression effect, as reported in white rice, edible nuts, and seeds by Arroyo-Manzanares et al. [18]. Compared to literature results, we obtained acceptable ME percentages except for AF B1 and B2. The mycotoxins NEO and DAS in wheat showed a strong signal enhancement. The signal suppression for the ENNs, especially for ENNB in rice, was more important than that for wheat and corn.

In conclusion, the matrix effects caused by different matrices were significant for most of the mycotoxins. The calibration curves help to reduce the MEs and, therefore, increase the extraction yields to improve the accuracy of the analysis.

2.5. Method Validation

2.5.1. Calibration Curves and Linearity

The calibration curves were constructed using a blank matrix with of the following concentrations: 0.1, 0.2, 0.5, 1, 2, 5, 10, 25, 50, 100, 200, 300, 400, and 500 $\mu g \cdot L^{-1}$. The peak area (Y) was plotted against the concentration of the analyte (X). The linear range of DON, 3-AcDON, 15-AcDON, FUS-X, DAS, FB1, FB2, and FB3 was from 0.5 to 500 $\mu g \cdot kg^{-1}$; the range of ZEN, OTA, VCG, SMC, T-2, HT-2, GLT, NEO, ENNA, ENNA1, ENNB, ENNB1, and BEA was from 0.1 to 200 $\mu g \cdot kg^{-1}$; and the range for AF B1, AF B2, AF G1, and AF G2 was from 0.1 to 50 $\mu g \cdot kg^{-1}$ (Table 2). A linear relation was considered to exist when the linear regression coefficient (r^2) was equal to or higher than 0.995.

2.5.2. Limit of Detection and Quantification

Limits of detection (LODs) and limits of quantification (LOQs) were calculated from spiked blank samples at the lowest spiking level. The lowest spiking levels were 3-fold and 10-fold lower than the signal-to-noise ratio (S/N) of the multiple reaction monitoring (MRM) chromatograms. The results showed that LODs were in the range of 0.1–5.0 $\mu g \cdot kg^{-1}$ except for AF B1, BEA, and ENNs, which were 0.03, 0.05, and 0.05 $\mu g \cdot kg^{-1}$, respectively, whereas the LOQs were in the range of 0.1–25.0 $\mu g \cdot kg^{-1}$ (Table 2). These results were far below the values for methanol extraction from cereals and derived products from a study in Tunisia, which reported LODs for ENNA, ENNA1, ENNB, ENNB1, and BEA of 215, 140, 145, 165, and 170 $\mu g \cdot kg^{-1}$, respectively, and LOQs of 600, 400, 400, 500, and 500 $\mu g \cdot kg^{-1}$, respectively [39]. The values for FBs were higher than those of the other mycotoxins due to the low signal responses and matrix effects. Nonetheless, when adding the values of FB1 and FB2, we obtain a value which below the maximum limits of European Union regulations (<1000 $\mu g \cdot kg^{-1}$; FB3 is not included in the regulations). The LODs of regulated mycotoxins (AF B1, AF B2, AF G1, AF G2, DON, ZEN, T-2, HT-2, and OTA) were lower than the maximum limits of the different analytes in regulated cereals per EU regulations [40].

2.5.3. Method Accuracy and Precision

The method accuracy was evaluated by determining the recovery of the standard mycotoxins that were spiked into the blank matrices at three different concentrations as shown in Table 3. The recovery values were in the range of 60.2%–115.8%, and the relative standard deviations (RSDs) of the values were between 0.7% and 19.6%. The results demonstrated that the method applied to the cereals was highly accurate and reliable.

In our study, we also use the certified reference material DON (PriboLab, Level # DW-163) for method verification. The reference material value was 0.5 ± 0.1 $\mu g \cdot kg^{-1}$, and the actual measured value was 0.44 $\mu g \cdot kg^{-1}$, which proved the state of instrument is stable and the method is feasible.

The precision of the method was evaluated by intra- and inter-day repeatability and by measuring six times. In the case of the wheat matrices, the intra-day repeatability values were within 2.4%–12.0%, and the inter-day repeatability ranged from 6.1% to 17.6% (Table 2). The results showed that the RSDs of inter-day repeatability test were higher than those for the intra-day repeatability test. Nonetheless, these values were still below 20% and within the allowable range.

Table 2. Fitting equations, linear ranges, correlation coefficients (r), limits of detection (LODs), and limits of quantifications (LOQs) of the 25 mycotoxins.

Compound	Calibration Curve	Linear Range ($\mu g \cdot kg^{-1}$)	r	LOD ($\mu g \cdot kg^{-1}$)	LOQ ($\mu g \cdot kg^{-1}$)	Intra-Day Precision ($n = 6$) RSD (%)	Inter-Day Precision ($n = 6$) RSD (%)
DON	Y = 98.62X + 169.00	5.0~500	0.9995	5.0	15.0	3.4	7.0
15-DON	Y = 313.95X + 1394.48	5.0~500	0.9997	2.0	5.0	3.6	7.1
3-DON	Y = 191.02X + 375.01	5.0~500	0.9999	2.0	5.0	3.3	7.3
AF B1	Y = 853.62X + 1079.28	0.1~50	0.9950	0.03	0.1	8.9	7.2
AF B2	Y = 1284.87X + 470.48	0.3~15	0.9974	0.1	0.3	5.3	9.2
AF G1	Y = 945.23X + 253.68	0.1~50	0.9999	0.1	0.3	8.1	12.2
AF G2	Y = 378.07X + 111.02	0.3~15	0.9960	0.1	0.3	8.8	16.9
ZEN	Y = 272.06X + 695.88	1.0~200	0.9993	1.0	4.0	6.4	6.5
SMC	Y = 2071.13X + 1706.52	0.4~200	0.9998	0.1	0.4	4.2	9.6
OTA	Y = 573.77X + 2398.65	0.4~200	0.9997	0.1	0.4	8.4	9.3
HT-2	Y = 174.01X + 181.35	0.4~200	0.9996	0.1	0.4	8.3	7.9
T-2	Y = 1144.87X + 2500.6	0.4~200	0.9996	0.1	0.4	6.1	7.2
VCG	Y = 581.24X + 1550.22	0.2~200	0.9997	0.1	0.4	8.2	7.4
ENNA	Y = 222.75X + 123.09	0.2~200	0.9998	0.05	0.2	7.8	12.6
ENNA1	Y = 6333.74X + 5213.22	0.2~200	0.9998	0.05	0.2	2.4	11.1
ENNB	Y = 6124.97X + 673.19	0.1~200	0.9998	0.05	0.2	5.6	12.9
ENNB1	Y = 4212.47X + 2487.88	0.1~200	0.9997	0.05	0.2	6.9	14.7
BEA	Y = 3170.66X + 659.08	0.1~200	0.9999	0.05	0.2	7.3	9.3
FUS-X	Y = 28.65X + 15.62	0.5~500	0.9996	10.0	25.0	6.9	9.4
GLT	Y = 473.91X + 2091.75	5.0~200	0.9993	5.0	10.0	4.6	6.1
NEO	Y = 97.93X + 1000.67	5.0~200	0.9993	5.0	10.0	7.0	17.6
DAS	Y = 97.93X + 1000.67	10.0~500	0.9993	10.0	25.0	5.6	14.3
FB1	Y = 83.71X + 323.38	25.0~500	0.9992	15.0	25.0	6.3	6.3
FB2	Y = 70.66X + 421.58	25.0~500	0.9950	15.0	25.0	12.0	8.5
FB3	Y = 60.71X + 2650.23	25.0~500	0.9995	15.0	25.0	7.4	7.8

Y: peak area, X: concentration in $\mu g \cdot L^{-1}$. RSD: relative standard deviation.

Table 3. Recovery and precision of 25 mycotoxins spiked in three different cereal matrices at three concentrations (%RSD of peak areas).

Mycotoxin	Wheat (n = 6)			Corn (n = 6)			Rice (n = 6)		
	Level 1	Level 2	Level 3	Level 1	Level 2	Level 3	Level 1	Level 2	Level 3
DON	115.8 (0.7)	97.9 (5.5)	98.8 (6.8)	98.4 (15.8)	106.6 (5.9)	101.2 (1.9)	97.0 (4.7)	116.9 (5.3)	106.5 (1.0)
15-ADON	91.0 (3.6)	100.8 (4.0)	96.8 (2.6)	88.9 (3.4)	95 (2.5)	97.9 (0.5)	85.8 (3.9)	113.1 (3.4)	93.0 (2.5)
3-ADON	93.4 (2.9)	104.7 (1.4)	96.8 (1.7)	85 (9.7)	93.8 (4.3)	102.5 (4.5)	73.4 (7.2)	100.6 (0.5)	80.9 (2.8)
AF B1	61.7 (10.8)	118.1 (1.7)	87.4 (12.2)	82.3 (9.3)	60.3 (18.5)	90 (5.4)	82.8 (7.0)	91.7 (17.1)	105.0 (5.1)
AF B2	114.0 (18.1)	63.0 (6.2)	71.0 (12.0)	60.6 (11.1)	108.3 (14.9)	91.6 (12.9)	67.9 (6.0)	98.3 (4.7)	79.4 (19.0)
AF G1	84.8 (10.5)	73.2 (12.8)	88.2 (14.8)	73.4 (18.8)	80.7 (8.5)	86.7 (3.4)	88.8 (8.1)	89.4 (19.2)	95.9 (18.6)
AF G2	75.7 (11.4)	91.5 (9.5)	84.5 (5.8)	114.7 (13.4)	89.9 (17.1)	63.7 (19.6)	62.5 (11.8)	103.3 (16.5)	100.4 (8.7)
ZEN	70.0 (6.4)	73.5 (8.5)	63.3 (12.2)	92 (4.4)	68.1 (7.3)	88.6 (5.3)	84.6 (17.1)	64.3 (10.5)	79.2 (2.1)
SMC	64.8 (10.0)	62.8 (5.8)	69.9 (8.4)	69.1 (4.7)	69.1 (3.6)	63.5 (9.1)	66.4 (10.5)	58.2 (1.5)	86.2 (6.2)
OTA	77.4 (1.4)	88.7 (11.2)	60.1 (2.0)	77.8 (7.0)	65.7 (2.7)	65.5 (7.7)	74.0 (8.7)	68.6 (5.9)	81.4 (9.5)
HT-2	77.5 (5.1)	73.0 (11.8)	84.1 (8.3)	73.6 (0.8)	70.8 (10.3)	89.7 (2.8)	83.8 (9.2)	103.6 (4.4)	86.1 (3.3)
T-2	62.4 (8.8)	67.9 (2.5)	69.9 (2.0)	77.0 (5.7)	69.4 (4.0)	85.9 (4.3)	82.2 (3.6)	122.4 (0.6)	85.0 (1.5)
VCG	61.4 (5.5)	106.5 (6.7)	62.4 (6.1)	67.6 (8.4)	71.9 (10.1)	113.2 (0.2)	92.7 (1.4)	79.0 (7.1)	89.5 (9.0)
ENNA	71.2 (9.1)	63.5 (1.5)	67.0 (9.2)	65.7 (6.9)	63.5 (4.0)	89.2 (0.6)	73.9 (5.2)	69.3 (10.6)	99.0 (1.1)
ENNA1	74.5 (3.5)	64.9 (5.3)	61.9 (8.9)	67.2 (4.4)	63.7 (4.9)	75.6 (0.7)	82.3 (11.7)	74.2 (4.4)	73.4 (6.7)
ENNB	67.4 (14.3)	69.7 (5.3)	70.2 (5.1)	71.4 (8.3)	96.6 (13.0)	81.7 (8.7)	104.2 (6.7)	79.3 (2.7)	87.7 (3.9)
ENNB1	67.7 (12.2)	65.0 (11.2)	60.2 (5.0)	64.9 (9.2)	73.1 (1.1)	76.2 (15.4)	80.0 (11.4)	77.3 (2.8)	79.1 (1.0)
BEA	65.4 (7.6)	71.7 (3.3)	60.3 (3.2)	66.5 (3.5)	74.8 (3.9)	65.1 (10.6)	72.8 (9.4)	67.9 (1.6)	62.5 (4.1)
FUS-X	97.5 (14.5)	107.9 (4.9)	119.3 (11.7)	101.4 (4.5)	87.2 (10)	111.2 (10.6)	76.5 (3.6)	110.4 (19.0)	92.1 (6.9)
GLT	67.7 (1.3)	74.3 (5.2)	82.4 (2.6)	75.7 (3.2)	75.7 (1.0)	79.3 (4.4)	69.9 (3.9)	86.3 (2.5)	74.9 (1.2)
NEO	63.7 (9.5)	69.3 (12.5)	64.6 (1.1)	63.2 (18.5)	95.3 (15.8)	62.6 (15.9)	64.6 (7.6)	108.7 (14.2)	67.2 (10.9)
DAS	77.0 (12.2)	84.0 (14.3)	100.7 (4.1)	70.8 (2.9)	88.1 (1.4)	87.8 (11.7)	78.8 (17.4)	72.6 (4.1)	88.4 (2.5)
FB1	109.4 (4.5)	107.1 (7.8)	73.9 (2.3)	118.9 (2.1)	115.4 (2.8)	83.7 (4.7)	107.8 (3.6)	113.7 (2.8)	80.2 (1.5)
FB2	102.3 (3.2)	102.4 (6.9)	74.7 (2.4)	107.4 (4.9)	110.8 (5.6)	87.8 (2.1)	100.9 (0.7)	99.5 (2.4)	105.7 (2.0)
FB3	115.6 (5.2)	102.8 (5.9)	89.6 (1.9)	115.5 (2.3)	114 (2.7)	89.1 (3.5)	111.2 (5.4)	107.4 (3.7)	91.3 (1.1)

Group 1: DON, 15-AcDON, 3-AcDON, FUS-X, GLT, DAS, FB1, FB2, and FB3 at 100, 200, and 500 µg·kg^{-1}; Group 2: ZEN, SMC, OTA, HT-2, T-2, VCG, ENNA, ENNA1, ENNB, ENNB1, BEA, and NEO at 20, 50, and 100 µg·kg^{-1}; Group 3: AF B1 and AF G1 at 5, 10, and 25 µg·kg^{-1}; Group 4: AF G2 and AF B2 at 1.5, 3.0, and 7.5 µg·kg^{-1}.

2.6. Application of the Developed Method on Real Samples

The developed and optimized method was applied for the analysis of 65 samples, comprising 26 wheat samples from different areas of China, 14 corn samples collected from two different years (2014 and 2015), and 25 various brands of rice obtained from the local supermarket. As shown in Figure 5, 90% of the samples were contaminated with mycotoxins. The detected mycotoxins were DON, 3-AcDON, 15-AcDON, ENNA, ENNA1, ENNB, ENNB1, FB1, FB2, FB3, T-2, OTA, BEA, ZEN, and SMC. The other mycotoxins (AF B1, AF B2, AF G1, AF G2, HT-2, VCG, NEO, FUS-X, and DAS) were not detected.

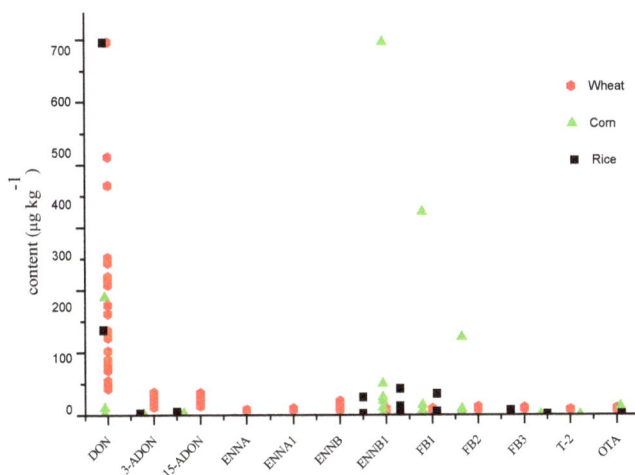

Figure 5. Multiple mycotoxins determination of cereal samples collected from farmland and local supermarkets in China.

In contaminated samples, all the wheat samples contained DON within concentrations ranging from 42.1 to 718.0 $\mu g \cdot kg^{-1}$, but the values are still below the state limit standards. Nevertheless, we should pay attention to the addition of the concentrations of 3-AcDON and 15-AcDON reaching values ranging from 6.1 to 32.8 $\mu g \cdot kg^{-1}$. It is worth mentioning that 19%, 27%, 69%, and 35% of the total wheat samples were contaminated with ENNA, ENNA1, ENNB, and ENNB1, respectively. The detected concentrations ranged from 0.1 to 17.2 $\mu g \cdot kg^{-1}$. The concentrations of ENNB were similar to that of ENNA in the wheat sample, as reported in the case of wheat grains from Poland [30] and Morocco [32]. Moreover, the values of FBs (FB1+, FB2+, and FB3), T-2, and OTA ranged from 0.1 to 5.9 $\mu g \cdot kg^{-1}$, 0.2 to 0.7 $\mu g \cdot kg^{-1}$, and 0.2 to 3.7 $\mu g \cdot kg^{-1}$, respectively. These values are all far below the maximum residual limits authorized by the European Union regulations [40].

The main mycotoxins detected in corn (12 samples out of 14 were contaminated) were FBs with concentrations ranging from 0.1 to 1190.7 $\mu g \cdot kg^{-1}$. Samples from 2014 were more contaminated than the ones from 2015, indicating that storage conditions should be better controlled. The quantity of detected mycotoxins in rice was lower than that of wheat and corn. The concentrations of mycotoxins FBs, OTA, BEA, and SMC ranged from 0.1 to 11.0 $\mu g \cdot kg^{-1}$, and are below the maximum residual limits, as well as the mycotoxins that are not mentioned in the regulations, such as ENNs and BEA. We should strengthen supervision and try to identify the potential risks and the effect of toxin combinations.

3. Conclusions

The parameters of the QuEChERS purification combined with ultrahigh-performance liquid chromatography tandem mass spectrometry were optimized for the simultaneous quantification of 25 mycotoxins in different cereals (wheat, corn, and rice). The method is rapid, simple, and economical. The optimization was focused on the mobile phase type, elution gradient, the extraction solvent, and the matrix effect removal. The method provided a good linear relation, precision, LODs, LOQs, and recoveries. The matrix effect could be controlled by using a matrix-matched calibration. The experimental results also showed that the amino column NH_2 is suitable for the extraction of ENNs and BEA, whereas the C18 column is suitable for the FBs, which may provide useful suggestion for cleaning up some kinds of mycotoxins in future applications. The optimized method was applied to 26 wheat, 14 corn, and 25 rice samples. We could monitor and examine the potential risk of mycotoxins in cereals by a quick and reliable method.

4. Experimental Section

4.1. Materials

The reagents methanol, acetonitrile, and formic acid were of HPLC grade and were purchased from Thermo Fisher Scientific Inc. (Shanghai, China). The extraction kits QuEChERS (Part No: 5982-0650, Agilent Technologies Inc., Santa Clara, CA, USA), sorbent C18, GCB, and PSA were purchased from Sigma-Aldrich Corp. (St. Louis, MO, USA). The solid-phase extraction column C18 and the amino column NH_2 were purchased from Agilent Technologies Inc. (Santa Clara, CA, USA). Ultra-pure water was produced by a Mill-Q system (Millipore, Billerica, MA, USA). All other reagents were of analytical grade.

Solid mycotoxin standards of deoxynivalenol (DON), 3-acetyldeoxynivalenol (3-AcDON), 15-acetyldeoxynivalenol (15-AcDON), fusarenon-X (FUS-X), neosolaniol (NEO), 4,5-diacetoxyscirpenol (DAS), aflatoxin B1 (AF B1), aflatoxin B2 (AF B2), aflatoxin G1 (AF G1), aflatoxin G2 (AF G2), T-2 toxin, HT-2 toxin, fumonisin B1 (FB1), fumonisin B2 (FB2), fumonisin B3 (FB3), sterigmatocystin (SMC), verruculogen (VCG), ochratoxin A (OTA), zearalenone (ZEN), gliotoxin (GLT), enniatin A (ENNA), enniatin A1 (ENNA1), enniatin B (ENNB), enniatin B1 (ENNB1), and beauvericin (BEA) were purchased from Sigma-Aldrich (St. Louis, MO, USA).

4.2. Preparation of the Standard Solutions

The standards AF B1, AF B2, AF G1, and AF G2 were dissolved in methanol at concentrations of 200, 60, 200, and 60 µg·L^{-1}, respectively. Stock solutions of the other standards were prepared in methanol at a concentration of 100 mg/L. All the standard solutions were stored at $-18\,°C$ in darkness. The working standard solutions were freshly prepared before use by diluting the stock solution in order to obtain a 50:50 *v/v* ratio of methanol/water. The concentrations of the working mixed solutions were set accordingly to the response values of the mycotoxins by MS. The concentration of the working mixed standard solutions for DON, 3-AcDON, 15-AcDON, FUS-X, DAS, GLT, FB1, FB2, and FB3 were 200 µg·L^{-1}. In the case of ZEN, OTA, VCG, SMC, T-2, HT-2, NEO, ENNA, ENNA1, ENNB, ENNB1, and BEA, the concentration of the working mixed solutions was 50 µg·L^{-1}. Finally, the concentrations of the working mixed solutions for AF B1, AF B2, AF G1, and AF G2 were 10, 3, 10, and 3 µg·L^{-1}, respectively.

A blank substrate was measured with the matrix of a standard solution according to Section 2.3, with the same concentration as the working mixed standard solutions.

4.3. Sample Preparation

The sample (2 ± 0.05 g) was finely milled in a 50 mL centrifuge tube before extraction. In the case of spiked sample, the required volume of spiking standard solution was added prior to the extraction step. A mixture of acetonitrile/water (20 mL, 80:20, *v /v*) containing 0.1% (*v/v*) formic acid was added to the solution. The mixture was then vortexed for 30 s and shaken on an automatic thermostatic cultivation shaker (Yiheng Technology Co., Ltd., Shanghai, China) for 30 min. The QuEChERS extraction kit, containing sodium chloride (1.0 g), anhydrous magnesium sulfate (4.0 g), sodium citrate

(1.0 g), and sodium hydrogen citrate sesquihydrate (0.5 g), was added to the mixture and the tube was vigorously shaken by hand for 2 min. The sample was then centrifuged (3-18K, Sigma, St. Louis, MO, USA) at 8000 revolutions per minute (relative centrifugal force, RCF, $3500 \times g$) for 5 min at 10 °C. The supernatant (2 mL) containing the extracts was transferred to a centrifuge tube, and submitted to a stream of N_2 at 55 °C until complete dryness. The residue was sequentially dissolved in a mixture of methanol/water (0.5 mL, 1:1, *v/v*), vortexed for 1 min, then filtered through nylon filter (0.22 μm, Membrana GmbH, Wuppertal, Germany), and the extracts were finally recovered in a 200 μL microtube for UPLC–MS/MS analysis.

4.4. UPLC–MS/MS Conditions

The extracted mycotoxins were analyzed by ultrahigh-performance liquid chromatography coupled to tandem quadrupole mass spectrometry (UPLC–MS/MS, XEVO-TQ, Waters Corp., Milford, MA, USA). The data processing was performed using the MassLynx 4.0 software (Waters Corp., Milford, MA, USA).

The separation of the mycotoxins was performed using a CORTECS C18 column (100 × 2.1 mm, 1.6 μm, Waters Corp., Manchester, NH, USA) under a flow rate of 200 μL·min^{-1}. The volumes of the strong wash (90% methanol) and weak wash (10% methanol) solvents were 100 μL and 600 μL, respectively. The proportion of the mobile phase was set as described in Section 2.2. Mobile phase A was methanol and mobile phase B was a water solution containing 0.5% (*v/v*) formic acid. The mobile phase A was gradually eluted from 5% during 5.5 min, 850% during 5.8 min, 100% during 9.0 min, 5% during 10.0 min. The column was maintained at room temperature and the sample temperature was 20 °C. The injection volume was 5 μL.

Tandem mass spectrometry (MS/MS) detection was performed in a positive electrospray ionization mode (ESI+). For the infusion experiments, the mycotoxins standards (0.1 mg/L) were dissolved in methanol and a flow rate of 25 μL·min^{-1} was applied. The parameters were optimized by using the IntelliStart program supplied by Waters Corp. (Milford, MA, USA). The capillary voltage was set at 2.5 kV. Gaseous nitrogen was used as the cone, nebulizing, and desolvation gas. The source and desolvation temperature was 110 °C and 500 °C, respectively. The cone and desolvation gas flows were maintained at 20 L·h^{-1} and 800 L·h^{-1}, respectively. The collision gas flow was 0.17 mL·min^{-1}. The analysis of the mycotoxins was performed in a multiple reaction monitoring (MRM) mode. For each mycotoxin, at least one precursor ion and two fragment ions were monitored. The most abundant product ion was selected for quantification while the second-most intense ion was used for the qualitative analysis. The shape of the peaks was optimized by waiting 1 min before and after the retention time. The acquisition parameters for the 25 analyzed mycotoxins are summarized in Table 4.

Table 4. Ultrahigh-performance liquid chromatography tandem mass spectrometry (UPLC–MS/MS) acquisition parameters of the 25 mycotoxins.

Mycotoxins	Retention Time (min)	Precursor Ion (*m/z*)	Product Ion (*m/z*)	Dwell Time (s)	Cone Voltage (V)	Collision Energy (V)
DON	3.22	297.28	249.1 (Q) 203.1 (q)	0.005 0.005	20 20	10 14
15-ADON	4.38	339.3	137.1 (Q) 231.2 (q)	0.005 0.005	22 22	10 12
3-ADON	4.37	339.2	231.2 (Q) 213.2 (q)	0.005 0.005	20 20	14 22
AF B1	5.13	313.28	241.1 (Q) 285.0 (q)	0.005 0.005	46 46	38 22
AF B2	4.97	315.2	259.1 (Q) 287.1 (q)	0.005 0.005	40 40	30 26

Table 4. *Cont.*

Mycotoxins	Retention Time (min)	Precursor Ion (*m/z*)	Product Ion (*m/z*)	Dwell Time (s)	Cone Voltage (V)	Collision Energy (V)
AF G1	4.78	329.03	199.8 (Q)	0.005	40	40
			243.0 (q)	0.005	40	28
AF G2	4.63	331.29	189.1 (Q)	0.005	44	34
			245.1 (q)	0.005	44	28
ZEN	6.45	319.05	283.0 (Q)	0.005	18	22
			187.3 (q)	0.005	18	14
SMC	6.63	325.20	281.0 (Q)	0.005	45	36
			253.1 (q)	0.005	45	42
OTA	6.43	404.24	239.1 (Q)	0.005	26	36
			221.1 (q)	0.005	26	26
HT-2	5.75	447.30	345.2 (Q)	0.005	34	18
			285.1 (q)	0.005	34	22
T-2	6.08	489.31	245.1 (Q)	0.005	42	34
			387.2 (q)	0.005	42	22
VCG	6.73	534.32	392.2 (Q)	0.005	22	18
			191.1 (q)	0.005	22	24
ENNA	7.66	682.29	210.0 (Q)	0.005	48	26
			228.0 (q)	0.005	48	28
ENNA1	7.50	668.28	99.9 (Q)	0.005	48	60
			210.0 (q)	0.005	48	26
ENNB	7.30	640.18	195.9 (Q)	0.005	46	24
			213.9 (q)	0.005	46	26
ENNB1	7.45	654.26	99.9 (Q)	0.005	46	54
			195.9 (q)	0.005	46	24
BEA	7.36	784.24	243.9 (Q)	0.005	48	28
			262.0 (q)	0.005	48	26
FUS-X	3.74	354.84	174.8 (Q)	0.005	18	34
			136.8 (q)	0.005	18	34
GLT	5.22	326.81	262.9 (Q)	0.005	16	10
			244.9 (q)	0.005	16	16
NEO	3.82	399.93	214.9 (Q)	0.005	16	18
			304.9 (q)	0.005	16	14
DAS	5.18	383.94	306.9 (Q)	0.005	18	12
			246.9 (q)	0.005	18	16
FB1	5.56	722.35	334.2 (Q)	0.005	56	42
			352.2 (q)	0.005	56	36
FB2	6.10	706.36	318.3 (Q)	0.005	56	40
			336.3 (q)	0.005	56	38
FB3	5.87	706.48	336.3 (Q)	0.005	44	32
			95.0 (q)	0.005	44	48

Note: 25 mycotoxins standards were dissolved in methanol (0.1 mg/L). The protonated ion peaks (M + 1) (precursor ion, *m/z*) of the 25 mycotoxins were determined by full scan under flow injection mode. The fragment ions were obtained on the basis of the MS_2 (product ion scan) and two fragment ions with relatively higher peak intensities selected as the quantitative and qualitative ion, respectively. Automatic optimization of the mass parameters was conducted with the IntelliStart program from Waters Corp. Q: quantitative ion, q: qualitative ion.

Acknowledgments: This work was financially supported by the National Key Program on Quality and Safety Risk Assessment for Agro-products (2014 GJFP2014006) and The CAAS Agricultural Science and Technology Program for Innovation Team on Quality and Safety Risk Assessment of Cereal Products. We are grateful to all colleagues of our team and thank the reviewers of our paper for their valuable suggestions.

Author Contributions: B.W. and J.S. conceived and designed the experiments; J.S., W.L. performed the experiments; J.S. analyzed the data; J.S., W.L. and Y.Z. contributed reagents/materials/analysis tools; J.S. wrote the paper. B.W., X.H. and L.W. contributed suggestions and revision of the paper.

Conflicts of Interest: The authors declare no conflict of interest.

References

1. De Saeger, S.; Sibanda, L.; Desmet, A.; Van Peteghem, C. A collaborative study to validate novel field immunoassay kits for rapid mycotoxin detection. *Int. J. Food Microbiol.* **2002**, *75*, 135–142. [CrossRef]

2. Tamura, M.; Mochizuki, N. Development of a Multi-mycotoxin Analysis in Beer-based Drinks by a Modified QuEChERS Method and Ultra-High-Performance Liquid Chromatography Coupled with Tandem Mass Spectrometry. *Anal. Sci.* **2011**, *27*, 629–635. [CrossRef] [PubMed]

3. Tanaka, T.; Inoue, S.; Sugiura, Y.; Ueno, Y. Simultaneous determination of trichothecene mycotoxins and zaaralenone in cereal by gas chromatography mass spectrometry. *J. Chromatogr. A* **2000**, *882*, 23–28. [CrossRef]

4. Ferreira, I.; Fernandes, J.O.; Cunha, S.C. Optimization and validation of a method based in a QuEChERS procedure and gas chromatography–mass spectrometry for the determination of multi-mycotoxins in popcorn. *Food Control* **2012**, *27*, 188–193. [CrossRef]

5. Ndube, N.; van der Westhuizen, L.; Green, I.R.; Shephard, G.S. HPLC determination of fumonisin mycotoxins in maize: A comparative study of naphthalene-2,3-dicarboxaldehyde and *o*-phthaldialdehyde derivatization reagents for fluorescence and diode array detection. *J. Chromatogr. B* **2011**, *879*, 2239–2243. [CrossRef] [PubMed]

6. Chan, D.; MacDonald, S.J.; Boughtflower, V.; Brereton, P. Simultaneous determination of aflatoxins and ochratoxin A in food using a fully automated immunoaffinity column clean-up and liquid chromatography–fluorescence detection. *J. Chromatogr. A* **2004**, *1059*, 13–16. [CrossRef] [PubMed]

7. Romero-Gonzalez, R.; Garrido Frenich, A.; Martinez Vidal, J.L.; Prestes, O.D.; Grio, S.L. Simultaneous determination of pesticides, biopesticides and mycotoxins in organic products applying a quick, easy, cheap, effective, rugged and safe extraction procedure and ultra-high performance liquid chromatography-tandem mass spectrometry. *J. Chromatogr. A* **2011**, *1218*, 1477–1485. [CrossRef] [PubMed]

8. Zachariasova, M.; Lacina, O.; Malachova, A.; Kostelanska, M.; Poustka, J.; Godula, M.; Hajslova, J. Novel approaches in analysis of Fusarium mycotoxins in cereals employing ultra performance liquid chromatography coupled with high resolution mass spectrometry. *Anal. Chim. Acta* **2010**, *662*, 51–61. [CrossRef] [PubMed]

9. Arroyo-Manzanares, N.; Huertas-Pérez, J.F.; García-Campaña, A.M.; Gámiz-Gracia, L. Simple methodology for the determination of mycotoxins in pseudocereals, spelt and rice. *Food Control* **2014**, *36*, 94–101. [CrossRef]

10. Malachova, A.; Sulyok, M.; Beltran, E.; Berthiller, F.; Krska, R. Optimization and validation of a quantitative liquid chromatography-tandem mass spectrometric method covering 295 bacterial and fungal metabolites including all regulated mycotoxins in four model food matrices. *J. Chromatogr. A* **2014**, *1362*, 145–156. [CrossRef] [PubMed]

11. Škrbić, B.; Đurišić-Mladenović, M.G.N.; Živančev, A.J. Multi-mycotoxin Analysis by UHPLC-HESI-MSMS: A Preliminary Survey of Serbian Wheat Flour A. *Agron. Res.* **2011**, *9*, 461–468.

12. Vaclavik, L.; Zachariasova, M.; Hrbek, V.; Hajslova, J. Analysis of multiple mycotoxins in cereals under ambient conditions using direct analysis in real time (DART) ionization coupled to high resolution mass spectrometry. *Talanta* **2010**, *82*, 1950–1957. [CrossRef] [PubMed]

13. Rubert, J.; Dzuman, Z.; Vaclavikova, M.; Zachariasova, M.; Soler, C.; Hajslova, J. Analysis of mycotoxins in barley using ultra high liquid chromatography high resolution mass spectrometry: Comparison of efficiency and efficacy of different extraction procedures. *Talanta* **2012**, *99*, 712–719. [CrossRef] [PubMed]

14. Desmarchelier, A.; Tessiot, S.; Bessaire, T.; Racault, L.; Fiorese, E.; Urbani, A.; Chan, W.C.; Cheng, P.; Mottier, P. Combining the quick, easy, cheap, effective, rugged and safe approach and clean-up by immunoaffinity column for the analysis of 15 mycotoxins by isotope dilution liquid chromatography tandem mass spectrometry. *J. Chromatogr. A* **2014**, *1337*, 75–84. [CrossRef] [PubMed]

15. Sun, J.; Li, W.; Zhang, Y.; Sun, L.; Dong, X.; Hu, X.; Wang, B. Simultaneous determination of twelve mycotoxins in cereals by ultra-high performance liquid chromatography-tandem mass spectrometry. *Acta Agron. Sin.* **2014**, *40*, 691–701. [CrossRef]

16. Kirinčič, S.; Škrjanc, B.; Kos, N.; Kozolc, B.; Pirnat, N.; Tavčar-Kalcher, G. Mycotoxins in cereals and cereal products in Slovenia Official control of foods in the years 2008–2012. *Food Control* **2015**, *50*, 157–165. [CrossRef]

17. Myresiotis, C.K.; Testempasis, S.; Vryzas, Z.; Karaoglanidis, G.S.; Papadopoulou-Mourkidou, E. Determination of mycotoxins in pomegranate fruits and juices using a QuEChERS-based method. *Food Chem.* **2015**, *182*, 81–88. [CrossRef] [PubMed]

18. Arroyo-Manzanares, N.; Huertas-Perez, J.F.; Gamiz-Gracia, L.; Garcia-Campana, A.M. A new approach in sample treatment combined with UHPLC-MS/MS for the determination of multiclass mycotoxins in edible nuts and seeds. *Talanta* **2013**, *115*, 61–67. [CrossRef] [PubMed]

19. Fernandes, P.J.; Barros, N.; Câmara, J.S. A survey of the occurrence of ochratoxin A in Madeira wines based on a modified QuEChERS extraction procedure combined with liquid chromatography–triple quadrupole tandem mass spectrometry. *Food Res. Int.* **2013**, *54*, 293–301. [CrossRef]

20. Rubert, J.; Leon, N.; Saez, C.; Martins, C.P.; Godula, M.; Yusa, V.; Manes, J.; Soriano, J.M.; Soler, C. Evaluation of mycotoxins and their metabolites in human breast milk using liquid chromatography coupled to high resolution mass spectrometry. *Anal. Chim. Acta* **2014**, *820*, 39–46. [CrossRef] [PubMed]

21. Liu, Y.; Han, S.; Lu, M.; Wang, P.; Han, J.; Wang, J. Modified QuEChERS method combined with ultra-high performance liquid chromatography tandem mass spectrometry for the simultaneous determination of 26 mycotoxins in sesame butter. *J. Chromatogr. B* **2014**, *970*, 68–76. [CrossRef] [PubMed]

22. Sirhan, A.Y.; Tan, G.H.; Wong, R.C.S. Method validation in the determination of aflatoxins in noodle samples using the QuEChERS method (Quick, Easy, Cheap, Effective, Rugged and Safe) and high performance liquid chromatography coupled to a fluorescence detector (HPLC–FLD). *Food Control* **2011**, *22*, 1807–1813. [CrossRef]

23. Frenich, A.G.; Romero-Gonzalez, R.; Gomez-Perez, M.L.; Vidal, J.L. Multi-mycotoxin analysis in eggs using a QuEChERS-based extraction procedure and ultra-high-pressure liquid chromatography coupled to triple quadrupole mass spectrometry. *J. Chromatogr. A* **2011**, *1218*, 4349–4356. [CrossRef] [PubMed]

24. Belen Serrano, A.; Capriotti, A.L.; Cavaliere, C.; Piovesana, S.; Samperi, R.; Ventura, S.; Lagana, A. Development of a Rapid LC-MS/MS Method for the Determination of Emerging Fusarium mycotoxins Enniatins and Beauvericin in Human Biological Fluids. *Toxins* **2015**, *7*, 3554–3571. [CrossRef] [PubMed]

25. Rubert, J.; Soler, C.; Mañes, J. Application of an HPLC–MS/MS method for mycotoxin analysis in commercial baby foods. *Food Chem.* **2012**, *133*, 176–183. [CrossRef]

26. Capriotti, A.L.; Cavaliere, C.; Foglia, P.; Samperi, R.; Stampachiacchiere, S.; Ventura, S.; Lagana, A. Multiclass analysis of mycotoxins in biscuits by high performance liquid chromatography-tandem mass spectrometry. Comparison of different extraction procedures. *J. Chromatogr. A* **2014**, *1343*, 69–78. [CrossRef] [PubMed]

27. Azaiez, I.; Giusti, F.; Sagratini, G.; Mañes, J.; Fernández-Franzón, M. Multi-mycotoxins Analysis in Dried Fruit by LC/MS/MS and a Modified QuEChERS Procedure. *Food Anal. Methods* **2014**, *7*, 935–945. [CrossRef]

28. Koesukwiwat, U.; Sanguankaew, K.; Leepipatpiboon, N. Evaluation of modified QuEChERS method for analysis of mycotoxins in rice. *Food Chem.* **2014**, *153*, 44–51. [CrossRef] [PubMed]

29. Dzuman, Z.; Zachariasova, M.; Veprikova, Z.; Godula, M.; Hajslova, J. Multi-analyte high performance liquid chromatography coupled to high resolution tandem mass spectrometry method for control of pesticide residues, mycotoxins, and pyrrolizidine alkaloids. *Anal. Chim. Acta* **2015**, *863*, 29–40. [CrossRef] [PubMed]

30. Bryla, M.; Waskiewicz, A.; Podolska, G.; Szymczyk, K.; Jedrzejczak, R.; Damaziak, K.; Sulek, A. Occurrence of 26 Mycotoxins in the Grain of Cereals Cultivated in Poland. *Toxins* **2016**, *8*, 160. [CrossRef] [PubMed]

31. Arroyo-Manzanares, N.; Huertas-Perez, J.F.; Gamiz-Gracia, L.; Garcia-Campana, A.M. Simple and efficient methodology to determine mycotoxins in cereal syrups. *Food Chem.* **2015**, *177*, 274–279. [CrossRef] [PubMed]

32. Blesa, J.; Moltó, J.-C.; El Akhdari, S.; Mañes, J.; Zinedine, A. Simultaneous determination of Fusarium mycotoxins in wheat grain from Morocco by liquid chromatography coupled to triple quadrupole mass spectrometry. *Food Control* **2014**, *46*, 1–5. [CrossRef]

33. Ren, Y.; Zhang, Y.; Shao, S.; Cai, Z.; Feng, L.; Pan, H.; Wang, Z. Simultaneous determination of multi-component mycotoxin contaminants in foods and feeds by ultra-performance liquid chromatography tandem mass spectrometry. *J. Chromatogr. A* **2007**, *1143*, 48–64. [CrossRef] [PubMed]

34. Wang, R.; Su, X.; Chen, F.; Wang, P.; Fan, X.; Zhang, W. Determination of 26 Mycotoxins in Feedstuffs by Multifunctional Cleanup Column and Liquid Chromatography-Tandem Mass Spectrometry. *Chin. J. Anal. Chem.* **2015**, *43*, 264–270. [CrossRef]

35. Njumbe Ediage, E.; Van Poucke, C.; De Saeger, S. A multi-analyte LC-MS/MS method for the analysis of 23 mycotoxins in different sorghum varieties: The forgotten sample matrix. *Food Chem.* **2015**, *177*, 397–404. [CrossRef] [PubMed]

36. Ren, Y.; Zhang, Y.; Lai, S.; Han, Z.; Wu, Y. Simultaneous determination of fumonisins B1, B2 and B3 contaminants in maize by ultra high-performance liquid chromatography tandem mass spectrometry. *Anal. Chim. Acta* **2011**, *692*, 138–145. [CrossRef] [PubMed]

37. Bolechová, M.; Benešová, K.; Běláková, S.; Čáslavský, J.; Pospíchalová, M.; Mikulíková, R. Determination of seventeen mycotoxins in barley and malt in the Czech Republic. *Food Control* **2015**, *47*, 108–113. [CrossRef]

38. Yogendrarajah, P.; Van Poucke, C.; De Meulenaer, B.; De Saeger, S. Development and validation of a QuEChERS based liquid chromatography tandem mass spectrometry method for the determination of multiple mycotoxins in spices. *J. Chromatogr. A* **2013**, *1297*, 1–11. [CrossRef] [PubMed]

39. Oueslati, S.; Meca, G.; Mliki, A.; Ghorbel, A.; Mañes, J. Determination of Fusarium mycotoxins enniatins, beauvericin and fusaproliferin in cereals and derived products from Tunisia. *Food Control* **2011**, *22*, 1373–1377. [CrossRef]

40. The Commission of the European Communities. Commission Regulation (EC) No 1881/2006 of 19 December 2006 setting maximum levels for certain contaminants in foodstuffs. *Off. J. Eur. Unione* **2006**, *L364*, 5–24.

© 2016 by the authors. Licensee MDPI, Basel, Switzerland. This article is an open access article distributed under the terms and conditions of the Creative Commons Attribution (CC BY) license (http://creativecommons.org/licenses/by/4.0/).

toxins

MDPI

Article

Harmonized Collaborative Validation of Aflatoxins and Sterigmatocystin in White Rice and Sorghum by Liquid Chromatography Coupled to Tandem Mass Spectrometry

Hyun Ee Ok [1], Fei Tian [1], Eun Young Hong [1], Ockjin Paek [2], Sheen-Hee Kim [2], Dongsul Kim [2] and Hyang Sook Chun [1,*]

[1] Advanced Food Safety Research Group, BK21 Plus, School of Food Science and Technology, Chung-Ang University, Anseong 17546, Korea; ohe0904@hanmail.net (H.E.O.); TianFei@cau.ac.kr (F.T.); eyhongs@gmail.com (E.Y.H.)

[2] Food Contaminants Division, National Institute of Food & Drug Safety Evaluation, Osong 28159, Korea; ojpaek92@korea.kr (O.P.); cinee@korea.kr (S.-H.K.); dongsul@korea.kr (D.K.)

* Correspondence: hschun@cau.ac.kr; Tel.: +82-10-8946-9273

Academic Editor: Aldo Laganà
Received: 5 October 2016; Accepted: 6 December 2016; Published: 13 December 2016

Abstract: An interlaboratory study was performed in eight laboratories to validate a liquid chromatography–tandem mass spectrometry (LC/MS/MS) method for the simultaneous determination of aflatoxins and sterigmatocystin (STC) in white rice and sorghum (*Sorghum bicolor*). Fortified samples (at three different levels) of white rice and sorghum were extracted, purified through a solid-phase extraction (SPE) column, and then analyzed by LC/MS/MS. The apparent recoveries (ARs) ranged from 78.8% to 95.0% for aflatoxins and from 85.3% to 96.7% for STC. The relative standard deviations for repeatability (RSD_r) and reproducibility (RSDR) of aflatoxins were in the ranges 7.9%–33.8% and 24.4%–81.0%, respectively. For STC, the RSD_r ranged from 7.1% to 40.2% and the RSDR ranged from 28.1% to 99.2%. The Horwitz ratio values for the aflatoxins and STC ranged from 0.4 to 1.2 in white rice and from 0.3 to 1.0 in sorghum, respectively. These results validated this method for the simultaneous determination of aflatoxins and STC by LC/MS/MS after SPE column cleanup. The percentages of satisfactory Z-score values ($|Z| \leq 2$) were the following: for white rice, 100% for aflatoxins and STC; for sorghum, 100%, except in data from two laboratories for STC (0.3 µg/kg). This validated that the LC/MS/MS method was successfully applied for the determination of aflatoxins and STC in 20 white rice and 20 sorghum samples sourced from Korean markets.

Keywords: aflatoxins; sterigmatocystin; simultaneous determination; liquid chromatography–tandem mass spectrometry

1. Introduction

Concern about human exposure to aflatoxins has tended inevitably to focus on high-risk commodities such as corn, nuts, and dried fruit, where levels of aflatoxins can be both variable and relatively high [1,2]. Although it is known that cereals can be contaminated with aflatoxins, the research emphasis, e.g., for wheat, has tended towards the monitoring and control of *Fusarium* toxins, where toxins such as deoxynivalenol commonly and frequently occur at mg/kg levels [3]. Although rice is not immediately thought of as a high-risk commodity in terms of contamination levels of aflatoxins, there is substantial evidence indicating endemic low concentration (mg/kg) occurrence of aflatoxin B_1 contamination in rice [4,5]. Because rice is a staple food worldwide, low-level contamination can be of concern because it can lead to long-term exposure at above recommended levels. Sorghum is another important cereal crop worldwide. It is used in food items such as cookies, cakes, porridge, unleavened bread, and beverages [6]. Sorghum contains significant amounts of tannins in the testa, low amounts of phenolic acids in the grain [7], and abundant polyphenols, which enhance its resistance to pests and microbial infestation [8]. However, sorghum is vulnerable to fungal contamination. It has been estimated that economic losses in Asia and Africa due to fungal infestation were >$130 million annually [9]. *Aspergillus, Fusarium, Penicillium,* and *Alternaria* species are commonly detected in contaminated sorghum, and mycotoxigenic strains of these fungal species have been isolated from different sorghum varieties [9].

Aflatoxins is a family of structurally related mycotoxins that includes aflatoxin B_1 (AFB$_1$), aflatoxin B_2 (AFB$_2$), aflatoxin G_1 (AFG$_1$), and aflatoxin G_2 (AFG$_2$) [10]. These are the most important mycotoxins detected in food [11] and have been classified in group 1 as human carcinogens by the International Agency for Research on Cancer [12]. Many strains of *Aspergillus*, such as *Aspergillus flavus, Aspergillus parasiticus, Aspergillus bombycis, Aspergillus ochraceoroseus, Aspergillus nomius,* and *Aspergillus pseudotamari* can produce aflatoxins [13]. Maximum levels of aflatoxin contamination have been established by different countries to protect public health. In particular, the European Commission regulated 2 µg/kg for AFB$_1$ and 4 µg/kg for total aflatoxins in cereals and derived products [14], and the Korea Ministry of Food and Drug Safety imposed limits of AFB$_1$ < 10 µg/kg and total aflatoxins < 15 µg/kg [15].

Sterigmatocystin (STC) is a mycotoxin produced by the fungi of many different *Aspergillus* species. Other species such as *Bipolaris, Chaetomium,* and *Emericella* are also able to produce STC. STC-producing fungi have frequently been isolated from different foodstuffs. STC has been detected regularly in grains, corn, bread, cheese, spices, coffee beans, soybeans, pistachio nuts, animal feed, and silage [16]. STC exhibits various toxicological, mutagenic, and carcinogenic effects in animals, and it has been recognized as a 2B carcinogen (possible human carcinogen) by the International Agency for Research on Cancer [17]. Recent reviews described the occurrence of STC in a variety of foodstuffs [18–20]. The European Food Safety Authority (EFSA) Panel on Contaminants in the Food Chain (CONTAM Panel) analyzed a total of 1259 samples of cereal grains, cereal products, and nuts, collected between August 2013 and November 2014 in nine European countries (and originating from 45 countries), for the presence of STC. In cereal grains, STC was detected in 2%–6% of the wheat, rye, maize, and barley samples, mostly at levels < 1.5 µg/kg. A higher incidence and higher levels of contamination (14 samples, 1.5–6 µg/kg; 1 sample, 33 µg/kg) were observed in rice (in virtually all unprocessed rice and 21% of processed rice from the EU) and oats (22%) [21]. Sorghum is also commonly detected with STC contamination. Queslati et al. [22] analyzed 60 sorghum samples from Tunisian markets; 33% of them were detected with STC contamination at the mean level of 20.5 µg/kg. Chala et al. [23] detected similar levels of contamination (21.2 µg/kg) in 34% of the sorghum samples (70 samples) collected from Ethiopian markets; the highest concentration was 323 µg/kg.

Analytical methods involving chromatography have been developed for STC and aflatoxins. Thin-layer chromatography [24], high-performance liquid chromatography (HPLC) [25,26], and gas chromatography [27] have been used. More recently, liquid chromatography–mass spectrometry (LC/MS) methods were reported [28].

Liquid chromatography–tandem mass spectrometry (LC/MS/MS) has been shown to be suitable for the analysis of mycotoxins in cereals [29–31]; it enables simultaneous qualification and quantification. According to the EFSA [32], among the analytical methods available for the determination of STC, LC/MS methods demonstrate the limit of detection (LOD). However, the sensitivity of these methods depends on the matrices and methods used.

In the present study, we validated the method for quantifying aflatoxins and STC in white rice and sorghum (*Sorghum bicolor*) through an interlaboratory study. The method uses a single solid phase extract column for cleanup, and simultaneous determination by LC/MS/MS for the toxins.

2. Results

2.1. Collaborative Validation of LC/MS/MS Determination of Aflatoxins and STC in White Rice

The results of the interlaboratory study of aflatoxins and STC in white rice are tabulated in Table 1. The mean apparent recoveries (ARs) at all spike levels (0.3–10 μg/kg) were within the acceptable limits (70%–110%). However, the mean AR of aflatoxins (89%) was below the ARs of STC (94%) and the ARs of AFG_1 and AFG_2 (82%–88%) were noticeably lower in comparison with AFB_1 and AFB_2 (91%–93%). No outliers of aflatoxins or STC were observed for white rice. The relative standard deviations for repeatability (RSDr) and reproducibility (RSDR) for aflatoxins were 5.7%–18.6% and 26.0%–44.1%, respectively. A Horwitz ratio (HorRat) value in the range 0.5–2.0 was confirmed at all the spiked levels. Furthermore, values for RSDr, RSDR, and HorRat for STC were 5.2%–18.9%, 15.3%–21.9%, and 0.4 (<0.5), respectively.

2.2. Collaborative Validation Results of LC/MS/MS Determination of Aflatoxins and STC in Sorghum

The results of the interlaboratory study of aflatoxins and STC in sorghum are tabulated in Table 2. The mean ARs for aflatoxins and STC were in the ranges 79%–99% and 85%–97%, respectively. One outlier was observed in the sorghum sample spiked with 1.0 μg/kg of aflatoxins. The RSDr and RSDR for aflatoxins were in the ranges 8.1%–24.4% and 13.6%–42.5%, respectively. The HorRat values of aflatoxins ranged from 0.3 to 1.0. For STC, the AR, RSDr, and RSDR were in the ranges 85%–97%, 14.2%–29.0%, and 32.0%–54.7%, respectively. The HorRat values of STC ranged from 0.8 to 1.0.

2.3. Calculation of the Z-Scores

Eight Korean laboratories participated in the collaborative validation of aflatoxins and STC in white rice and sorghum. Calculation of the Z-scores to evaluate each laboratory's proficiency was based on the results provided by the participating laboratories (Tables 3 and 4). Eight laboratories completed the interlaboratory comparison for aflatoxins and STC successfully. Mean Z-scores were 0.58 and 0.49 for white rice and sorghum, respectively. All Z-scores for aflatoxins in both white rice and STC, which varied from –1.91 to 1.33, were within the acceptable limits ($|Z| \leq 2$). The Z-scores reported for STC were within the range –2.19 to 2.82, with two unsatisfactory data (–2.19 and 2.82) at the 0.3 μg/kg level.

2.4. Application of the Validated LC/MS/MS Method to Market Samples

The validated LC/MS/MS method was used to determine aflatoxins and STC in 20 white rice and 20 sorghum samples sourced from Korean markets. Results are tabulated in Table 5. No rice samples were found to be contaminated with AFB_1, AFB_2, AFG_1 or AFG_2.

In sorghum samples, AFB_1 was detected in 10% (2/20) of samples with values above the LOD. In one sorghum sample, AFB_1 was detected between the LOD and limit of quantification (LOQ), and one sample was above the LOQ (detection range 0.1–1.0 μg/kg). AFG_2 was detected in 15% (3/20) of sorghum samples with values above the LOD. In three samples, it was detected between the LOD and LOQ. No sorghum samples were detected with AFB_2 or AFG_1.

Regarding STC, it was detected in 70% (14/20) of white rice samples with values below the LOQ, while in 30% (6/20) it was detected between the LOD and LOQ. In contrast, STC was detected in only 10% (2/20) of sorghum samples. In two sorghum samples, STC and AFB_1 were detected, but their concentrations were low (0.1–1.0 μg/kg). Furthermore, in one sorghum sample, STC, AFB_1, and AFG_2 were detected simultaneously.

Table 1. Collaborative validation results of liquid chromatography–tandem mass spectrometry (LC/MS/MS) determination of aflatoxins and sterigmatocystin (STC) in white rice.

Laboratory	AFB$_1$ [a]			AFB$_2$ [b]			AFG$_1$ [c]			AFG$_2$ [d]			STC [e]		
							Spiked Samples (μg/kg)								
	1.0	2.5	5.0	1.0	2.5	5.0	1.0	2.5	5.0	2.0	5.0	10.0	0.30	0.75	1.50
1	1.09	2.54	5.02	1.08	2.55	5.13	1.07	2.54	5.03	2.00	5.06	10.78	0.30	0.71	1.55
2	0.69	1.51	3.24	0.87	1.67	3.58	0.62	1.56	3.13	1.41	3.01	6.48	0.26	0.61	1.19
3	0.76	1.96	3.96	0.76	1.89	3.85	0.63	1.80	3.71	1.38	3.23	7.04	0.29	0.69	1.37
4	0.95	2.56	4.68	0.98	2.57	4.99	0.91	2.14	3.86	1.73	3.05	7.65	0.22	0.65	1.24
5	0.99	2.43	4.76	1.04	2.54	4.90	0.95	2.48	4.98	2.10	5.26	9.93	0.28	0.68	1.35
6	1.01	2.63	5.02	0.95	2.58	5.14	1.04	2.72	5.40	2.03	5.17	10.37	0.29	0.70	1.38
7	0.69	1.78	3.91	0.49	1.53	3.74	0.60	1.58	3.77	1.28	2.72	6.94	0.34	0.75	1.55
8	1.13	2.84	5.65	1.11	2.79	5.70	1.08	2.71	5.27	2.03	5.26	11.10	0.33	0.80	1.53
Mean (μg/kg) [f]	0.91	2.28	4.53	0.91	2.27	4.63	0.86	2.19	4.39	1.75	4.10	8.79	0.29	0.70	1.39
AR (%) [f]	91.0	91.2	90.6	91.0	90.8	92.6	86.0	87.6	87.8	87.5	82.0	87.9	96.7	93.3	92.7
Outlier [g]	0	0	0	0	0	0	0	0	0	0	0	0	0	0	0
RSD$_r$ (%) [h]	16.4	5.7	9.3	16.9	13.4	7.3	18.1	15.6	9.4	18.6	13.3	10.9	15.9	18.9	5.2
RSDR (%) [i]	31.2	31.2	26.5	35.9	33.7	26.0	39.1	35.4	30.3	32.1	44.1	34.0	21.9	18.4	15.3
HorRat	0.7	0.8	0.7	0.8	0.9	0.7	0.9	0.9	0.9	0.8	1.2	1.1	0.4	0.4	0.4

[a] AFB$_1$: aflatoxin B$_1$; [b] AFB$_2$: aflatoxin B$_2$; [c] AFG$_1$: aflatoxin G$_1$; [d] AFG$_2$: aflatoxin G$_2$; [e] STC: sterigmatocystin; [f] AR: apparent recovery; [g] Outlier: cochran and single Grubbs parameters; [h] RSD$_r$: relative standard deviation of repeatability; [i] RSDR: relative standard deviation of reproducibility.

Table 2. Collaborative validation results of LC/MS/MS determination of aflatoxins and STC in sorghum.

Laboratory	AFB₁[a]			AFB₂[b]			AFG₁[c]			AFG₂[d]			STC[e]		
	1.0	2.5	5.0	1.0	2.5	5.0	1.0	2.5	5.0	2.0	5.0	10.0	0.30	0.75	1.50
1	1.00	2.44	4.64	0.89	2.04	4.79	0.95	2.24	4.77	1.81	4.26	9.50	0.21	0.51	1.06
2	1.03	1.69	2.87	0.74	1.55	2.94	0.79	1.71	2.98	2.22	3.50	6.03	0.14	0.35	0.76
3	1.10	2.39	4.56	0.93	2.22	4.23	1.01	2.25	4.31	0.82	3.38	7.67	0.27	0.65	1.45
4	0.62	1.84	3.49	0.71	1.86	4.46	0.58	2.46	4.24	1.28	3.96	9.18	0.25	0.66	1.35
5	0.90	2.35	4.88	0.93	2.19	4.88	0.89	2.23	4.68	1.87	4.31	9.36	0.30	0.72	1.46
6	1.04	2.62	5.40	1.07	2.57	5.42	1.01	2.55	5.68	2.01	4.75	10.34	0.32	0.76	1.56
7	0.97	1.94	3.96	0.93	1.83	4.04	0.50	1.43	3.63	1.54	3.04	7.10	0.47	0.72	1.23
8	0.92	2.37	3.95	0.92	2.42	3.46	0.91	2.32	3.48	1.65	4.31	6.38	0.33	0.77	1.39
Mean (µg/kg)[f]	0.99	2.21	4.22	0.89	2.08	4.28	0.83	2.15	4.22	1.65	3.94	8.19	0.29	0.64	1.28
AR (%)[f]	99.0	88.4	84.4	89.0	83.2	85.6	83.0	86.0	84.4	82.5	78.8	81.9	96.7	85.3	85.3
Outlier[g]	1	0	0	0	0	0	0	0	0	0	0	0	0	0	0
RSDr (%)[h]	11.9	14.6	8.1	18.3	17.4	9.6	21.6	14.6	17.4	19.3	24.4	12.8	29.0	24.2	14.2
RSDR (%)[i]	13.6	24.8	29.4	23.0	26.9	28.8	38.5	28.6	32.7	42.5	28.1	30.8	54.7	38.0	32.0
HorRat	0.3	0.6	0.8	0.5	0.7	0.8	0.9	0.7	0.9	1.0	0.8	1.0	1.0	0.8	0.8

[a] AFB₁: aflatoxin B₁; [b] AFB₂: aflatoxin B₂; [c] AFG₁: aflatoxin G₁; [d] AFG₂: aflatoxin G₂; [e] STC: sterigmatocystin; [f] AR: apparent recovery; [g] Outlier: cochran and single Grubbs parameters; [h] RSDr: relative standard deviation of repeatability; [i] RSDR: relative standard deviation of reproducibility.

Table 3. Results of mycotoxin analysis and relevant scoring in white rice.

Toxins	Assigned Values (µg/kg)	Laboratory							
		1	2	3	4	5	6	7	8
AFB$_1$ [a]	1	0.79	−1.03	−0.70	0.16	0.34	0.43	−1.00	0.99
	2.5	0.47	−1.40	−0.57	0.50	0.27	0.63	−0.91	1.02
	5	0.45	−1.17	−0.52	0.13	0.21	0.45	−0.56	1.02
AFB$_2$ [b]	1	0.76	−0.18	−0.70	0.32	0.61	0.19	−1.91	0.92
	2.5	0.52	−1.08	−0.69	0.56	0.50	0.58	−1.34	0.96
	5	0.46	−0.95	−0.71	0.33	0.25	0.46	−0.80	0.97
AFG$_1$ [c]	1	0.95	−1.12	−1.05	0.23	0.40	0.82	−1.20	0.97
	2.5	0.64	−1.14	−0.71	−0.09	0.52	0.96	−1.12	0.95
	5	0.58	−1.15	−0.62	−0.48	0.54	0.91	−0.57	0.79
AFG$_2$ [d]	2	0.58	−0.77	−0.82	−0.04	0.80	0.64	−1.05	0.65
	5	0.87	−0.98	−0.78	−0.95	1.06	0.98	−1.25	1.06
	10	0.91	−1.05	−0.80	−0.51	0.52	0.72	−0.84	1.05
STC [e]	0.3	0.17	−0.40	−0.05	−0.98	−0.09	−0.01	0.79	0.57
	0.75	0.04	−0.54	−0.03	−0.32	−0.10	0.02	0.31	0.63
	1.5	0.48	−0.61	−0.08	−0.46	−0.15	−0.05	0.47	0.40

[a] AFB$_1$: aflatoxin B$_1$; [b] AFB$_2$: aflatoxin B$_2$; [c] AFG$_1$: aflatoxin G$_1$; [d] AFG$_2$: aflatoxin G$_2$; [e] STC: sterigmatocystin.

Table 4. Results of mycotoxin analysis and relevant scoring in sorghum.

Toxins	Assigned Values (µg/kg)	Laboratory							
		1	2	3	4	5	6	7	8
AFB$_1$ [a]	1	0.26	0.38	0.69	−1.50	−0.23	0.41	0.11	−0.12
	2.5	0.43	−0.93	0.34	−0.66	0.26	0.75	−0.48	0.30
	5	0.38	−1.22	0.31	−0.66	0.60	1.07	−0.24	−0.25
AFB$_2$ [b]	1	0.02	−0.68	0.17	−0.80	0.17	0.80	0.16	0.15
	2.5	−0.09	−0.97	0.24	−0.41	0.20	0.88	−0.46	0.60
	5	0.47	−1.22	−0.04	0.17	0.54	1.04	−0.21	−0.74
AFG$_1$ [c]	1	0.54	−0.19	0.84	−1.14	0.27	0.82	−1.52	0.38
	2.5	0.17	−0.79	0.18	0.57	0.15	0.73	−1.31	0.31
	5	0.50	−1.13	0.08	0.02	0.42	1.33	−0.54	−0.67
AFG$_2$ [d]	2	0.36	1.31	−1.89	−0.84	0.50	0.82	−0.25	0.00
	5	0.29	−0.40	−0.51	0.01	0.34	0.74	−0.82	0.34
	10	0.59	−0.98	−0.24	0.45	0.53	0.98	−0.50	−0.82
STC [e]	0.3 [f]	−1.17	−2.19	−0.32	−0.49	0.25	0.51	2.82	0.60
	0.75	−0.81	−1.79	0.06	0.12	0.47	0.71	0.44	0.79
	1.5	−0.67	−1.57	0.50	0.19	0.55	0.84	−0.17	0.33

[a] AFB$_1$: aflatoxin B$_1$; [b] AFB$_2$: aflatoxin B$_2$; [c] AFG$_1$: aflatoxin G$_1$; [d] AFG$_2$: aflatoxin G$_2$; [e] STC: sterigmatocystin; [f] Two laboratories had $2 < |Z| \leq 3$.

Table 5. Application of validated LC/MS/MS for determination of aflatoxins and sterigmatocystin in rice and sorghum samples sourced from Korean markets.

Myotoxin	Product	Sample	Number of Samples			Concentration Range [h] (µg/kg)	
			<LOD [f]	LOD–LOQ [g]	>LOQ	0.1~1.0	1.0~2.0
AFB$_1$ [a]	Rice	20	20 (100) [i]	-	-	-	-
	Sorghum	20	18 (90)	1 (5)	1 (5)	2 (10)	-
AFB$_2$ [b]	Rice	20	20 (100)	-	-	-	-
	Sorghum	20	20 (100)	-	-	-	-
AFG$_1$ [c]	Rice	20	20 (100)	-	-	-	-
	Sorghum	20	20 (100)	-	-	-	-
AFG$_2$ [d]	Rice	20	20 (100)	-	-	-	-
	Sorghum	20	17 (85)	3 (15)	-	-	-
STC [e]	Rice	20	14 (70)	6 (30)	-	-	-
	Sorghum	20	-	18 (90)	2 (10)	2 (10)	-

[a] AFB$_1$: aflatoxin B$_1$; [b] AFB$_2$: aflatoxin B$_2$; [c] AFG$_1$: aflatoxin G$_1$; [d] AFG$_2$: aflatoxin G$_2$; [e] STC: sterigmatocystin; [f] LOD: limit of detection; [g] LOQ: limit of quantification; [h] AF concentration was not corrected for the AR result; [i] Figure in parenthesis indicates percentage.

3. Discussion

This study describes a harmonized collaborative validation of multiple mycotoxin detection in white rice and sorghum using LC/MS/MS. We found that the LC/MS/MS method possessed the performance characteristics required to obtain accurate results. In the case of white rice spiked with 1.0–10.0 µg/kg aflatoxins, the mean AR (82%–91%), and the values obtained for RSDr (5.7%–18.6%) and RSDR (26.5%–44.1%) were considered acceptable, taking into consideration the performance criteria (AR 70%–110% and RSD$_R$ ≤ 45.2%) suggested by EU guidelines for aflatoxins [33]. The RSDR value (44.1%) of AFG$_2$ spiked samples (5.0 µg/kg) was close to the maximum value stated by the EU guidelines. It might be caused by the sample-to-sample variation of matrix effects or the inter-laboratory variability result from different technical expertise and established workflows for individual laboratories. The important expression of the method's precision is the HorRat value, which is calculated as a ratio of %RSDR to the predicted reproducibility RSD, %PRSDR. The %PRSDR is a function of the analyte concentration and is expressed as $2C^{-0.1505}$, where C is the estimated mean concentration [24]. In comparison with AOAC guidelines [24], a HorRat value in the range 0.5–2.0 was confirmed for white rice at all the spiked levels. This indicated that the presented method was reproducible for the determination of aflatoxins contained in white rice. HorRat values of STC were 0.4 (<0.5). Consistent deviations from the ratio that were on the low side (values < 0.5) may indicate unreported averaging or excellent training and experience [34].

For spiked sorghum, all parameters of aflatoxins satisfied the criteria mentioned [33] above, although one outlier (0.62 µg/kg) was observed in the AFB$_1$ result at the 1.0 µg/kg level. Results also showed a HorRat value of 0.3 at the 1.0 µg/kg level of AFB$_1$ in sorghum. However, no problems were recognized throughout this interlaboratory study, suggesting that outliers may have been due to random errors. It was also noticeable that the ARs of aflatoxins and STC in sorghum were slightly lower than they were in white rice, perhaps due to the matrix suppression effect of sorghum.

Calculation of the Z-scores suggested that all eight laboratories completed the interlaboratory comparison successfully. Unsatisfactory Z-scores for STC from individual laboratories were mainly a consequence of low or high ARs.

In summarizing information on analytical methods for STC in foodstuffs, Veršilovskis and de Saeger [19] concluded that LC/MS/MS should be used in efforts to develop sensitive methods and that sample preparation should be improved. This was realized in some new methods developed for the determination of STC in various grains and cheese by Veršilovskis et al. (LOD 0.03–0.15 µg/kg) [20,35] and in the multi-mycotoxin method reported by Sulyok et al. (LOD 0.4 µg/kg) [29]. Furthermore, Goto and co-workers [36] developed a new method for analyzing STC in grains using an immunoaffinity column (IAC) and LC/MS. This method is effective for STC analysis in grains and holds potential for a new application of a commercial IAC in STC analysis of aflatoxins. Based on EU guidelines, the AR, repeatability, and reproducibility levels were acceptable. Hence, this method should be useful for future studies on the occurrence and risk attached to aflatoxins and STC.

In our study, the validated LC/MS/MS method was successfully applied for the analysis of natural samples sourced from Korean markets. The results showed a low occurrence of aflatoxins in both white rice and sorghum samples, with no sample exceeding the EU maximum limits for aflatoxins (<2 µg/kg) [17,18]. AFB_1 and AFG_2 with concentrations > LOD values were only detected in sorghum samples. In contrast, the incidence of STC with concentrations < LOD values was high in sorghum samples, but their concentrations were low (0.1–1.0 µg/kg). Of particular importance was the simultaneous detection of STC, AFB_1, and AFG_2 in a sorghum sample.

4. Conclusions

Harmonized collaborative validation of aflatoxins and STC in white rice and sorghum was carried out using LC/MS/MS. The HorRat value was <1.5, which indicates that the method described in this study is reproducible for the determination of aflatoxins and STC contained in both white rice and sorghum at concentrations in the range 0.3–10.0 µg/kg. Currently, no country has legislation for STC. As aflatoxins and STC are naturally occurring substances, it is important to gather further information regarding suitable analytical methods for their detection. Our results indicate that the validated LC/MS/MS method described here can be successfully applied for the determination of aflatoxins and STC that naturally contaminate white rice and sorghum.

5. Materials and Methods

5.1. Reagents and Chemicals

Formic acid (LC/MS grade) was purchased from Fisher Scientific (Geel, Belgium). Water, acetonitrile, and methanol (HPLC grade) were purchased from J.T. Baker (Phillipsburg, NJ, USA). Mycotoxin solid standards AFB_2 (5 mg, Lot No. A9887), AFG_1 (1 mg, Lot No. A0138), and AFG_2 (5 mg, Lot No. A0263) were purchased from Sigma-Aldrich (St. Louis, MO, USA). AFB_1 (2.01 µg/mL in acetonitrile, Lot No. L15153A) and STC (50.2 µg/mL in acetonitrile, Lot No. L14424S) were purchased from Romer Labs (Tulln, Austria). For each aflatoxin and STC standard, a stock solution of 100 ng/mL was prepared in acetonitrile and stored at −20 °C. The aflatoxin solutions were calibrated spectrophotometrically at 350 nm, using the method published by the association of analytical communities (AOAC) [37]. Aflatoxins and STC working standard solutions were prepared by evaporating an exact volume of stock solution under nitrogen gas and re-dissolving the residue in acetonitrile to reach a final concentration of 10 ng/mL for AFB_1, AFB_2 and AFG_1, 20 ng/mL for AFG_2, and 3 ng/mL for STC, respectively.

5.2. Interlaboratory Study Design

Eight laboratories in Korea participated in this interlaboratory study to validate the method. The same aflatoxin and STC free samples of white rice and sorghum that were previously examined by a laboratory were spiked with aflatoxin and STC standard solutions to reach a final concentration of 1, 2.5 and 5 µg/kg for AFB_1, AFB_2, AFG_1 and AFG_2, and 0.3, 0.75 and 1.5 µg/kg for STC, respectively. All spiked samples were prepared by the organizer's laboratory and sent to each participating laboratory for analysis. Each participant was supplied with three bottles of blank white rice and sorghum samples, and three bottles of spiked (3 levels) white rice and sorghum samples.

5.3. Sample Extraction

A spiked 5 g sample was placed in a 50 mL conical tube with 20 mL of extract solution (acetonitrile/water, 50:50, *v*/*v* containing 0.1% formic acid) and placed on a vortex shaker for 5 min. After extraction, the sample was centrifuged at 3500 rpm for 5 min and then filtered through glass microfiber filter paper (GF/A grade, 110 mm; Whatman, Little Chalfont, UK). A 4 mL volume of filtrate extract was diluted with 16 mL water. An SPE column (ISOLUTE® Myco, 60 mg/3 mL; Biotage, Cardiff, UK) was preconditioned with 2 mL acetonitrile and then 2 mL water. Five milliliters of dilute solution were applied to the SPE column, followed by washing with 2 mL water and then 2 mL acetonitrile/water (10:90, *v*/*v*). The toxins that remained on the column were eluted with 4 mL acetonitrile. The collected eluate was evaporated with N_2 gas under vacuum at 50 °C. The residue was dissolved in 1 mL acetonitrile/water (50:50, *v*/*v*) and then filtered through a polyvinylidene fluoride syringe filter (0.2 µm; Whatman, Pittsburgh, PA, USA).

5.4. Calibration Solutions

Aflatoxins and STC free samples of white rice and sorghum were used for the manufacture of matrix-matched calibration curves. Sample solutions were prepared using the same method described in sample extraction. Standard solution of mycotoxin was added into sample solution to reach different concentrations (0.5, 1.0, 2.5, 5.0, 10.0 µg/kg for AFB_1, AFB_2, and AFG_1; 1.0, 2.0, 5.0, 10.0, 20.0 µg/kg for AFG_2; and 0.15, 0.30, 0.75, 1.50, 3.00 µg/kg for STC, respectively). Calibration solutions were prepared by the organizer's laboratory and sent to each participating laboratory for analysis. The concentrations of aflatoxins and STC in the sample solutions were calculated from calibration curves. The LOD and LOQ were calculated using the slope (*b*) of the matrix calibration curve and the residual standard error ($s_{y/x}$) with the following equations [38]: LOD = 3.3 $s_{y/x}$/*b*; LOQ = 10 $s_{y/x}$/*b*. The LOQ for AFB_1, AFB_2, AFG_1, AFG_2, and STC was 0.28, 0.35, 0.42, 0.90, and 0.52 µg/kg, respectively.

5.5. LC/MS/MS Analysis

Each laboratory determined the aflatoxins and STC concentrations by LC/MS/MS. Two microliters of the final solution were loaded on an XBbridge C18 column (1.7 µm, 2.1 × 100 mm; Waters, Milford, MA, USA) at 35 °C. Gradient elution was carried out with mixtures of acetonitrile and water (see Table 6). Electrospray ionization in the positive mode was used. Multiple reactions monitoring (MRM) mode was used here with LC/MS/MS. All other experimental conditions were set by the respective participating laboratories. LC/MS/MS instruments and parameters used in the eight participating laboratories are tabulated in Table 7. Supplementary material, describing chromatograms of each mycotoxin detected by LC/MS/MS in MRM mode, is available (Figure S1).

Table 6. Analysis of aflatoxins and sterigmatocystin residues by LC/MS/MS.

Time, min [a]	A, % [b]	B, % [c]
0.0	90	10
3.0	90	10
10.0	30	70
10.1	10	90
12.0	10	90
12.1	90	10
15.0	90	10

[a] Flow rate, 0.2 mL/min; [b] Mobile phase A, 0.1% formic acid in water; [c] Mobile phase B, 0.1% formic acid in acetonitrile.

Table 7. LC/MS/MS instruments and parameters used in the eight participating laboratories.

Lab.	LC-MS/MS Equipment	AFB$_1$[a] Precursor Ion (m/z)	Product Ion (m/z)[f]	AFB$_2$[b] Precursor Ion (m/z)	Product Ion (m/z)	AFG$_1$[c] Precursor Ion (m/z)	Product Ion (m/z)	AFG$_2$[d] Precursor Ion (m/z)	Product Ion (m/z)	STC[e] Precursor Ion (m/z)	Product Ion (m/z)
1	ThermoAccela TSQ Quantum Ultra (Thermo Scientific, Waltham, MA, USA)	313.0 [M+H]+	284.9	315.0 [M+H]+	287.0	329.1 [M+H]+	311.0	331.1 [M+H]+	313.0	325.1 [M+H]+	310.0
2	HPLC coupled with TSQ Quantum (Thermo Scientific, Waltham, MA, USA)		241.1		259.0		199.8		245.0		310.0
3	ThermoAccela TSQ Quantum Ultra (Thermo Scientific, Waltham, MA, USA)		284.9		287.0		311.0		313.0		310.0
4	Acquity I-class UPLC Xevo TQ-S (Waters, Milford, MA, USA)		241.1		259.0		199.8		245.0		310.0
5	Acquity I-class UPLC Xevo TQ-2 (Waters, Milford, MA, USA)		284.9		287.0		243.0		245.0		281.0
6	ThermoAccela TSQ Quantum Ultra (Thermo Scientific, Waltham, MA, USA)		284.9		287.0		311.0		313.0		310.0
7	Acquity I-class UPLC Xevo TQ-S (Waters, Milford, MA, USA)		241.1		259.0		199.8		245.0		310.0
8	Acquity I-class UPLC Xevo TQ-2 (Waters, Milford, MA, USA)		241.1		259.0		199.8		245.0		310.0

[a] AFB$_1$: aflatoxin B$_1$; [b] AFB$_2$: aflatoxin B$_2$; [c] AFG$_1$: aflatoxin G$_1$; [d] AFG$_2$: aflatoxin G$_2$; [e] STC: sterigmatocystin; [f] In addition to the product ion selected in each laboratory for quantification.

5.6. Statistical Analysis

The results received from the participating laboratories were initially evaluated for evidence of outliers using the following statistical tests: the Cochran test and the single value and pair value Grubbs tests (between laboratory means) [39]. The RSD_r and RSDR, and the HorRat value (the ratio of the RSDR to the predicted RSDR) were obtained using an analysis of variance according to AOAC guidelines [24]. The criteria for analytical methods mentioned in Commission Regulation (EC) No. 401/2006 were used for evaluation of these parameters [33]. The Z-score compares the analytical result to the assigned value and can be used to describe the comparability of results. The Z-score was derived from the results of each participant according to the following equation [40]:

$$Z\text{-score} = \frac{X_{lab} - X_{ref}}{\sigma_p},$$

where X_{lab} is the result reported by the participant, expressed as a dimensionless mass ratio, e.g., $1\ \mu g/kg = 10^{-9}$; X_{ref} is the assigned value expressed as a dimensionless ratio, e.g., $1\ \mu g/kg = 10^{-9}$; and σ_P is the target standard deviation (SD) for proficiency assessment. The target SD (σ_p) was calculated from the modified Horwitz equation:

$$\sigma_p = \frac{0.22 \times c}{mr},$$

where c is the concentration of the assigned value, expressed as a dimensionless mass ratio, e.g., $1\ \mu g/kg = 10^{-9}$; and mris the dimensionless mass ratio, $1\ \mu g/kg = 10^{-9}$. The Z-score classification was as follows: $|Z| \leq 2$, acceptable; $2 < |Z| \leq 3$, questionable; $|Z| > 3$, unacceptable.

Supplementary Materials: The following are available online at www.mdpi.com/2072-6651/8/12/371/s1, Figure S1: Chromatograms of multimycotoxins: (A) aflatoxin B1; (B) aflatoxin B2; (C) aflatoxin G1; (D) aflatoxin G2; and (E) STC.

Acknowledgments: This study was financially supported by the Ministry of Food and Drug Safety (15161MFDS012). We are grateful for the collaboration of the following participants in the interlaboratory study: Nongshim Co. Ltd. (Cheong-Tae Kim), Pulmuone Co. Ltd. (Sang Woo Cho), Korea Textile Inspection & Testing Institute (Jang-Hyuk Ahn), Chung-Ang University (Chan Lee), Kyungpook National University (Sung-Eun Lee), and Wonkwang University (Hoon Choi).

Author Contributions: Designed experiments and interpreted data: H.S.C. Performed experiments: H.E.O., F.T. Statistical analysis: E.Y.H., S.H.K. Prepared and distributed samples: O.P., D.K. Wrote the manuscript: H.S.C., F.T.

Conflicts of Interest: The authors declare no conflict of interest.

References

1. Andrade, P.; de Mello, M.H.; França, J.; Caldas, E. Aflatoxins in food products consumed in Brazil: A preliminary dietary risk assessment. *Food Addit. Contam. Part A* **2013**, *30*, 127–136. [CrossRef] [PubMed]
2. Sugita-Konishi, Y.; Sato, T.; Saito, S.; Nakajima, M.; Tabata, S.; Tanaka, T.; Norizuki, H.; Itoh, Y.; Kai, S.; Sugiyama, K. Exposure to aflatoxins in Japan: Risk assessment for aflatoxin B1. *Food Addit. Contam.* **2010**, *27*, 365–372. [CrossRef] [PubMed]
3. Montes, R.; Segarra, R.; Castillo, M.-Á. Trichothecenes in breakfast cereals from the Spanish retail market. *J. Food Compos. Anal.* **2012**, *27*, 38–44. [CrossRef]
4. Trucksess, M.; Abbas, H.; Weaver, C.; Shier, W. Distribution of aflatoxins in shelling and milling fractions of naturally contaminated rice. *Food Addit. Contam. Part A* **2011**, *28*, 1076–1082. [CrossRef] [PubMed]
5. Rahmani, A.; Soleimany, F.; Hosseini, H.; Nateghi, L. Survey on the occurrence of aflatoxins in rice from different provinces of Iran. *Food Addit. Contam. Part B* **2011**, *4*, 185–190. [CrossRef] [PubMed]
6. Njumbe Ediage, E.; Van Poucke, C.; De Saeger, S. A multi-analyte LC-MS/MS method for the analysis of 23 mycotoxins in different sorghum varieties: The forgotten sample matrix. *Food Chem.* **2015**, *177*, 397–404. [CrossRef] [PubMed]
7. Hussaini, A.M.; Timothy, A.G.; Olufunmilayo, H.A.; Ezekiel, A.S.; Godwin, H.O. Fungi and some mycotoxins found in mouldy sorghum in Niger state, Nigeria. *World J. Agric. Sci.* **2009**, *5*, 5–17.

8. Audilakshmi, S.; Stenhouse, J.; Reddy, T.; Prasad, M. Grain mould resistance and associated characters of sorghum genotypes. *Euphytica* **1999**, *107*, 91–103. [CrossRef]

9. Chandrashekar, A.; Bandyopadhyay, R.; Hall, A.J.; Chandrashekar, A.; Bandyopadhyay, R. Technical and institutional options for sorghum grain mold management. In Proceedings of the International Consultation, ICRISAT, Patancheru, India, 18–19 May 2000.

10. Sweeney, M.J.; Dobson, A.D. Mycotoxin production by aspergillus, fusarium and penicillium species. *Int. J. Food Microbiol.* **1998**, *43*, 141–158. [CrossRef]

11. Marin, S.; Ramos, A.; Cano-Sancho, G.; Sanchis, V. Mycotoxins: Occurrence, toxicology, and exposure assessment. *Food Chem. Toxicol.* **2013**, *60*, 218–237. [CrossRef] [PubMed]

12. International Agency for Research on Cancer (IARC). Agents Classified by the IARC Monographs, Volumes 1–116. Available online: https://monographs.iarc.fr/ENG/Classification/ClassificationsAlpha Order.pdf (accessed on 16 July 2016).

13. Bennett, J.; Klich, M. Chotoxins. C lin. *Microbiol. Rev* **2003**, *16*, 497–516. [CrossRef]

14. European Commission. Commission regulation (EU) no 165/2010 of 26 February 2010 amending regulation (EC) no 1881/2006 setting maximum levels for certain contaminants in foodstuffs as regards aflatoxins (text with eea relevance). *Off. J. Eur. Union* **2010**, *038*, 244–248.

15. Korea Food and Drug Administration (KFDA). *Food Code, KFDA Notification No 2011-42*; Korea Food and Drug Administration: Seoul, Korea, 2011.

16. Veršilovskis, A.; De Saeger, S. Sterigmatocystin: Occurrence in foodstuffs and analytical methods—An overview. *Mol. Nutr. Food Res.* **2010**, *54*, 136–147. [CrossRef] [PubMed]

17. World Health Organization. International agency for research on cancer (IARC) working group on the evaluation of carcinogenic risks to humans. In *Overall Evaluations of Carcinogenicity: An Updating of IARC Monographs Volumes 1 to 42*; World Health Organization: Geneva, Switzerland, 1987; Volume 7.

18. Sengun, I.; Yaman, D.B.; Gonul, S. Mycotoxins and mould contamination in cheese: A review. *World Mycotoxin J.* **2008**, *1*, 291–298. [CrossRef]

19. Zastempowska, E.; Grajewski, J.; Twarużek, M. Food-borne pathogens and contaminants in raw milk—A review. *Ann. Anim. Sci.* **2016**. [CrossRef]

20. Veršilovskis, A.; Bartkevičs, V.; Mikelsone, V. Sterigmatocystin presence in typical Latvian grains. *Food Chem.* **2008**, *109*, 243–248. [CrossRef] [PubMed]

21. Mo, H.G.J.; Pietri, A.; MacDonald, S.J.; Anagnostopoulos, C.; Spanjere, M. Survey on sterigmatocystin in food. *EFSA Support. Publ.* **2015**, *12*. [CrossRef]

22. Oueslati, S.; Blesa, J.; Moltó, J.C.; Ghorbel, A.; Mañes, J. Presence of mycotoxins in sorghum and intake estimation in Tunisia. *Food Addit. Contam. Part A* **2014**, *31*, 307–318. [CrossRef] [PubMed]

23. Chala, A.; Taye, W.; Ayalew, A.; Krska, R.; Sulyok, M.; Logrieco, A. Multimycotoxin analysis of sorghum (*Sorghum bicolor* L. Moench) and finger millet (*Eleusine coracana* L. Garten) from Ethiopia. *Food Control* **2014**, *45*, 29–35. [CrossRef]

24. Association of Official Analytical Chemists (AOAC) International. *Official Methods of Analysis of AOAC International*; AOAC International: Gaithersburg, MD, USA, 2005.

25. Veršilovskis, A.; De Saeger, S.; Mikelsone, V. Determination of sterigmatocystin in beer by high performance liquid chromatography with ultraviolet detection. *World Mycotoxin J.* **2008**, *1*, 161–166. [CrossRef]

26. Ok, H.E.; Chung, S.H.; Lee, N.; Chun, H.S. Simple high-performance liquid chromatography method for the simultaneous analysis of aflatoxins, ochratoxin A, and zearalenone in dried and ground red pepper. *J. Food Prot.* **2015**, *78*, 1226–1231. [CrossRef] [PubMed]

27. Tanaka, K.; Sago, Y.; Zheng, Y.; Nakagawa, H.; Kushiro, M. Mycotoxins in rice. *Int. J. Food Microbiol.* **2007**, *119*, 59–66. [CrossRef] [PubMed]

28. Veršilovskis, A.; Van Peteghem, C.; De Saeger, S. Determination of sterigmatocystin in cheese by high-performance liquid chromatography-tandem mass spectrometry. *Food Addit. Contam.* **2009**, *26*, 127–133. [CrossRef] [PubMed]

29. Sulyok, M.; Berthiller, F.; Krska, R.; Schuhmacher, R. Development and validation of a liquid chromatography/tandem mass spectrometric method for the determination of 39 mycotoxins in wheat and maize. *Rapid Commun. Mass Spectrom.* **2006**, *20*, 2649–2659. [CrossRef] [PubMed]

30. Sulyok, M.; Krska, R.; Schuhmacher, R. A liquid chromatography/tandem mass spectrometric multi-mycotoxin method for the quantification of 87 analytes and its application to semi-quantitative screening of moldy food samples. *Anal. Bioanal. Chem.* **2007**, *389*, 1505–1523. [CrossRef] [PubMed]

31. Frenich, A.G.; Vidal, J.L.M.; Romero-González, R.; del Mar Aguilera-Luiz, M. Simple and high-throughput method for the multimycotoxin analysis in cereals and related foods by ultra-high performance liquid chromatography/tandem mass spectrometry. *Food Chem.* **2009**, *117*, 705–712. [CrossRef]

32. Böhm, J.; De Saeger, S.; Edler, L.; Fink-Gremmels, J.; Mantle, P.; Peraica, M.; Stetina, R.; Vrabcheva, T. Scientific opinion on the risk for public and animal health related to the presence of sterigmatocystin in food and feed. *Eur. Food Saf. Auth. J.* **2013**, *11*, 1–81.

33. European Commission. No 401/2006 of 23 February 2006. Laying down the methods of sampling and analysis for the official control of the levels of mycotoxins in foodstuffs. *Off. J. Eur. Union L* **2006**, *70*, 12–34.

34. Horwitz, W.; Albert, R. The horwitz ratio (horrat): A useful index of method performance with respect to precision. *J. AOAC Int.* **2006**, *89*, 1095–1109. [PubMed]

35. Veršilovskis, A.; Bartkevičs, V.; Mikelsone, V. Analytical method for the determination of sterigmatocystin in grains using high-performance liquid chromatography–tandem mass spectrometry with electrospray positive ionization. *J. Chromatogr. A* **2007**, *1157*, 467–471. [CrossRef] [PubMed]

36. Sasaki, R.; Hossain, M.; Abe, N.; Uchigashima, M.; Goto, T. Development of an analytical method for the determination of sterigmatocystin in grains using LCMS after immunoaffinity column purification. *Mycotoxin Res.* **2014**, *30*, 123–129. [CrossRef] [PubMed]

37. Association of official analytical chemists (AOAC). Natural Toxins: Aflatoxins. In *Official Methods of Analysis*, 18th ed.; International Association of Official Analytical Chemists: Gaithersburg, MD, USA, 2005; Volume Chapter 49; p. 4.

38. Miller, J.C.; Miller, J.N. *Statistics for Analytical Chemistry*; John Wiley and Sons: New York, NY, USA, 1988.

39. Horwitz, W. Protocol for the design, conduct and interpretation of method-performance studies: Revised 1994 (technical report). *Pure Appl. Chem.* **1995**, *67*, 331–343. [CrossRef]

40. Thompson, M.; Ellison, S.L.; Wood, R. The international harmonized protocol for the proficiency testing of analytical chemistry laboratories (IUPAC technical report). *Pure Appl. Chem.* **2006**, *78*, 145–196. [CrossRef]

© 2016 by the authors. Licensee MDPI, Basel, Switzerland. This article is an open access article distributed under the terms and conditions of the Creative Commons Attribution (CC BY) license (http://creativecommons.org/licenses/by/4.0/).

Review

Recent Advances and Future Challenges in Modified Mycotoxin Analysis: Why HRMS Has Become a Key Instrument in Food Contaminant Research

Laura Righetti [1], Giuseppe Paglia [2], Gianni Galaverna [1] and Chiara Dall'Asta [1,*]

[1] Department of Food Science, University of Parma, Parco Area delle Scienze 95/A, Parma 43124, Italy;
laura.righetti1@studenti.unipr.it (L.R.); gianni.galaverna@unipr.it (G.G.)

[2] Center of Biomedicine, European Academy of Bolzano/Bozen, Via Galvani 31, Bolzano 39100, Italy;
beppepaglia@gmail.com

* Correspondence: chiara.dallasta@unipr.it; Tel.: +39-0521-905-431

Academic Editor: Aldo Laganà
Received: 30 October 2016; Accepted: 25 November 2016; Published: 2 December 2016

Abstract: Mycotoxins are secondary metabolites produced by pathogenic fungi in crops worldwide. These compounds can undergo modification in plants, leading to the formation of a large number of possible modified forms, whose toxicological relevance and occurrence in food and feed is still largely unexplored. The analysis of modified mycotoxins by liquid chromatography–mass spectrometry remains a challenge because of their chemical diversity, the large number of isomeric forms, and the lack of analytical standards. Here, the potential benefits of high-resolution and ion mobility mass spectrometry as a tool for separation and structure confirmation of modified mycotoxins have been investigated/reviewed.

Keywords: modified mycotoxins; high resolution mass spectrometry; ion mobility spectrometry

1. Introduction

The presence of food and feed contaminants, in particular secondary fungal metabolites, has become of increasing concern for consumers and producers. Indeed, more than 400 mycotoxins with widely different chemical structures have been identified so far, and their number is expected to increase further due to climate changes [1]. Extreme weather conditions are increasingly affecting the mycotoxin map in Europe and also world-wide, leading to an unpredictability of the range of mycotoxins occurring in food crops. In addition, it is already well known [2] that plants and other living organisms (i.e., fungi, bacteria, mammals) can alter the chemical structure of mycotoxins as part of their defense against xenobiotics, and thus contribute to further increase the wide spectrum of possible occurring contaminants. Mycotoxins, indeed, may undergo [2,3] phase-I, and phase II metabolism, involving in the former, chemical reactions such as oxidation, reduction and hydrolysis, and in the latter conjugation with amino acids, glucoses, sulfate groups and glutathione. All these modifications significantly change the chemical structure of the parent compounds.

According to the recent EFSA opinion [3], which aimed to harmonize the terminology across the scientific community, all these metabolites are referred to as "modified mycotoxins" [4], being structurally altered forms of the parent mycotoxins. The definition also covers the metabolites originating after thermal/ process degradation. Modified mycotoxins may co-occur as contaminants in addition to parent compounds in food and feed; so far modified forms for trichothecenes, zearalenone (ZEN), fumonisins, *Alternaria* toxins and ochratoxin A (OTA) have been identified. However, up to now their occurrence in naturally infected cereals has been exclusively confirmed for deoxynivalenol (DON), ZEN and fumonisins [5–7]. In fact, the lack of analytical standards and reference materials, has substantially complicated their identification, partially restraining research progress in the field as well.

The metabolic conjugation with polar molecules is commonly considered as an inactivation reaction, because the aglycone usually loses its biological activity. However, the possible hydrolysis of modified mycotoxins back to their toxic parents during mammalian digestion raises toxicological concerns [8] and thus the requirement for their detection. Modified mycotoxins were originally considered "masked" [9] since they may elude conventional analysis because of impaired extraction efficiency caused by increased polarity when a less polar solvent is used for the extraction of non-modified mycotoxins. Moreover, the changed physicochemical properties of their molecules lead to modified chromatographic behavior and due to the lack of analytical standards they are currently not routinely screened.

All of these effects may lead to a potential underestimation of the total mycotoxin content of the sample. Therefore, monitoring the presence of these potentially hazardous metabolites remains one of the main tasks for ensuring food safety and human/animal health.

In this frame, liquid chromatography coupled with mass spectrometry has represented the golden standard for at least a decade. This review aims, therefore, at pointing out the possible advantage of innovative MS techniques in mycotoxin analysis, and to highlight the improvements still needed to meet the future challenges in the field.

2. From Targeted LC-MS/MS Determination to Untargeted HR-MS Analysis

In the field of residues and contaminants, analytical methods used for surveillance purposes must ensure optimal sensitivity and accuracy. For this reason, chromatographic analysis coupled to fluorescence or UV detection was the reference techniques for many decades. Over the last two decades, the LC-MS/MS platform became the method of choice, in particular for its ability to allow the development of multi-residue and multi-class methods [10–14]. First attempts were aimed at quantifying a single mycotoxin, but later on the research moved to the simultaneous determination of multiple mycotoxins, leading to the development of the so-called multi-toxin methods for quantitative as well as screening purposes [13–15].

The current "golden standard" in routine food safety control is represented by unit resolution tandem mass spectrometric analyzers such as triple quadrupole (QqQ) [16,17], mainly because this technique ensures analytical parameters that easily meet quality criteria required by law [18–20]. Multiple reaction monitoring (MRM) has been traditionally selected for mycotoxin analysis, monitoring in parallel quantitative and qualitative ion transitions, providing both sensitivity and selectivity. Achieved limits of quantification (LOQs) and detection (LODs) usually match regulatory requirements for the official control method for contaminants. The first low-resolution multi-toxins method was developed in 2006 [13] for the quantitative determination of 39 parent and modified mycotoxins. Among the modified mycotoxins, the methods included 3Ac-DON, 15Ac-DON, DON3Glc, ZEN14Glc, ZEN14Sulf, and hydrolysed FB1. More recently [21] a method for the simultaneous quantification of both parent and modified *Alternaria* mycotoxins in cereals based food, including alternariol (AOH), AOH3Sulf, AOH3Glc, alternariol-methyl ether (AME), AME3Sulf, and AME3Glc was developed. *Alternaria* modified mycotoxins were in-house synthesized, since they are not commercially available yet. In addition, the applicability of the developed methods has been demonstrated by analysis of a variety of naturally contaminated cereals and real-life samples purchased on the market [7,13,21–23].

Despite having become a well-established technique, the QqQ method set-up is time-consuming when aimed at determining a large number of substances. Likewise, this technique presents limitation on the number of compounds that can be analyzed in one run. In addition, only targeted analytes can be detected, making necessary the use of an analytical standard, which is a critical issue in the modified mycotoxins field.

Thus, with the introduction of benchtop high resolution mass spectrometers (HRMS), such as Time-of-Flight (ToF) and Orbitrap, full-scan techniques started to be investigated as a complementary approach for the triple-quadrupole-based methods on the basis of increased resolution power and detectability. In addition, the use of LC coupled to HRMS offers some advantages over QqQ, since the acquisition of high resolution full scanned data permits the combination of target analysis with screening of non-target compounds, novel compound identification, and retrospective data analysis.

The increasing popularity of HRMS is mainly due to the advantages of using the full-scan acquisition mode [10,24,25] with high sensitivity, combined with high resolving power (up to 100,000 FWHM) and accurate mass measurement (<5 ppm). Moving from low resolution MS to high resolution should improve in principle specificity, although this is difficult to be transferred to a superior performance of the target analysis when moving from LC-MS/MS to HRMS methods, since ion suppression/enhancement phenomena due to the matrix may occur in both approaches. A possible approach to cope with the matrix effect is represented by stable isotope dilution assay (SIDA) [26]. Multi-mycotoxins methods applying this technique have been developed [27,28], also thanks to in-house synthesis of labeled isotopologue mycotoxin standards, including isotopologues of modified forms (i.e., DON3Glc, 3Ac-DON, 15Ac-DON) [29]. Authenticity and method performance have been demonstrated by analyzing naturally contaminated samples, such as malt, beer and maize [30]. Despite ion suppression phenomena, the enhanced selectivity and sensitivity provided by HRMS allow the development of methods that cover a wide range of compounds with different physicochemical properties, as demonstrated by Dzuman and coworkers [31] who developed an LC-HRMS method for the detection of 323 pesticides, 55 mycotoxins, and 11 plant toxins.

The major advantage of using HRMS over MS/MS techniques is actually due to the possibility to perform retrospective data analysis [32], thus enabling the possibility to reconsider analytical results for stored data. The measurement of accurate MS and MS/MS spectra (resolution <5 ppm) allows the determination of compounds without previous compound-specific tuning, to carry out retrospective analysis of data, and to perform structural elucidation of unknown or suspected compounds. This is particularly worth noting when modified mycotoxins are considered, especially if combined toxicological effects are in the pipeline.

3. Use of HRMS Methods for Targeted Quantification of Natural Toxins

From a quantitative point of view, as main advantage, a full-scan technique allows the extraction from the HRMS full scan data of a theoretically unlimited number of analytes without any compromise regarding the resulting detectability. Generic tuning setting can be used, without the need for optimizing parameters for each analyte.

Although the use of HRMS in food is very recent, over the last few years there was a significant increase in the number of studies reporting LC-HRMS-based approaches for targeted quantitative analysis of residues and contaminants in complex food matrices [10,24,33,34], most of them using a multi-contaminant approach [11,12,31,32].

Among the classes of food contaminants, the possible use of HRMS-based methods for natural toxins was successfully applied to the analysis of a large variety of samples from the market, as recently reviewed by Senyuva et al. [35]. In general, the collection of exact m/z values, HRMS/MS spectra and retention time allow the build up of (myco)toxins spectral libraries potentially sharable between Q-Orbitrap instruments. As an example, Ates et al. [36] created a database containing empirical formulae, polarity, fragment ions (up to five), and retention times for 670 plant and fungal metabolites. The library was validated by correct identification of known mycotoxins in proficiency test materials, and then applied to the screening of emerging mycotoxins in cereal samples from the market.

While the use of HRMS for the quantification of one or few analytes does not pose any significant advantage compared to MRM-based methods, multi-contaminant methods seem to be the most promising approach for food and feed surveillance in the coming years. It must be underlined, however, that the validation steps required for HRMS-based methods does not differ from those applied for QqQ methods. Moreover, sample preparation (i.e., extraction, enrichment, clean up, chromatographic separation) still remains a crucial step to reduce ion suppression phenomena, and ensure the required specificity and detectability.

Concerning mycotoxins, the possible set up of HRMS multi-toxin methods has been increasingly exploited in recent years, especially for monitoring the co-occurrence of regulated and emerging mycotoxins. In this frame, the potential capability of HRMS to return a full picture of the pool of modified mycotoxins in a selected food may represent the basis for future studies of combined toxicity.

However, it should be mentioned that only five modified mycotoxins have been included so far in multi-toxin HRMS quantitative methods, since analytical standards are available on the market only for DON3Glc, 3Ac-DON, 15Ac-DON, as well as α and β zearalenol (ZEL) [31]. Other modified forms have been included in semi-quantitative or screening methods, based on in-house prepared reference compounds, usually obtained by chemical/enzymatic synthesis or isolation from natural sources. In other studies, conjugates are semi-quantified on the base of the parent compounds. Although inaccurate, in case of novel compounds, when neither commercial calibrants nor in-house synthesized standards are available, this assumption allows for a rough semi-quantitation of the novel conjugate compared to the parent form. For instance, zearalenone biotransformation to zearalenone malonyl-glucoside in wheat was estimated by assuming that both the parent and the modified form had the same response during MS ionization [37]. As an alternative, the formation of DON-oligoglucosides during malting and brewing processes was expressed as their molar ratio to DON [38].

In consideration of the lack of commercial standards, HRMS actually provides more qualitative than quantitative benefits to modified mycotoxin analysis.

4. Use of HRMS Non-Targeted Screening Methods for Natural Toxins

In consideration of the possible collection of full-scan spectra, a theoretically unlimited number of compounds from different classes may be detected simultaneously by HRMS at low concentration level. Therefore, HRMS is often used for non-targeted screening of unknowns, since compound to be monitored should not be established in advance. As a general remark, when natural toxins are considered, it must be noted that unknown compounds should be better defined as "expected unknowns" and "unexpected unknowns", the former being modified forms of natural compounds that can be anticipated on the basis of biological pathways, and the latter novel compounds never described before. In addition, known compounds may occur in unexpected biological matrices, thus representing an unexpected known analyte. Examples of this definition are collected in Table 1.

In a more general meaning, a "non-targeted" analysis could be described as a screening against a large database of compounds, or a retrospective analysis of a dataset for compounds not specifically anticipated. This approach usually leads to a list of potential contaminants occurring in a sample that should be further confirmed by targeted analysis. The applicability of HRMS as a non-targeted approach indeed is based on the screening of an accurate mass of both precursor and fragments ions in one single run, by using data-independent analysis (DIA), without monitoring any preselected parent ions and based on general settings. This permits retrospective data analysis from the recorded HR-full-scan spectra; consequently, the presence of 'newly discovered' mycotoxins can be investigated with the data of prior-analyzed samples without the need for analytical standards.

As can be seen below (Figure 1), the untargeted workflow usually moves from seeking the exact masses of a list of compounds in the full scan spectra, to generating molecular formula, and comparing theoretical and experimental isotopic patterns and fragment spectra with those collected in the reference library, or available in on-line databases. Therefore, the full process can be described as the search of a limited number of compounds (those reported in the library) in an unlimited dataset (the stored data).

Table 1. Examples of parent and modified mycotoxins classified in known/unknown categories, according to the above mentioned definition.

Known/Unknown Categories	Modified Mycotoxin	Matrix	MS Equipment	Identification Based on	Analytical Standard
Expected knowns	Aflatoxin M1	Cheese [39]	Q-Trap	authentic standards	commercially available
	DON3Glc	Wheat and maize [22]	Q-Trap	authentic standards	commercially available
	3/15Ac-DON	Wheat [23]	Q-Trap	authentic standards	commercially available
Unexpected knowns	Aflatoxin M1	Feed [40]	QqQ	authentic standards	commercially available
	enniatins, alternaria toxins, T-2/HT-2 toxins	Dietary supplements [41]	Q-Trap	authentic standards	commercially available
	FB2	Culture media [42]	QqQ	authentic standards	commercially available
Expected unknowns	T2-Glc	Wheat and oats [43]	LTQ Orbitrap	HRMS	in-house synthesized
	15Ac-DON-Glc	Wheat [44]	LTQ Orbitrap XL	authentic standards	in-house synthesized
	DON-oligoglycoside	Malt and Beer [38]	Exactive Orbitrap	HRMS	n.a.
	NIV-Glc	Wheat [45]	LTQ Orbitrap	HRMS/MS	n.a.
	Desmethyl Enn B1	Human liver [46]	Q-Tof	HRMS/MS	n.a.
	ZEN-MalGlc	Wheat [47]	Q-trap	MS/MS	n.a.
Unexpected unknowns	Feruloyl-T2	Barley [48]	Exactive Plus Orbitrap	HRMS/MS	n.a.
	DON-2H-glutathione	Wheat [44]	LTQ Orbitrap XL	HRMS/MS	n.a.
	Pentahydroxyscirpene (PHS)	Barley [49]	Q-Tof	MS/MS	in-house synthesized
	DON-3-Glc lactone	Wheat [50]	Exactive Orbitrap	HRMS	n.a.

Table columns: MS equipment = mass spectrometry instrument; identification based on = mycotoxin identification based on matching retention time, m/z and MS/MS fragment with that of authentic standards, or based on accurate mass (HRMS) and accurate mass fragments (HRMS/MS); analytical standard = mycotoxin standard commercially available, not available (n.a.) or in-house synthesized by research group.

Screening methods are usually based on a generic sample preparation (i.e., QuEChERS extraction procedure) [15,31], and thus data collected allow in principle the retrospective review of any potential compounds of interest. Nonetheless, the final detection and quantification of targeted compounds may be negatively affected by interferences from the matrix. In some cases, co-elution may impact on the mass accuracy by causing ion suppression. This may lead to the missing of suppressed compounds during the automated filtering process. Therefore, the improvement of sample preparation as well as the fit-for-purpose adjustment of software parameters are of great relevance for the analysis.

A comprehensive description of the systematic workflow for quantitative target analysis, targeted screening of listed compounds, and untargeted screening of unknowns was reported by Krauss et al. [25].

Considering the large number of data generated by using HRMS, reliable bioinformatics tools for processing untargeted data are needed, as well as software packages for an automated compound detection. Therefore the development of a reliable bioinformatics workflow, providing a software solution from features extraction to unknown identification data is fundamental. The selection of relevant compounds from a large data set still represents the bottleneck [51]. Recently developed approaches show that statistical analyses, in combination with untargeted screening, are useful methods to preselect relevant compounds [52]. In addition, there are different software tools that allow peak deconvolution and the removal of background noise [25] by comparing different chromatograms. As an example, the use of a blank matrix as a control sample can substantially reduce the number of compounds to be screened by the software algorithm.

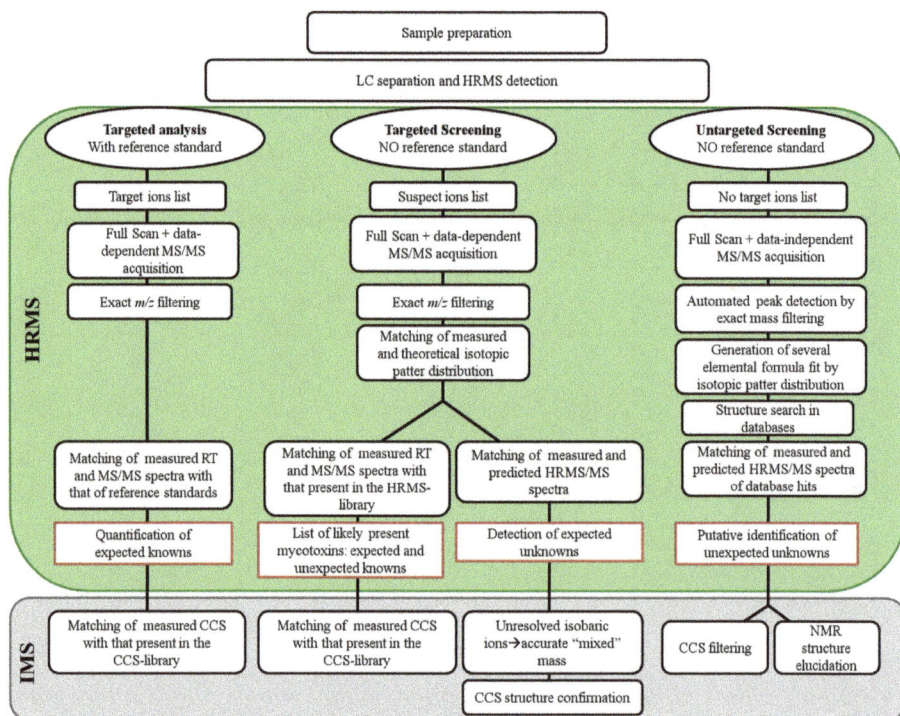

Figure 1. Workflow for targeted analysis with mycotoxin reference standards, targeted screening without analytical standards, and untargeted screening for unexpected unknowns (adapted from Krauss et al. [25]). CCS = Collision Cross Section.

At this point, several elemental composition formula may be predicted from an accurate mass measurement using MS trademark packages. Subsequently, molecular formula candidates and MS/MS pathway may be jointly investigated by structure elucidation software. Nowadays, HRMS qualitative software tools are linked with on-line databases (i.e., Pubchem, Metlin) providing a comparison of accurate mass, isotopic pattern, and MS/MS fragment ions. In the end, the users may find a percentage of matches, which are interconnected with the unknown compound. However, it should be noted that the search for unknowns in on-line databases is limited to the recorded spectra of available reference standards. In this frame, a big step forward has been achieved with computer-based tools based on in silico strategies [51,52], since the unknown chemical structures may be putatively confirmed by matching measured and computational predicted MS/MS fragmentation.

5. Use of HRMS Methods for Structural Identification of Unknowns

Another relevant application of HRMS is the structural identification of unknown compounds, i.e., novel compounds identified for the first time in the considered matrix. Although the unequivocal structural elucidation of compounds still requires ^{1}H- or ^{13}C NMR spectroscopy, in most cases there is already sufficient information to tentatively annotate and identify the unknowns. This is the case, for instance, of expected unknowns, modified mycotoxins where the possible modification pattern carried out by plants or microbes is well-established. As an example, the identification of new trichothecene conjugates such as FUSX-glucoside and NIV-glucoside [45], or acetyl-T2 [48,53] was an extrapolation of existing knowledge. The tentative identification of mycotoxin gluco-conjugates based on accurate mass, isotopic pattern distribution, and MS/MS fragmentation is indeed feasible. Accurate mass can be theoretically calculated and MS/MS fragments may be quite easily predicted since they generally lead to the loss of the glucosidic unit [M−H−glucose]$^-$ and thus the detection of the aglyconic form. More challenging is the elucidation of the binding position of the sugar unit. As suggested by Dall'Asta et al. [54] the binding position in the DON molecule distinctly influences the stability of the molecular ion, thus leading to different fragments. In particular, the ion corresponding to [M−H−CH$_2$O]$^-$ is reported to be characteristic of all 3-substituted trichothecenes; however that is not true for the other classes of mycotoxins. By following this approach, also DON-oligoglycosides [38] were identified in malt and beer samples. Therefore, the molecular formula and exact masses of DON-di, tri, and tetra glucosides can be easily calculated by adding glucose units corresponding to 162.0523 Da, as summarized in Table 2, and thus many putative structures can be predicted. As a result, nine possible molecular structures were hypothesized for DON-di-Glc, corresponding both to di-glucoside conjugates, with one molecule of glucose conjugated to each of the hydroxyl groups of DON or oligosaccharides. Although chromatographic separation was optimized changing from a reverse phase to an HILIC column, isobar separation was not achieved. In addition, also considering the peaks that are not baseline-resolved, no more than four out of nine peaks were detected both in malt and beer samples, as is depicted in Figure 2.

As well as gluco-conjugates, also mycotoxins conjugation with malonyl-glucoside frequently occurs. So far, occurrence of T2, HT2, ZEN, α and β ZEL, and DON malonyl-glucosyl (MalGlc) derivatives have been reported in artificially infected samples [44,47,48,53]. The extracted ion chromatogram (EIC) of tentative HT2-malonyl-glucoside showed two peaks that may result from the conjugation of malonic acid to different hydroxyl groups of glucose (four possible positions) or from conjugation of malonyl-glucoside to the two different position of the HT2 hydroxyl groups [39]. As an example, the ZEL-MalGlc HRMS putative identification workflow is depicted in Figure 3.

Thus, also in these examples, it was not possible to achieve a high degree of certainty for the identification of expected unknowns and thus the authors concluded that NMR analysis was required. In addition, putative co-eluting isomers having the same m/z value can be present in the sample, considering the high number of theoretical structure compared to the detected peaks.

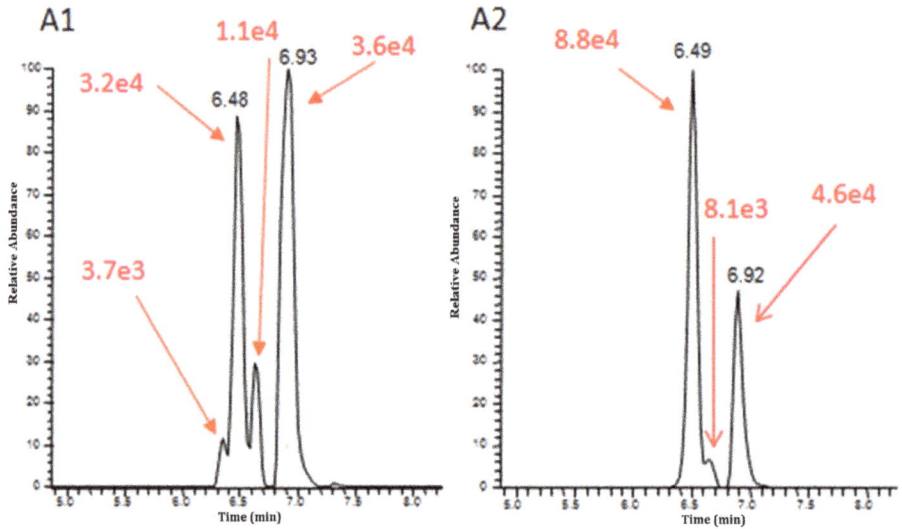

Figure 2. Extracted ion chromatogram for DON di glucosides determined in malt (**A1**) and beer (**A2**) by using HILIC phase chromatography coupled to HRMS (Orbitrap). (Reproduced with permission from [29], copyright (2016) American Chemical Society).

As far as modified mycotoxins, the metabolic modification occurring in plants usually follows phase I and phase II patterns. Therefore, the possible modification can be theoretically anticipated, and a list of expected unknowns can be used for peak annotation. As an example, a list of possible phase I and phase II modifications is reported in Table 2.

Table 2. Summary of calculated exact masses for putative phase I and phase II mycotoxin modifications.

Modification	Mass Change (Da)	Molecular Formula Change
Hydrogenation	2.0151	H_2
Hydroxylation	15.9944	O
Methylation	14.0151	CH_2
Acetylation	42.0100	C_2H_2O
Glycine	57.0209	C_2H_3NO
Sulfate	79.9563	SO_3
Sulfonation	102.9460	SO_3Na
Ferulic acid	176.0468	$C10H_8O_3$
Cysteine	119.0036	$C_3H_5NO_2S$
Acetyl-cysteine	161.0141	$C_5H_7NO_3S$
Glucose	162.0523	$C_6H_{10}O_5$
Cysteine-glycine	176.0250	$C_5H_8N_2O_3S$
Glucuronic acid	176.0315	$C_6H_8O_6$
Acetyl-glucoside	203.0550	$C_8H_{11}O_6$
Malonyl glucoside	248.0527	$C_9H_{12}O_8$
Glutathione	305.0682	$C_{10}H_{15}N_5O_6S$
Di-glucoside	324.1051	$C_{12}H_{20}O_{10}$
Malonyl di-glucoside	410.1055	$C_{15}H_{22}O_{13}$
Tri-glucoside	486.1579	$C_{18}H_{30}O_{15}$
Di malonyl-di glucoside	497.1137	$C_{18}H_{25}O_{16}$
Tetra-glucoside	648.2107	$C_{24}H_{40}O_{20}$

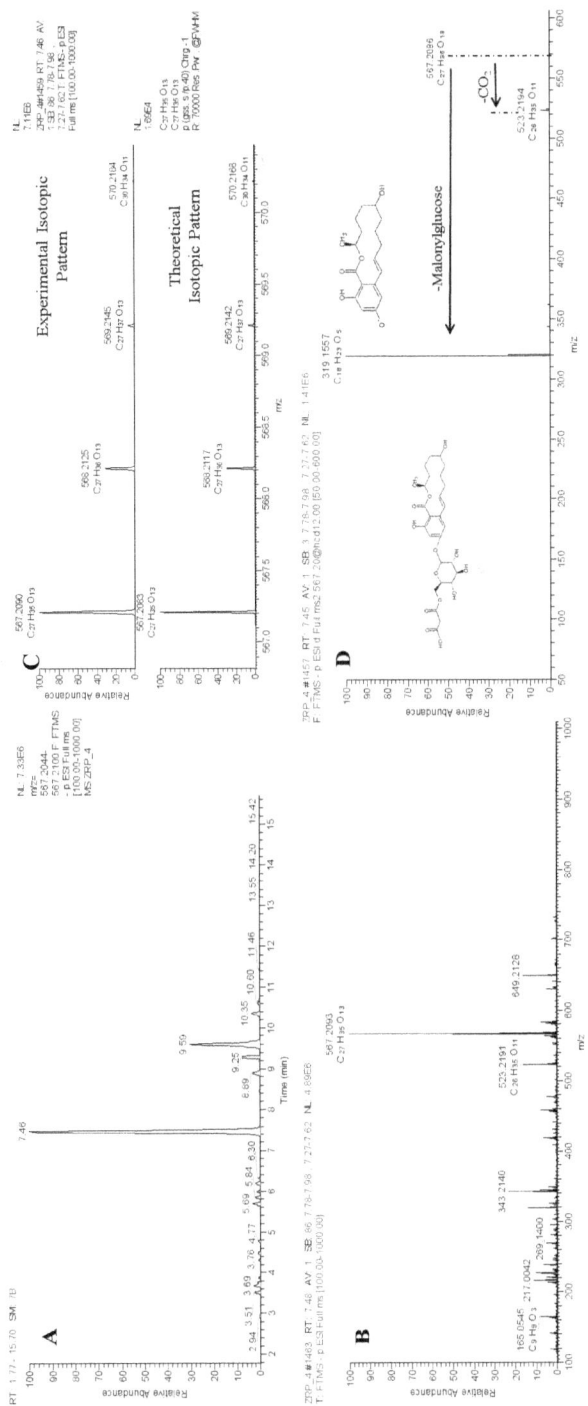

Figure 3. ZEL-MalGlc putative identification steps: UHPLC-Q-Exactive (**A**) full scan extracted ion chromatogram (resolving power 70,000 FWHM, extraction window 5 ppm); (**B**) molecular formula assignment of parent ion; theoretical and experimental isotopic pattern comparison (**C,D**) high resolution fragmentation pathways obtained by using DDA acquisition.

Several expected unknowns of mycotoxins have been tentative identified so far thanks to the use of HRMS [55]. When only one isomer configuration is possible, the recorded HRMS/MS spectra allow a satisfactory compound structure elucidation. However it should be stated that many mycotoxins are characterized by more than one hydroxyl group, that can be further increased by phase I metabolism. As a result, conjugation reactions may take place on several substituent groups and thus separation and characterization of the different isomers still remain challenging.

However, although it is frequently based on the search of expected unknowns, structural elucidation in complex samples is limited by the non-specific nature of the electrospray ionization process. It has been stated, indeed, that full scan spectra may contain up to 90% of noise compared to low concentration metabolites [56]. In such cases, the use of stable isotope labeling (SIL) may successfully assist metabolic profiling [57]. The software-driven correction of isotope pattern abundance errors resulted in better identification rates of the molecular formulas. In particular, artifacts generated by solvents, matrix background or noise, can be filtered out, enabling better detection and annotation of novel and unexpected compounds.

Recently, the in vivo SIL approach was exploited for the identification of *Fusarium* toxin metabolites in grains, based on the comparison of natural and ^{13}C-labelled patterns of metabolites showing identical chromatographic behavior but different (shifted) MS spectra [48,58]. The spectra comparison and metabolite identification was supported by the use of a dedicated software [59]. This powerful approach may simplify the identification of novel unexpected unknown compounds (i.e., feruloyl-T2 [48]) and successfully support the metabolic fingerprint.

6. Advantages and Challenges of HRMS in Modified Mycotoxin Analysis

When natural toxins are considered, the advantages derived from a qualitative use of HRMS overcome those obtained from a quantitative analysis. Retrospective data analysis, expected unknown screening, and novel compound identification are actually features that may strongly improve research in this field in the very near future. This is especially interesting in the field of emerging and modified mycotoxins, when analytical standards, calibrants, and reference materials are not commercially available. In addition, the urgent trend to base risk assessment on the combined toxicity derived from a pool of contaminants instead of a single compound, makes necessary the collection of (co)occurrence data. In this frame, the information generated by HRMS identification of novel compounds can be added to parent toxins libraries, and further used for retrospective analysis of full-scan data.

This can usefully concur to acquire qualitative occurrence data for better food quality controls, and more importantly, to ensure food safety.

Besides the advantages described above, HRMS still presents some critical points to be considered for its application in contaminants analysis. Among the possible disadvantages, the cost of instrumentation and the difficulties in big data management are often reported, but the improvement of software and storage systems as well as the introduction on the market of bench-top instruments favor broader diffusion LC-HRMS based methodologies.

However, from a chemical point of view, issues such as isobars co-elution, and unknown molecule identification have to be solved to allow an effective use of HRMS for food safety purposes. The innovative technique of ion mobility spectrometry (IMS) can be used to ascertain complementary information about analytes, adding a third dimension of separation based on size, shape, and charge of ions. The coupling of the two strategies, IMS and LC-HRMS platform, may act as a powerful tool to (i) improve the quality of mass spectral information obtained thanks to a filtered background; (ii) to increase the peak capacity to separate isomeric/isobaric compounds which can neither be resolved by MS and sometimes nor by UHPLC; (iii) to enhance analyte identification thanks to structural information (size and shape) based on collision cross-section measurement.

7. The Potential Benefits of Ion Mobility Mass Spectrometry in Mycotoxin Analysis

In the first section, the many possibilities of HRMS as a tool for modified mycotoxin analysis were reviewed. However, it should be stated that confirmation and structure elucidation of new unknown molecules and the unambiguous identification of isomers still remains challenging when using conventional mass spectrometry methods.

In comparison, NMR spectroscopy has been successfully employed with the goal of recognizing modified mycotoxins structure [22,60]. The NMR-based approach is efficient for evaluating isomeric heterogeneity, and for structural elucidation, but has the limitation of needing a considerable amount of analytes and obtaining a single molecular species following the purification steps.

Ion mobility spectrometry (IMS) is a promising approach that can overcome the above mentioned HRMS and NMR limitations, making it an ideal candidate for improving confidence in the identification and separation of structurally closely related isomers. IMS is a gas-phase electrophoretic technique that provides a new dimension (3D) of separation based on size, shape, and charge of ions. The ion mobility spectrometer consists of four main components that can be identified as: the sample introduction system (SIS), the ionization source, the drift tube region for separation and selection of ions, and the detection area [61,62]. In this review we mainly focus on the separation region, detailed information on the other components can be found elsewhere [61]. Once ionized, ions are directed to the drift tube region that contains an electric field, drift gas separates them according to their mobility. Ions moving in a gas phase medium and in the presence of an electric field are accelerated due to coulomb forces and slowed down due to collisions with molecules of the gas medium [61]. Thus small and compact ions travel faster and will reach the detector before large and heavy ions. In this way isobars are separated in the mobility spectrum, where the ion current is plotted as a function of the drift time or the compensation voltage [50].

Eight different types of IMS have been recently reviewed [61]. However, it should be noted that not all IMS devices are stand-alone instruments and for the purpose of the present review, only the IMS hyphenated with a mass spectrometer are discussed. A common hyphenated technique includes coupling IMS with MS (IMS-MS) in which IMS works as a pre-filter by confirming ion identities for the MS system. In addition, since IM separation typically occurs in a millisecond timeframe and MS detection, in a typical TOF instrument, takes only microseconds, additional separation techniques such as liquid chromatography (LC) can be hyphenated without compromising the speed of MS detection. So far, four major IMS-MS separation approaches are currently commercially available coupled with MS: drift-time IMS (DT-IMS) [63], traveling-wave IMS (TW-IMS) [64,65], high field asymmetric waveform IMS (FAIMS) [66], also known as differential-mobility spectrometry (DMS), and trapped IMS (TIMS) [67–69].

In DTIMS and TWIMS all the ions pass through the mobility cell ions and are separated based on the time it takes to traverse the cell. Such devices are generally used for untargeted screening experiments. FAIMS/DMS devices separate ions by varying voltages, filtering ions in a space-dispersive fashion. TIMS-MS separates ions based upon differences in mobility, after trapping and selectively ejecting them. Readers interested in the principles behind these IMS technologies can refer to previous reviews [70].

Applications in the food analysis field, especially when analyzing contaminants, have been mainly addressed applying FAIMS technologies [71–73] probably because of the advantages offered by the filtering effect as well as for the possibility of the device being moved and placed at the front-end of the mass spectrometer. On the other hand, TWIMS applications are rapidly growing [74–76] because, enabling CCS values measurement, it has found application as both a separation device and a structural elucidation tool.

Considering the complexity of food and feed samples, the use of LC-IMS-MS hyphenated methods is starting to be considered vital for versatile applications. Indeed, LC-IMS-MS potentially provides three major benefits to modified mycotoxins detection compared to traditional approaches. First, LC-IMS-MS improves the peak capacity and signal-to-noise ratio of traditional analytical approaches providing cleaner mass spectra obtained from the filtered background [71,77,78]. Second, it allows the separation of co-eluting isobaric metabolites according to their mobility [79], simplifying the interpretation of mass spectra. Third, IMS enhances confidence in analyte identification thanks to the measurements of the collision cross section (CCS) [80], a physicochemical measure related to the conformational structure of ions (size, shape, and charge).

7.1. Peak Capacity and Signal-to-Noise Ratio Improvement

The coupling of IM with liquid chromatography (LC) and HRMS gives a degree of orthogonality to both techniques by separating co-eluting LC compounds in mobility space before mass analysis. Hence, the overall peak capacity of the method is increased [81,82] making IM-MS highly suitable for food safety control analysis. In this frame, a broad range of food contaminants such as herbicides [83], pesticides [75,84], mycotoxins [71,77,78], and veterinary drug residues [85] have been successfully detected by IMS.

Although mycotoxins have been scarcely analyzed by IMS-MS, three applications using low resolution mass spectrometry have been reported so far [71,77,78]. An unusual type of IMS, corona discharge ion mobility spectrometry (CD-IMS) was applied to determine aflatoxin B1 (AFB1) and B2 (AFB2) in pistachio samples aiming to monitor spoilage status [77]. Sample extracts were directly introduced into the corona discharged ionization region via a headspace (HS) device without any chromatographic separation. As expected by the authors, IMS was not able to distinguish between AFB1 and AFB2 due to their similar chemical structure and their very close molecular weight. Thus they measured the total AFBs since it was demonstrated that their IMS response factors were identical [77]. The resulting LOQ and LOD (0.5 and 0.1 ng·mL^{-1}, respectively) were in line with those reported in literature obtained using different chromatographic and spectrometric techniques. In addition, pistachio samples were analyzed to demonstrate the capability of the method in detecting aflatoxins in real samples. The same authors applied the proposed approach some years later [78] for the analysis of ochratoxin in licorice roots. Following extraction and purification by passage through an immuno-affinity column, the achieved LOD in real samples was compliant with established concentration limits for licorice roots (20 µg/Kg).

A significant improvement of the detection limits was also measured for ZEN and its metabolites α-ZEL, β-ZEL, and β-zearalanol (ZAL) in maize using FAIMS technology and direct infusion or flow injection [71]. In fact, compared to ESI-MS or ESI-MSMS, a five-fold improvement in the signal to noise ratio was reported. This result was attributed to the ability of FAIMS equipment to operate as an ion filter, focusing ions and reducing the chemical background attributable to the matrix. The achieved LODs for ZEN, α-ZEL, β-ZEL, and β-ZAL were 0.4 ng·mL^{-1}, 3.2 ng·mL^{-1} and 3.1 ng·mL^{-1}, respectively. Thus, FAIMS filter step prior to ESI-MS analysis was able to selectively resolve and quantify species that otherwise cannot be selectively analyzed by ESI-MS alone. Additionally, since analytes are separated on account of their compensation voltages, reducing the time required for each sample run to about 1 min, the authors suggested that FAIMS might allow overstepping of the chromatographic separation [71].

LC-ESI-FAIMS was also used to develop a quantitative method for the determination of marine toxins in mussel tissue [73]. They investigated in depth how to improve the method sensitivity in relation to the number of CV values monitored at a given time. In fact, one of the limiting factors for analytical sensitivity is the duty cycle of the FAIMS device, which has a switching time between different CV values of about 100 ms, the time required to empty the device of ions that experience a particular CV. This means that limiting the number of CV values simultaneously monitored is an effective way of limiting sensitivity losses observed when using FAIMS in combination with LC. Two approaches were investigated for limiting the number of monitored CVs. The first one was developed using time periods with a limited number of optimized CV values at a given retention time, the second one reducing the number of monitored CVs to three values, which provided coverage of all analytes close to, but not at their optimal CV. The latter method proved to be more robust but less sensitive due to the fact that toxins were not detected at their optimum CV values, but was more suitable for analysis of large sample sets where RT could be expected to drift slightly.

The above mentioned IMS-MS approaches are pioneering, since they represent the first application of ion mobility spectrometry in the mycotoxin field. On the whole, the achieved limits of detection for these applications were in agreement with those required from the regulatory authorities, and confirming the method applicability to real samples and then the fitness for purpose, led to the expectation that IMS may help trace analysis control compliance. Thus, the development of further IMS-MS methods to extend the number of monitored parent and modified mycotoxins are encouraged.

In this regard, more recently, a novel approach to screening multi-class pesticides by TW ion mobility time-of-flight mass spectrometry detection was successfully developed [75]. The authors demonstrated that combining full scan and mobility XIC (extracted ion chromatogram) it was possible to detect the mass spectrum of indoxacarb, that was masked by other co-eluting compounds in the scan spectra. This example demonstrated how drift times give a higher level of selectivity to the overall method as no interfering compound resulted at the same retention time, drift time, and measured exact mass. In addition, once the pesticide has been identified using its retention time, exact mass, and drift time, the resulting cleaned mass spectrum facilitates the identification process. In line with these findings, the cleaning effect due to ion mobility separation on MS/MS spectra was demonstrated [86]. Applying the DIA mode for the MS analysis of complex extracts could result in MS/MS spectra containing a mixture of product ions derived from co-eluting precursors, complicating interpretation of the spectra and the overall identification process (Figure 4). Combining IMS-MS with DIA might allow the separation of co-eluted precursor ions before fragmentation, resulting in a drift-time correlation of product ions with their respective precursor ions and thus cleaner MS/MS product-ion spectra.

This phenomenon might offer a straight benefit when analyzing mycotoxins in complex food and feed matrices and in particular modified mycotoxins, considering that they are usually present at a low concentration level. Despite the HRMS improvement, quite significant discrepancies when comparing relative intensities of fragment ions measured in pure solvent with those measured in matrix were reported [31]. Thus, these clean MS and MS/MS spectra might, in turn, facilitate compound identification and reduce false-positive assignments in complex food matrices.

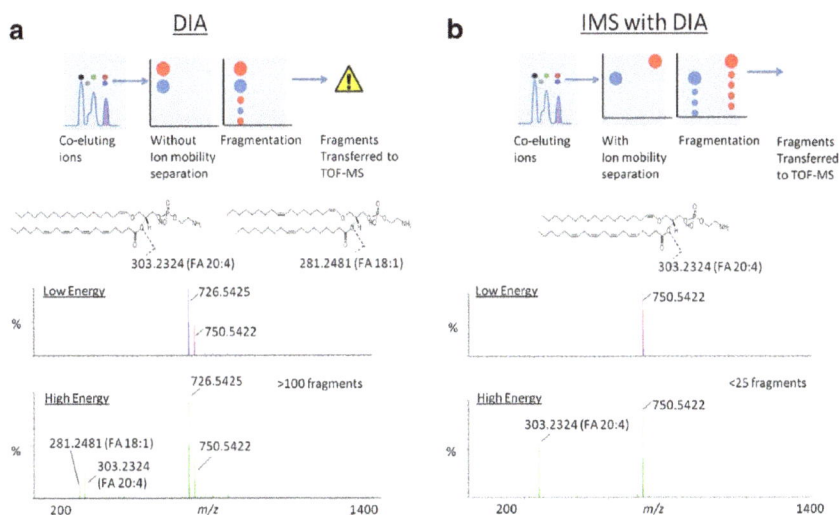

Figure 4. MS and MS/MS cleaner mass spectra obtained by using LC-IM-QTOF (**b**) compared to those obtained by LC-QTOF (**a**) (from Paglia et al. [86]).

7.2. LC-IMS-MS Enables the Separation of Isobar Molecules

Filtering out interferences, LC-IMS-MS may also allow the separation of co-eluting isobars and isomers that are difficult to separate by traditional LC-MS. Isobar co-elution may occur when expected unknowns are tentatively identified by LC-HRMS, as described in some examples reported in the previous sections. This issue is of a great relevance for mycotoxin analysis, above all when co-eluting modified forms differ in their toxicological profile, as for the acetylated derivatives of DON. With regard to intestinal toxicity, 3Ac-DON was found to be less toxic than DON, which was less toxic than 15Ac-DON [87]. As a result, a precise quantification of the different isoforms has to be performed. However, considering that they differ only in the position of the acetyl group and thus similar polarities of these two mycotoxins, it has not been possible to achieve chromatographic separation so far [88]. Hence, different strategies have been developed to reach a correct quantification. By MS-single quadrupole detection, Biancardi et al. [89] calibrated the response for 15Ac-DON and 3Ac-DON separately and reported the results as the sum of 3- and 15Ac-DON (Ac-DONs). Afterwards, thanks to the selective detection power of MS/MS, separate identification was performed due to the difference in characteristic daughter ions in MRM mode, identified as m/z 339.5 >137.2 and 339.5 >321.1, and m/z 339.5 >213.0 and 339.5 >230.9 for 15Ac-DON and 3Ac-DON, respectively [88]. In a recently published multi-mycotoxin method [15], the two acetylated derivatives were detected by taking advantage of the detection polarity. In fact, 15Ac-DON was detected in positive mode (m/z 356.1, $[M+NH_4]^+$) and 3Ac-DON in negative mode as acetate adduct (m/z 397.1, $[M+CH_3COO]^-$).

Hence, it is evident that the potential of LC-IMS-MS to enhance isomer separation, would overcome challenges associated with modified mycotoxin isomers analysis that will not otherwise be achieved. In addition, the proper separation and then quantification of 3-Ac-DON and 15-Ac-DON is essential in order to collected further data and better characterize their potential contribution to the total exposure to DON.

Even more challenging is the case of modified mycotoxins, whose analytical standards are not available, since the traditional quantification methods, represented by tandem MS, are not applicable. The detection and identification of these unknown modified forms is permitted by taking advantage of the accurate mass and the isotopic pattern distribution provided by HRMS. However, no information about putative co-eluting isomers having the same m/z value can be obtained with only LC-HRMS. When analyzing oligo-glycosides mycotoxins, indeed, information about binding position and configuration (α or β) between the sugar moiety and the mycotoxin, and the linkages, $\alpha/\beta 1$-4, $\alpha/\beta 1$-6, cannot be achieved. Regarding binding configuration, information available in the literature is quite contradictory. McCormick et al. [60] reported the occurrence of both T2-α-Glc and T2-β-Glc in naturally contaminated wheat and oat samples. By contrast, Meng-Reiter et al. [48], stated that since the UDP-glucosyltransferase is an inverting enzyme, the detection of the α-glucosyl isomer should be unexpected. In agreement, Zachariasova et al. [38] reported an increase of free DON after DON oligoglucosides incubation with fungal β-glycosidase, that was obviously caused by its release from the β-bound forms. However, also stating the glucosylic bound for DON-oligo-Glc, the number of possible theoretical isomeric structures was higher than the number of chromatographic peaks detected (see Figures 2 and 3) [38]. Thanks to the enhancement in isomer separation offered by applying DTIMS coupled with quadrupole time-of-flight spectrometer (Q-TOF), one more DON-di-Glc and two more DON-tri-Glc peaks separated by their drift time were detected [79]. These additional peaks could be due to the linkages, 1–4 or 1–6, between the sugar moieties and the mycotoxin, since the bounding position was confirmed by HRMS-MS/MS. Therefore, IMS-MS allowed the detection and subsequently the characterization of new isomeric DON-oligo-Glc, also increasing confidence in results.

The same approach could be applied to resolve other modified mycotoxins such as olygo-glycosides forms of ZEN or α/β ZEL, that have already been detected [38] but whose structures have not yet been elucidated, giving new insight into the mycotoxin biotransformation that may occur in plants and/or animals.

It is evident from the examples given that a strong synergy arises between IMS and MS. IMS-MS can act as a tool to separate complex mixtures, to resolve ions that may be indistinguishable by mass spectrometry alone. This is vital in the modified mycotoxins field, since many different types of isomers (diastereoisomers, epimers, anomers, protomers) may occur and need to be separated to achieve a correct quantification and subsequently to perform a reliable risk assessment. As for ADON isomers, also α/β-ZEL diastereoisomers present different toxicity. α/β-ZEL may undergo phase I and phase II metabolism [47], both in plants and in humans, leading to a wide range of metabolites having different configurations and potentially different toxicities. Thus, the separation and then elucidation of the binding configuration of the conjugated metabolites is highly advisable also for the toxicological outlook.

7.3. CCS Value: A New Unambiguous Molecular Descriptor

In addition to signal-to-noise improvement and enhancement in separation of co-eluting isobar molecules, IMS-MS has also been applied to more high-throughput analytical approaches for confirming compound identity, providing molecular structural and conformational information.

Once the drift time is recorded, this can be converted into a CCS value, which represents the effective area for the interaction between an individual ion and the neutral gas through which it is travelling. Thus CCS is an important distinguishing characteristic of an ion in the gas phase and, being related to its chemical structure and three-dimensional conformation, can provide specific information on ionic configuration and potential structural confirmation.

Nowadays, CCS values can be routinely measured as an integrated part of the LC-HRMS experiment. In DTIMS instruments, CCS can be directly derived from the drift time. In TWIMS [90–92] instruments, CCS can be experimentally derived by using IMS calibration performed using compounds of known CCS under defined conditions (i.e., gas type and pressure, travelling wave speed or height). This allows CCS to be used alongside the traditional molecular identifiers of precursor ion accurate mass, fragment ions, isotopic pattern, and retention time as a confirmation of compound identity [93]. Indeed, Goscinny and colleagues [75] developed a TWIMS approach to screening multi-class pesticides and suggested the inclusion of the pesticide CCS values as a new identification point (IP) (Commission Decision 2002/657/EC [94]), to increase confidence in the results. Overall, they measured CCS values for 150 pesticides, using standard solutions, building an in-house CCS library with associated retention times, accurate masses, and diagnostic fragments. Thus, the CCS value may be included in the contaminants screening workflow, as reported in Figure 1, and it can be used as an additional means of filtering the screening data to significantly reduce the proportion of false positive and false negative detections. Therefore, the CCS tolerance of ±2% in combination with the traditional confirmation threshold filters of m/z (±5/10 ppm) and retention time (±2.5%) will lead to a more definitive identification of the species of interest. A further dataset providing CCS values for 200 pesticides has been recently reported by Regueiro et al. [84].

Population of databases with CCS values for pesticides and mycotoxins is pivotal in order to support the inclusion of CCS values as a new identification point (IP).

In addition, since CCS measurements are undertaken in the gas phase, remotely from the ion source, their values are not affected by sample matrix and are consistent between instruments and across a range of experimental conditions [86,93]. Taking into account the analytical effort made in recent years for validating extraction and detection procedure depending on the sample matrix, the great advantage offered by the CCS measurement is evident. Moreover, it has been demonstrated that the concentration of the compound had no significant effect on the drift time values and thus on the CCSs [75]. This will help in avoiding false negative assignments in the screening confirmation procedure, mainly when analyzing contaminants close to method LOQs, since matching with HRMS/MS in libraries could be hard due to the low intensity of the fragmentation pattern [31].

In agreement, Paglia and co-workers [95] also confirmed the high reproducibility of CCS measurements of lipids classes in varying matrices. They created a CCS database for lipids that includes 244 CCS values, aiming to implement the ion-mobility derived CCS in routine lipidomics workflow.

These findings raised the possibility that CCS can be used to help the identification process of targeted compounds, and, as for pesticides and lipids, CCS can be inserted in a routine workflow for parent and modified mycotoxin screening and used as an identification parameter. Data bases of mycotoxins can be created using CCS obtained by running standard compounds, providing an additional coordinate to support mycotoxin identification, and reducing the number of false positive and false negatives of the targeted analysis. In addition, at the end of the untargeted screening process (see Figure 1), CCS can be used to confirm the structure of expected unknowns by matching the theoretical and the experimental CCS values.

In TWIMS devices, poly-DL-alanine is often used as IMS calibrant for deriving CCS. Since peptides have unique physical properties and gas-phase conformations, it might not be ideal to calculate the accurate CCS values for all metabolites and lipids classes. For instance, alternative calibrants have been proposed for specific lipids, which better reflect their chemical structure [96].

CCS values may also be estimated computationally if the 3D structure is known. A comparison of the theoretical and experimentally derived collision cross-sections can be utilized for the accurate assignment of isomeric metabolites. Recently, the CCS areas were used to elucidate the α and β epimeric forms of glycosylated T-2 and HT-2 toxins [80]. The two isomeric forms had already been separated by UHPLC-MS/MS [48] however, thanks to additional information provided by the CCS value, it was possible to confirm the bounding configuration between the toxin and the sugar moiety [80].

The aforementioned application is the only one developed so far in the field of mycotoxins exploiting CCS potential; however IMS-MS, as a tool to gain insight into structural information, would expect to rise rapidly, offering a unique means of characterization. New modified forms, i.e., expected unknown mycotoxins, may be discovered and unequivocally characterized by matching theoretical and experimental rotationally averaged cross-sectional areas, despite the lack of analytical standards. Regarding unexpected unknown mycotoxins, even though HR-IMS reduces the number of possible candidates due to accurate mass, isotopic pattern, MS/MS fragment ions, and CCS values, the identification might still be challenging. In these particular cases, the use of NMR still represents the only approach for identification.

8. Conclusions and Future Trends

In the last few years, a significant increase in the number of studies reporting HRMS-based approaches for food contaminant analysis has been reported. The analytical potential of high resolving power, accurate mass, and acquisition in full scan permits a retrospective analysis using extensive databases of hundreds of analytes and enabling the investigation of 'newly discovered' mycotoxins in the data of prior-analyzed samples. Therefore, HRMS is undoubtedly going to redefine LC-MS workflow since targeted and routine quantification as well as qualitative research analysis can be performed with the same instrument. This situation is also facilitated by the launch from many MS companies of the latest generation of HRMS instrumentation designed for routine analysis and equipped with user-friendly dedicated data processing software.

On the other hand, ion mobility spectrometry is starting to be successfully employed in mycotoxin trace analysis with the aim of increasing signal-to-noise ratio, gaining higher sensitivity, and with longer dynamic range [71,77,78]. However, applications of IMS in separation and structure confirmation of mycotoxins has not been explored adequately so far, even though it offers great potential for gaining insight into the formation and characterization of new modified forms. In particular, the CCS values may be added in targeted and untargeted screening workflow, providing an additional coordinate to support mycotoxin identification, reducing the number of false positive and false negatives and confirming the structure of expected unknowns. In conclusion, all evidence points towards future growth in the number of applications of HRMS and HR-IMS in food safety, as the power of this technique gains wider recognition.

Author Contributions: C.D. and G.G. conceived the review; C.D. and L.R. wrote the manuscript; G.P. and G.G. critically revised the manuscript.

Conflicts of Interest: The authors declare no conflict of interest.

References

1. Battilani, P.; Toscano, P.; Van der Fels-Klerx, H.J.; Moretti, A.; Camardo Leggieri, M.; Brera, C.; Rortais, A.; Goumperis, T.; Robinson, T. Aflatoxin B1 contamination in maize in Europe increases due to climate change. *Sci. Rep.* **2016**, *6*. [CrossRef] [PubMed]
2. Dall'Asta, C.; Berthiller, F. *Masked Mycotoxins in Food: Formation, Occurrence and Toxicological Relevance*, 1st ed.; Royal Society of Chemistry: Cambridge, UK, 2016; Volume 24.
3. EFSA Panel on Contaminants in the Food Chain (CONTAM). Scientific Opinion on the risks for human and animal health related to the presence of modified forms of certain mycotoxins in food and feed. *EFSA J.* **2014**, *12*. [CrossRef]
4. Rychlik, M.; Humpf, H.U.; Marko, D.; Danicke, S.; Mally, A.; Berthiller, F.; Klaffke, H.; Lorenz, N. Proposal of a comprehensive definition of modified and other forms of mycotoxins including "masked" mycotoxins. *Mycotoxin Res.* **2014**, *30*, 197–205. [CrossRef] [PubMed]
5. Berthiller, F.; Crews, C.; Dall'Asta, C.; Saeger, S.D.; Haesaert, G.; Karlovsky, P.; Oswald, I.P.; Seefelder, W.; Speijers, G.; Stroka, J. Masked mycotoxins: A review. *Mol. Nutr. Food Res.* **2013**, *57*, 165–186. [CrossRef] [PubMed]
6. De Boevre, M.; Jacxsens, L.; Lachat, C.; Eeckhout, M.; Di Mavungu, J.D.; Audenaert, K.; Maene, P.; Haesaert, G.; Kolsteren, P.; De Meulenaer, B.; et al. Human exposure to mycotoxins and their masked forms through cereal-based foods in Belgium. *Toxicol. Lett.* **2013**, *218*, 281–292. [CrossRef] [PubMed]
7. Falavigna, C.; Lazzaro, I.; Galaverna, G.; Battilani, P.; Dall'Asta, C. Fatty acid esters of fumonisins: First evidence of their presence in maize. *Food Addit. Contam. Part A* **2013**, *30*, 1606–1613. [CrossRef] [PubMed]
8. Dall'Erta, A.; Cirlini, M.; Dall'Asta, M.; Del Rio, D.; Galaverna, G.; Dall'Asta, C. Masked mycotoxins are efficiently hydrolyzed by human colonic microbiota releasing their aglycones. *Chem. Res. Toxicol.* **2013**, *26*, 305–312. [CrossRef] [PubMed]
9. Gareis, M.; Bauer, J.; Thiem, J.; Plank, G.; Grabley, S.; Gedek, B. Cleavage of zearalenone-glycoside, a "masked" mycotoxin, during digestion in swine. *Zentralbl Veterinarmed B* **1990**, *37*, 236–240. [CrossRef] [PubMed]
10. Kaufmann, A.; Butcher, P.; Maden, K.; Walker, S.; Widmer, M. Comprehensive comparison of liquid chromatography selectivity as provided bytwo types of liquid chromatography detectors (high resolution mass spectrometry and tandem mass spectrometry): "Where is the crossover point". *Anal. Chim. Acta* **2010**, *673*, 60–72. [CrossRef] [PubMed]
11. Mol, H.G.J.; Van Dam, R.C.J.; Zomer, P.; Mulder, P.P.J. Screening of plant toxins in food, feed and botanicals using full-scan high-resolution (Orbitrap) mass spectrometry. *Food Addit. Contam.* **2011**, *28*, 1405–1423. [CrossRef] [PubMed]
12. Martínez-Domínguez, G.; Romero-González, R.; Arrebola, F.J.; Garrido Frenich, A. Multi-class determination of pesticides and mycotoxins in isoflavones supplements obtained from soy by liquid chromatography coupled to Orbitrap high resolution mass spectrometry. *Food Control* **2015**, *59*, 218–224. [CrossRef]
13. Sulyok, M.; Berthiller, F.; Krska, R.; Schuhmacher, R. Development and validation of a liquid chromatography/tandem mass spectrometric method for the determination of 39 mycotoxins in wheat and maize. *Rapid Commun. Mass Spectrom.* **2006**, *20*, 2649–2659. [CrossRef] [PubMed]
14. Malachová, A.; Sulyok, M.; Beltrán, E.; Berthiller, F.; Krska, R. Optimization and validation of a quantitative liquid chromatography–tandem mass spectrometric method covering 295 bacterial and fungal metabolites including all regulated mycotoxins in four model food matrices. *J. Chrom. A* **2014**, *1362*, 145–156. [CrossRef] [PubMed]

15. Dzuman, Z.; Zachariasova, M.; Lacina, O.; Veprikova, Z.; Slavikova, P.; Hajslova, J. A rugged high-throughput analytical approach for the determination and quantification of multiple mycotoxins in complex feed matrices. *Talanta* **2014**, *121*, 263–272. [CrossRef] [PubMed]

16. Cirlini, M.; Dall'Asta, C.; Galaverna, G. Hyphenated chromatographic techniques for structural characterization and determination of masked mycotoxins. *J. Chromatog. A* **2012**, *1255*, 145–152. [CrossRef] [PubMed]

17. Berthiller, F.; Brera, C.; Crews, C.; Iha, M.H.; Kraska, R.; Lattanzio, V.M.T.; MacDonald, S.; Malone, R.J.; Maragos, C.; Solfrizzo, M.; et al. Development in mycotoxin analysis: An update for 2014–2015. *World Mycotoxin J.* **2016**, *9*, 5–29. [CrossRef]

18. Regulation No 882/2004/EC of 29 April 2004 of the European Parliament and of the Council on Official Controls Performed to Ensure the Verification of Compliance with Feed and Food Law, Animal Health and Animal Welfare Rules. Available online: http://data.europa.eu/eli/reg/2004/882/oj (accessed on 18 November 2016).

19. ISO/IEC 17025:2005 General Requirements for the Competence of Testing and Calibration Laboratories. Available online: http://www.iso.org/iso/catalogue_detail?csnumber=39883 (accessed on 18 November 2016).

20. Thompson, M.; Ellison, S.L.R.; Wood, R. Harmonized guidelines for single-laboratory validation of methods of analysis, IUPAC Technical report. *Pure Appl. Chem.* **2002**, *74*, 835–855. [CrossRef]

21. Walravens, J.; Mikula, H.; Rychlik, M.; Asam, S.; Ediage, E.M.; Di Mavungu, J.D.; Van Landschoot, A.; Vanhaecke, L.; De Saeger, S. Development and validation of an ultra-high-performance liquid chromatography tandem mass spectrometric method for the simultaneous determination of free and conjugated *Alternaria* toxins in cereal-based foodstuffs. *J. Chromatog. A.* **2014**, *1372*, 91–101. [CrossRef] [PubMed]

22. Berthiller, F.; Dall'Asta, C.; Schuhmacher, R.; Lemmens, M.; Adam, G.; Krska, R. Masked Mycotoxins: Determination of a Deoxynivalenol Glucoside in Artificially and Naturally Contaminated Wheat by Liquid Chromatography—Tandem Mass Spectrometry. *J. Agric. Food Chem.* **2005**, *53*, 3421–3425. [CrossRef] [PubMed]

23. Vend, O.; Berthiller, F.; Crews, C.; Krska, R. Simultaneous determination of deoxynivalenol, zearalenone, and their major masked metabolites in cereal-based food by LC-MS-MS. *Anal. Bioanal. Chem.* **2009**, *395*, 1347–1354. [CrossRef] [PubMed]

24. Kaufmann, A. The current role of high-resolution mass spectrometry in food analysis. *Anal. Bioanal. Chem.* **2012**, *403*, 1233–1249. [CrossRef] [PubMed]

25. Krauss, M.; Singer, H.; Hollender, J. LC-high resolution MS in environmental analysis: From target screening to the identification of unknowns. *Anal. Bional. Chem.* **2010**, *397*, 943–951. [CrossRef] [PubMed]

26. Rychlik, M.; Asam, S. Stable isotope dilution assays in mycotoxin analysis. *Anal. Bioanal. Chem.* **2008**, *390*, 617–628. [CrossRef] [PubMed]

27. Asam, S.; Rychlik, M. Quantitation of type B-trichothecene mycotoxins in foods and feeds by a multiple stable isotope dilution assay. *Eur. Food Res. Technol.* **2007**, *224*, 769–783. [CrossRef]

28. Habler, K.; Rychlik, M. Multi-mycotoxin stable isotope dilution LC-MS/MS method for Fusarium toxins in cereals. *Anal. Bioanal. Chem.* **2016**, *408*, 307–317. [CrossRef] [PubMed]

29. Habler, K.; Frank, O.; Rychlik, M. Chemical synthesis of deoxynivalenol-3-β-D-[$^{13}C_6$]-glucoside and application in stable isotope dilution assay. *Molecules* **2016**, *21*. [CrossRef] [PubMed]

30. Varga, E.; Glauner, T.; Köppen, R.; Mayer, K.; Sulyok, M.; Schuhmacher, R.; Krska, R.; Berthiller, F. Stable isotope dilution assay for the accurate determination of mycotoxins in maize by UHPLC-MS/MS. *Anal. Bional. Chem.* **2012**, *402*, 2675–2686. [CrossRef] [PubMed]

31. Dzuman, Z.; Zachariasova, M.; Veprikova, Z.; Godula, M. Multi-analyte high performance liquid chromatography coupled to high resolution tandem mass spectrometry method for control of pesticide residues, mycotoxins, and pyrrolizidine alkaloids. *Anal. Chim. Acta* **2015**, *863*, 29–40. [CrossRef] [PubMed]

32. León, N.; Pastor, A.; Yusà, V. Target analysis and retrospective screening of veterinary drugs, ergot alkaloids, plant toxins and other undesirable substances in feed using liquid chromatography–high resolution mass spectrometry. *Talanta* **2016**, *149*, 43–52. [CrossRef] [PubMed]

33. De Boevre, M.; Vanheule, A.; Audenaert, K.; Bekaert, B.; Di Mavungu, J.D.; Werbrouck, S.; Haesaert, G.; De Saeger, S. Detached leaf in vitro model for masked mycotoxin biosynthesis and subsequent analysis of unknown conjugates. *World Mycotoxin J.* **2014**, *7*, 305–312. [CrossRef]

34. Rochat, B. From targeted quantification to untargeted metabolomics: Why LC-high-resolution-MS will become a key instrument in clinical labs. *Trend Analyt Chem.* **2016**, *84*, 151–164. [CrossRef]

35. Senyuva, H.Z.; Gökmen, V.; Sarikaya, E.A. Future perspectives in Orbitrap™-high-resolution mass spectrometry in food analysis: A review. *Food Addit. Contam. Part A* **2015**, *32*, 1568–1606. [CrossRef] [PubMed]

36. Ates, E.; Godula, M.; Stroka, J.; Senyuva, H. Screening of plant and fungal metabolites in wheat, maize and animal feed using automated on-line clean-up coupled to high resolution mass spectrometry. *Food Chem.* **2014**, *142*, 276–284. [CrossRef] [PubMed]

37. Kovalsky, M.P.; Schweiger, W.; Hametner, C.; Stuckler, R.; Muehlbauer, G.J.; Varga, E.; Krska, R.; Berthiller, F.; Adam, G. Zearalenone-16-O-glucoside: A new masked mycotoxin. *J. Agric. Food Chem.* **2014**, *62*, 1181–1189. [CrossRef] [PubMed]

38. Zachariasova, M.; Vaclavikova, M.; Lacina, O.; Vaclavik, L.; Hajslova, J. Deoxynivalenol oligoglycosides: New "masked" *fuarium* toxins occurring in malt, beer, and breadstuff. *J. Agric. Food Chem.* **2012**, *60*, 9280–9291. [CrossRef] [PubMed]

39. Cavaliere, C.; Foglia, P.; Guarino, C.; Marzioni, F.; Nazzari, M.; Samperi, R.; Laganà, A. Aflatoxin M1 determination in cheese by liquid chromatography–tandem mass spectrometry. *J. Chromatogr. A* **2006**, *1135*, 135–141. [CrossRef] [PubMed]

40. Sahin, H.Z.; Celik, M.; Kotay, S.; Kabak, B. Aflatoxins in dairy cow feed, raw milk and milk products from Turkey. *Food Addit. Contam. Part B* **2016**, *9*, 152–158. [CrossRef] [PubMed]

41. Veprikova, Z.; Zachariasova, M.; Dzuman, Z.; Zachariasova, A.; Fenclova, M.; Slavikova, P.; Vaclavikova, M.; Mastovska, K.; Hengst, D.; Hajslova, J. Mycotoxins in Plant-Based Dietary Supplements: Hidden Health Risk for Consumers. *J. Agric. Food Chem.* **2015**, *63*, 6633–6643. [CrossRef] [PubMed]

42. Frisvad, J.C.; Larsen, T.O.; de Vries, R.; Meijer, M.; Houbraken, J.; Cabañes, F.J.; Ehrlich, K.; Samson, R.A. Secondary metabolite profiling, growth profiles and other tools for species recognition and important Aspergillus mycotoxins. *Stud. Mycol.* **2007**, *59*, 31–37. [CrossRef] [PubMed]

43. Lattanzio, V.M.T.; Visconti, A.; Haidukowski, M.; Pascale, M. Identification and characterization of new Fusariummasked mycotoxins, T2 and HT2 glycosyl derivatives, in naturally contaminated wheat and oats by liquid chromatography–high-resolution mass spectrometry. *Rapid Commun. Mass Spectrom.* **2012**, *47*, 466–475. [CrossRef] [PubMed]

44. Kluger, B.; Bueschl, C.; Lemmens, M.; Michlmayr, H.; Malachova, A.; Koutnik, A.; Maloku, I.; Berthiller, F.; Adam, G.; Krska, R.; et al. Biotransformation of the mycotoxin deoxynivalenol in fusarium resistant and susceptible near isogenic wheat lines. *PLoS ONE* **2015**, *10*. [CrossRef] [PubMed]

45. Nakagawa, H.; Ohmichi, K.; Sakamoto, S.; Sago, Y.; Kushiro, M.; Nagashima, H.; Yoshida, M.; Nakajima, T. Detection of a new Fusarium masked mycotoxin in wheat grain by high-resolution LC–OrbitrapTM MS. *Food Addit. Contam.* **2011**, *28*, 1447–1456. [CrossRef] [PubMed]

46. Ivanova, L.; Kruse Fæste, C.; Uhlig, S. In vitro phase I metabolism of the depsipeptide enniatin B. *Anal. Bioanal. Chem.* **2011**, *400*, 2889–2901. [CrossRef] [PubMed]

47. Berthiller, F.; Werner, U.; Suliok, M.; Kraska, R.; Hauser, M.T.; Schuhmacher, R. Liquid chromatography coupled to tandem mass spectrometry (LC-MS/MS) determination of phase II metabolites of the mycotoxin zearalenone in the model plant Arabidopsis thaliana. *Food Addit. Contam.* **2006**, *23*, 1194–1200. [CrossRef] [PubMed]

48. Meng-Reiterer, J.; Varga, E.; Nathanail, A.V.; Bueschl, C.; Rechthaler, J.; McCormick, S.P.; Michlmayr, H.; Malachová, A.; Fruhmann, P.; Adam, G.; et al. Tracing the metabolism of HT-2 toxin and T-2 toxin in barley by isotope-assisted untargeted screening and quantitative LC-HRMS analysis. *Anal. Bioanal. Chem.* **2015**, *407*, 8019–8033. [CrossRef] [PubMed]

49. Fruhmann, P.; Mikula, H.; Wiesenberger, G.; Varga, E.; Lumpi, D.; Stöger, B.; Häubl, G.; Lemmens, M.; Berthiller, F.; Krska, R.; et al. Isolation and Structure Elucidation of Pentahydroxyscirpene, a Trichothecene Fusarium Mycotoxin. *J. Nat. Prod.* **2014**, *77*, 188–192. [CrossRef] [PubMed]

50. Kostelanska, M.; Dzuman, Z.; Malachova, A.; Capouchova, I.; Prokinova, E.; Skerikova, A.; Hajslova, J. Effects of Milling and Baking Technologies on Levels of Deoxynivalenol and its Masked Form Deoxynivalenol-3-Glucoside. *J. Agric. Food Chem.* **2011**, *59*, 9303–9312. [CrossRef] [PubMed]

51. Kind, T.; Fiehn, O. Advances in structure elucidation of small molecules using mass spectrometry. *Bioanal. Rev.* **2010**, *2*, 23–60. [CrossRef] [PubMed]

52. Zedda, M.; Zwiener, C. Is nontarget screening of emerging contaminants by LC-HRMS successful? A plea for compound libraries and computer tools. *Anal. Bioanal. Chem.* **2012**, *403*, 2493–2502. [CrossRef] [PubMed]

53. Nathanail, A.V.; Varga, E.; Meng-Reiterer, J.; Bueschl, C.; Michlmayr, H.; Malachova, A.; Fruhmann, P.; Jestoi, M.; Peltonen, K.; Adam, G.; et al. Metabolism of the *Fusarium* Mycotoxins T-2 Toxin and HT-2 Toxin in Wheat. *J. Agric. Food Chem.* **2015**, *63*, 7862–7872. [CrossRef] [PubMed]

54. Dall'Asta, C.; Berthiller, F.; Schuhmacher, R.; Adam, G.; Lemmens, M.; Krska, R. DON-glycosides: Characterisation of synthesis products and screening for their occurrence in DON treated wheat samples. *Mycotoxin Res.* **2005**, *21*, 123–127. [CrossRef] [PubMed]

55. De Boevre, M.; Ediage, E.N.; Van Poucke, C.; De Sager, S. Untargeted analysis of modified mycotoxins using high-resolution mass spectrometry. In *Masked Mycotoxins in Food: Formation, Occurrence and Toxicological Relevance*, 1st ed.; Dall'Asta, C., Berthiller, F., Eds.; Royal Society of Chemistry: Cambridge, UK, 2016; Volume 24, pp. 50–72.

56. Keller, B.O.; Sui, J.; Young, A.B.; Whittal, R.M. Interferences and contaminants encountered in modern mass spectrometry. *Anal. Chim. Acta* **2008**, *627*, 71–81. [CrossRef] [PubMed]

57. Bueschl, C.; Kluger, B.; Lemmens, M.; Adam, G.; Wiesenberger, G.; Maschietto, V.; Marocco, A.; Strauss, J.; Bodi, S.; Thallinger, G.G.; et al. A novel stable isotope labelling assisted workflow for improved untargeted LC–HRMS based metabolomics research. *Metabolomics* **2014**, *10*, 754–769. [CrossRef] [PubMed]

58. Kluger, B.; Bueschl, C.; Lemmens, M.; Berthiller, F.; Häubl, G.; Jaunecker, G.; Adam, G.; Krska, R.; Schuhmacher, R. Stable isotopic labelling-assisted untargeted metabolic profiling reveals novel conjugates of the mycotoxin deoxynivalenol in wheat. *Anal. Bioanal. Chem.* **2013**, *405*, 5031–5036. [CrossRef] [PubMed]

59. Bueschl, C.; Kluger, B.; Berthiller, F.; Lirk, G.; Winkler, S.; Krska, R.; Schuhmacher, R. MetExtract: A new software tool for the automated comprehensive extraction of metabolite-derived LC/MS signals in metabolomics research. *Bioinformatics* **2012**, *28*, 736–738. [CrossRef] [PubMed]

60. McCormick, S.P.; Kato, T.; Maragos, C.M.; Busman, M.; Lattanzio, V.M.T.; Galaverna, G.; Dall-Asta, C.; Crich, D.; Price, N.P.J.; Kurtzman, C.P. Anomericity of T-2 Toxin-glucoside: Masked Mycotoxin in Cereal Crops. *J. Agric. Food Chem.* **2015**, *63*, 731–738. [CrossRef] [PubMed]

61. Cumeras, R.; Figueras, E.; Davis, C.E.; Baumbach, J.I.; Gràcia, I. Review on ion mobility spectrometry. Part 1: Current instrumentation. *Analyst* **2015**, *140*, 1376–1390. [CrossRef] [PubMed]

62. Kafle, G.K.; Khot, L.R.; Sankaran, S.; Bahlol, H.Y.; Tufariello, J.A.; Hill, H.H. State of ion mobility spectrometry and applications in agriculture: A review. *Eng. Agric. Environ. Food* **2016**, *9*, 346–357. [CrossRef]

63. Ibrahim, Y.M.; Baker, E.S.; Danielson, W.F.; Norheim, R.V.; Prior, D.C.; Anderson, G.A.; Belov, M.E.; Smith, R.D. Development of a New Ion Mobility (Quadrupole) Time-of-Flight Mass Spectrometer. *Int. J. Mass Spectrum.* **2015**, *377*, 655–662. [CrossRef] [PubMed]

64. Giles, K. Travelling wave ion mobility. *Int. J. Ion Mobil. Spectrom.* **2013**, *16*, 1–3. [CrossRef]

65. Shvartsburg, A.A.; Smith, R.D. Fundamentals of traveling wave ion mobility spectrometry. *Anal. Chem.* **2008**, *80*, 9689–9699. [CrossRef] [PubMed]

66. Shvartsburg, A.A. *Differential Ion Mobility Spectrometry: Nonlinear Ion Transport and Fundamentals of FAIMS*, 1st ed.; CRC Press: Boca Raton, FL, USA, 2008.

67. Michelmann, K.; Silveira, J.A.; Ridgeway, M.E.; Park, M.A. Fundamentals of trapped ion mobility spectrometry. *J. Am. Soc. Mass Spectrom.* **2015**, *26*, 14–24. [CrossRef] [PubMed]

68. Hernandez, D.R.; Debord, J.D.; Ridgeway, M.E.; Kaplan, D.A.; Park, M.A.; Fernandez-Lima, F. Ion dynamics in a trapped ion mobility spectrometer. *Analyst* **2014**, *139*, 1913–1921. [CrossRef] [PubMed]

69. Fernandez-Lima, F.; Kaplan, D.A.; Suetering, J.; Park, M.A. Gas-phase separation using a trapped ion mobility spectrometer. *Int. J. Ion Mobil. Spectrom.* **2011**, *14*, 1–10. [CrossRef] [PubMed]

70. Lapthorn, C.; Pullen, F.; Chowdhry, B.Z. Ion mobility spectrometry-mass spectrometry (IMS-MS) of small molecules: Separating and assigning structures to ions. *Mass Spectrom. Rev.* **2013**, *32*, 43–71. [CrossRef] [PubMed]

71. McCooeye, M.; Kolakowski, B.; Boison, J.; Mester, Z. Evaluation of high-field asymmetric waveform ion mobility spectrometry for the analysis of the mycotoxin zearalenone. *Anal. Chim. Acta* **2008**, *627*, 112–116. [CrossRef] [PubMed]

72. Karpas, Z. Applications of ion mobility spectrometry (IMS) in the field of foodomics. *Food. Res. Int.* **2013**, *54*, 1146–1151. [CrossRef]

73. Beach, D.G.; Melanson, J.E.; Purves, R.W. Analysis of paralytic shellfish toxins using high-field asymmetric waveform ion mobility spectrometry with liquid chromatography-mass spectrometry. *Anal. Bional. Chem.* **2015**, *407*, 2473–2484. [CrossRef] [PubMed]

74. Nielen, M.W.F.; Van Beek, T.A. Macroscopic and microscopic spatially-resolved analysis of food contaminants and constituents using laser-ablation electrospray ionization mass spectrometry imaging. *Anal. Bional. Chem.* **2014**, *406*, 6805–6815. [CrossRef] [PubMed]

75. Goscinny, S.; Joly, L.; De Pauw, E.; Hanot, V.; Eppe, G. Travelling-wave ion mobility time-of-flight mass spectrometry as an alternative strategy for screening of multi-class pesticides in fruits and vegetables. *J. Chromatogr. A* **2015**, *1405*, 85–93. [CrossRef] [PubMed]

76. Gonzales, G.B.; Smagghe, G.; Coelus, S.; Adriaenssens, D.; De Winter, K.; Desmet, T.; Raes, K.; Van Camp, J. Collision cross section prediction of deprotonated phenolics in a travelling-wave ion mobility spectrometer using molecular descriptors and chemometrics. *Anal. Chim. Acta* **2016**, *924*, 68–76. [CrossRef] [PubMed]

77. Sheibani, A.; Tabrizchi, M.; Ghaziaskar, H. Determination of aflatoxins B1 and b2 using ion mobility spectrometry. *Talanta* **2007**, *75*, 233–238. [CrossRef] [PubMed]

78. Khalesi, M.; Sheikh-Zeinoddin, M.; Tabrizchi, M. Determination of ochratoxin A in licorice root using inverse ion mobility spectrometry. *Talanta* **2011**, *83*, 988–993. [CrossRef] [PubMed]

79. Fenclová, M.; Lacina, O.; Zachariášová, M.; Hajšlová, J. Application of ion-mobility Q-TOF LC/MS platform in masked mycotoxins research. In Proceedings of the Recent Advances in Food Analysis, Prague, Czech Republic, 3–6 November 2015; Pulkrabová, J., Tomaniová, M., Nielen, M., Hajšlová, J., Eds.; UCT Prague Press: Prague, Czech Republic, 2015; p. 122.

80. Stead, S.; Joumier, J.M.; McCullagh, M.; Busman, M.; McCormick, S.; Maragos, C. Using ion mobility mass spectrometry and collision cross section areas to elucidate the α and β epimeric forms of glycosilated T-2 and HT-2 toxins. In Proceedings of the Recent Advances in Food Analysis, Prague, Czech Republic, 3–6 November 2015; Pulkrabová, J., Tomaniová, M., Nielen, M., Hajšlová, J., Eds.; UCT Prague Press: Prague, Czech Republic, 2015; p. 338.

81. Ruotolo, B.T.; Gillig, K.J.; Stone, E.G.; Russell, D.H. Peak capacity of ion mobility mass spectrometry: Separation of peptides in helium buffer gas. *J. Chromatogr. B* **2002**, *782*, 385–392. [CrossRef]

82. Ruotolo, B.T.; McLean, J.A.; Gillig, K.J.; Russell, D.H. Peak capacity of ion mobility mass spectrometry: The utility of varying drift gas polarizability for the separation of tryptic peptides. *J. Mass Spectrom.* **2004**, *39*, 361–367. [CrossRef] [PubMed]

83. Jafari, M.T.; Saraji, M.; Yousefi, S. Negative electrospray ionization ion mobility spectrometry combined with microextraction in packed syringe for direct analysis of phenoxyacid herbicides in environmental water. *J. Chromatogr. A* **2012**, *1249*, 41–47. [CrossRef] [PubMed]

84. Regueiro, J.; Negreira, N.; Berntssen, M.H.G. Ion mobility-derived collision cross section as an additional identification point for multi-residue screening of pesticides in fish feed. *Anal. Chem.* **2016**, *88*, 11169–11177. [CrossRef] [PubMed]

85. Jafari, M.T.; Khayamian, T.; Shaer, V.; Zarei, N. Determination of veterinary drug residues in chicken meat using corona discharge ion mobility spectrometry. *Anal. Chim. Acta* **2007**, *581*, 147–153. [CrossRef] [PubMed]

86. Paglia, G.; Kliman, M.; Claude, E.; Geromanos, S.; Astarita, G. Applications of ion-mobility mass spectrometry for lipid analysis. *Anal. Bional. Chem.* **2015**, *407*, 4995–5007. [CrossRef] [PubMed]

87. Pinton, P.; Tsybulskyy, D.; Lucioli, J.; Laffitte, J.; Callu, P.; Lyazhri, F.; Grosjean, F.; Bracarense, A.P.; Kolf-Clauw, M.; Oswald, I.P. Toxicity of deoxynivalenol and its acetylated derivatives on the intestine: Differential effects on morphology, barrier function, tight junction proteins, and mitogen-activated protein kinases. *Toxicol. Sci.* **2012**, *130*, 180–190. [CrossRef] [PubMed]

88. Monbaliu, S.; Van Poucke, C.; Van Peteghem, C.; Van Poucke, K.; Heungens, K.; De Saeger, S. Development of a multi-mycotoxin liquid chromatography/tandem mass spectrometry method for sweet pepper analysis. *Rapid Commun. Mass Spectrom.* **2009**, *23*, 3–11. [CrossRef] [PubMed]

89. Biancardi, A.; Gasparini, M.; Dall'Asta, C.; Marchelli, R. A rapid multiresidual determination of type A and type B trichothecenes in wheat flour by HPLC-ESIMS. *Food. Addit. Contam.* **2005**, *22*, 251–258. [CrossRef] [PubMed]

90. Campuzano, I.; Bush, M.F.; Robinson, C.V.; Beaumont, C.; Richardson, K.; Kim, H.; Kim, H.I. Structural characterization of drug-like compounds by ion mobility mass spectrometry: Comparison of theirethical and experimentally derived nitrogen collision cross sections. *Anal. Chem.* **2012**, *84*, 1026–1033. [CrossRef] [PubMed]

91. Fenn, L.S.; Kliman, M.; Mahsut, A.; Zhao, S.R.; McLean, J.A. Characterizing ion mobility-mass spectrometry conformation space for the analysis of complex biological samples. *Anal. Bioanal. Chem.* **2009**, *394*, 235–244. [CrossRef] [PubMed]

92. Bush, M.F.; Campuzano, I.D.; Robinson, C.V. Ion mobility mass spectrometry of peptide ions: Effects of drift gas and calibration strategies. *Anal. Chem.* **2012**, *84*, 7124–7130. [CrossRef] [PubMed]

93. McCullagh, M.; Cleland, G.; Hanot, V.; Stead, S.; Williams, J.; Goscinny, S. *Collision Cross Section a New Identification Point for a "Catch All" Non Targeted Screening Approach*; Waters Application Note; Waters and the Scientific Institute of Public Health: Brussels, Belgium, 2014; Available online: http://www.waters.com/waters/library.htm?lid=134803219&locale=it_IT (accessed on 29 November 2016).

94. Commission Decision No. 2002/657/EC of 12 August 2002 Implementing Council Directive 96/23/EC Concerning the Performance of Analytical Methods and the Interpretation of Results. Available online: http://data.europa.eu/eli/dec/2002/657/oj (accessed on 18 November 2016).

95. Paglia, G.; Angel, P.; Williams, J.P.; Richardson, K.; Olivos, H.J.; Thompson, J.W.; Menikarachchi, L.; Lai, S.; Walsh, C.; Moseley, A.; et al. Ion Mobility-Derived Collision Cross Section As an Additional Measure for Lipid Fingerprinting and Identification. *Anal. Chem.* **2015**, *87*, 1137–1144. [CrossRef] [PubMed]

96. Hines, K.M.; May, J.C.; McLean, J.A.; Xu, L. Evaluation of Collision Cross Section Calibrants for Structural Analysis of Lipids by Traveling Wave Ion Mobility-Mass Spectrometry. *Anal. Chem.* **2016**, *88*, 7329–7336. [CrossRef] [PubMed]

© 2016 by the authors. Licensee MDPI, Basel, Switzerland. This article is an open access article distributed under the terms and conditions of the Creative Commons Attribution (CC BY) license (http://creativecommons.org/licenses/by/4.0/).

Article

Metabolism of HT-2 Toxin and T-2 Toxin in Oats

Jacqueline Meng-Reiterer [1,2], Christoph Bueschl [1], Justyna Rechthaler [3], Franz Berthiller [1,4], Marc Lemmens [2] and Rainer Schuhmacher [1,*]

[1]   Center for Analytical Chemistry, Department of Agrobiotechnology (IFA-Tulln),
      University of Natural Resources and Life Sciences, Vienna (BOKU), Konrad-Lorenz-Str. 20, Tulln 3430,
      Austria; jacqueline.reiterer@boku.ac.at (J.M.-R.); christoph.bueschl@boku.ac.at (C.B.);
      franz.berthiller@boku.ac.at (F.B.)
[2]   Institute for Biotechnology in Plant Production, Department of Agrobiotechnology (IFA-Tulln),
      University of Natural Resources and Life Sciences, Vienna (BOKU), Konrad-Lorenz-Str. 20, Tulln 3430,
      Austria; marc.lemmens@boku.ac.at
[3]   Biotechnological Processes Campus-Tulln, University of Applied Sciences, Wr. Neustadt,
      Konrad-Lorenz-Str. 10, Tulln 3430, Austria; justyna.rechthaler@tulln.fhwn.ac.at
[4]   Christian Doppler Laboratory for Mycotoxin Metabolism, Department of Agrobiotechnology (IFA-Tulln),
      University of Natural Resources and Life Sciences, Vienna (BOKU), Konrad-Lorenz-Str. 20,
      Tulln 3430, Austria
*   Correspondence: rainer.schuhmacher@boku.ac.at; Tel.: +43-1-47654-97307

Academic Editor: Aldo Laganà
Received: 14 October 2016; Accepted: 24 November 2016; Published: 5 December 2016

Abstract: The *Fusarium* mycotoxins HT-2 toxin (HT2) and T-2 toxin (T2) are frequent contaminants in oats. These toxins, but also their plant metabolites, may contribute to toxicological effects. This work describes the use of ^{13}C-assisted liquid chromatography–high-resolution mass spectrometry for the first comprehensive study on the biotransformation of HT2 and T2 in oats. Using this approach, 16 HT2 and 17 T2 metabolites were annotated including novel glycosylated and hydroxylated forms of the toxins, hydrolysis products, and conjugates with acetic acid, putative malic acid, malonic acid, and ferulic acid. Further targeted quantitative analysis was performed to study toxin metabolism over time, as well as toxin and conjugate mobility within non-treated plant tissues. As a result, HT2-3-*O*-β-D-glucoside was identified as the major detoxification product of both parent toxins, which was rapidly formed (to an extent of 74% in HT2-treated and 48% in T2-treated oats within one day after treatment) and further metabolised. Mobility of the parent toxins appeared to be negligible, while HT2-3-*O*-β-D-glucoside was partly transported (up to approximately 4%) through panicle side branches and stem. Our findings demonstrate that the presented combination of untargeted and targeted analysis is well suited for the comprehensive elucidation of mycotoxin metabolism in plants.

Keywords: metabolomics; liquid chromatography–high-resolution mass spectrometry; stable isotope labelling; type A trichothecenes; masked mycotoxins; xenobiotics; cereals

1. Introduction

The type A trichothecene mycotoxins HT-2 toxin (HT2) and T-2 toxin (T2) are secondary metabolites of *Fusarium* species such as *F. sporotrichioides*, *F. poae*, *F. armeniacum*, and *F. langsethiae* [1]. Plants, especially small grain cereals, suffer from diseases that are related to the infection with trichothecene-producing fungi, for example *Fusarium* head blight (FHB) [2]. Several recent surveys of cereals and cereal-based foods across Europe have shown that HT2 and T2 are detected most frequently and at the highest concentrations in oats and oat products [1,3,4]. In mammals, T2 is rapidly transformed into HT2 by deacetylation during digestion of infested food or feed. Both T2 and HT2 are toxic to animals and humans, affecting the immune system and causing among other things apoptosis of proliferating cells as well as inhibition of protein synthesis [5,6]. The European Food Safety Authority (EFSA) published a tolerable daily intake (TDI) of 100 ng/kg body weight for the sum of HT2 and T2 [7], while indicative levels ranging from 15 to 2000 µg/kg for cereals and cereal products were established in Recommendation 2013/165/EU from the European Commission [8].

Plants use different strategies to detoxify harmful compounds. To enhance the polarity and reactivity of xenobiotics, phase I metabolism reactions occur such as hydroxylation, hydrolysis, or oxidation. On the other hand, covalent linkage with endogenous hydrophilic molecules such as sugars, malonic acid, glutathione, and amino acids is typical for phase II metabolism. All these reactions are catalysed by enzymes, which are present within plant cells. Phase I uses esterases, amidases, and cytochrome P450-dependent hydroxylases, for example, while phase II reactions include glucosyl-, malonyl-, or glutathione S-transferases. If xenobiotics already have active functional groups, as is the case for HT2 and T2, a direct conjugation with the hydrophilic substances mentioned above can take place [9,10]. A few authors have even reported that xenobiotics or formed xenobiotic conjugates moved out of the cells again and underwent long-range transport in plants [11,12].

While mycotoxin conjugation and subsequent sequestration into cell vacuoles or cell wall biopolymers (phase III metabolism [9,10]) are generally associated with detoxification for the affected plant, these plant-derived biotransformation products can exhibit increased as well as decreased toxicity for mammals [13–16]. Moreover, the potential reactivation of detoxified derivatives in the animal intestinal tract [17] underlines the importance of plant metabolite elucidation.

Analytical methods based on liquid chromatography coupled with mass spectrometry (LC-MS) are widespread for the determination of mycotoxins and their plant metabolites. Since many plant biotransformation products of mycotoxins are not included in conventional analytical methods, also because no regulated limits exist for them in food commodities, they are called masked mycotoxins [18,19]. LC-MS/MS techniques based on multiple reaction monitoring mode are frequently used for the analysis of various (masked) mycotoxins in food products and raw materials [20–22]. However, much more accurate mass-to-charge (m/z) values are obtained by high-resolution MS (HRMS) including Orbitrap or quadrupole time-of-flight (Q-TOF) instruments. Such measurements enable a higher degree of confidence for compound identification and more structural information can be gained, especially when product ion spectra (LC-HRMS/MS) are acquired [18]. Therefore, a high-resolving instrument is highly valuable for untargeted screening of metabolites in biological matrices. The combination with stable isotopic labelling (SIL) offers to differentiate between biologically derived metabolites and unspecific signals coming from solvents, reagents, biological matrix, or instrument noise [23,24]. SIL-assisted metabolomics workflows were recently developed to study the metabolism of endogenous or exogenous tracer compounds in biological systems [25].

Regarding the biotransformation of HT2 and T2 in plants, a few authors have reported the occurrence of glucoside (Glc) derivatives [26–31]. McCormick et al. have investigated the anomericity of T2-Glc (α- or β-Glc) in naturally contaminated wheat and oats. Two comprehensive metabolism studies of HT2 and T2 were performed in barley [32] as well as in wheat [33]. Beyond the formation of Glc derivatives, only limited information exists about the biotransformation process of HT2/T2 in oats, although this plant was shown to be the cereal crop that is most susceptible to HT2/T2 contamination.

The objective of this study was to elucidate the metabolism of HT2 and T2 in oats (*Avena sativa* L.) using a SIL-assisted metabolomics workflow for qualitative screening based on plant-treatment with a mixture of native and ^{13}C-labelled toxin, LC-Orbitrap-MS analysis in positive and negative ion mode and automated data processing by MetExtract II software [25]. Further structural information of the formed toxin derivatives was gained by LC-HRMS/MS measurements with a LC-Q-TOF-system. Within the scope of a time course experiment, targeted quantitative analysis was performed to investigate the kinetics of metabolite formation as well as the mobility of toxin and conjugate within the plant. In this study, we elucidated novel toxin derivatives, identified and quantified the main metabolites, and gained insights into metabolic processes in oats, which might be highly valuable for risk assessment of contaminated food and feed.

2. Results and Discussion

2.1. Overview of Annotated HT2 and T2 Metabolites in Oats

For qualitative screening of plant-derived HT2 and T2 metabolites, oat panicles were treated with a 1/1 mixture of non-labelled and uniformly ^{13}C-labelled toxin (termed $^{12}C/^{13}C$ sample from now on), extracted, analysed by LC-Orbitrap-MS and the data have been processed by the MetExtract II software. The automatically detected biotransformation product-derived LC-HRMS features were then manually inspected and verified, and false-positive results were removed from the data, recognisable by either implausible number of parent toxin-derived C-atoms, differences between labelled and corresponding non-labelled extracted ion chromatogram (EIC) peaks in retention time or peak shape, implausible isotope pattern, or the lack of characteristic parent toxin-derived fragments in LC-HRMS/MS spectra. In addition to automatically detected toxin derivatives, metabolites putatively missed by the strict filtering criteria of MetExtract II that had, however, been found in previous barley [32] and wheat [33] studies or as HT2 metabolites in oats, were searched for manually. Those toxin biotransformation products (M or ^{12}C and respective M' or ^{13}C ion signals) showed low abundances and were thus missing isotopologs (M + 1 and/or M' − 1) and were added to the lists when the $^{12}C/^{13}C$ signal intensity ratios were approximately 1/1 and retention times as well as accurate masses were comparable.

The resulting metabolites were annotated by the number of C-atoms derived from parent toxin. Structure units originating from the plant occurred as non-labelled moieties of the formed toxin conjugates and did not alter the $^{12}C/^{13}C$ mass shift between monoisotopic ^{12}C and ^{13}C-labelled metabolite ions. Further structure annotation and identification was performed on a LC-Q-TOF-system by LC-HRMS/MS measurements.

A total of 16 HT2 and 17 T2 metabolites were found in oat samples (Tables 1 and 2 and Figure 1), which were present as different ion species (mainly $[M+NH_4]^+$, $[M+Na]^+$, $[M+H]^+$, $[M+K]^+$ and $[M+HCOO]^-$, $[M-H]^-$, $[M+Cl]^-$) or in-source fragments.

In good agreement with previous studies on the metabolism of HT2 and T2 in barley [32] and wheat [33], most T2 metabolites matched those found for HT2 also in oats due to the rapid loss of the acetyl group (2 C-atoms) at C-4 position of T2. Only two metabolites were detected with intact T2 (24 C-atoms) backbone, namely 3-acetyl-T2 (18) and feruloyl-T2 (19). Taking HT2 (22 C-atoms) as a starting point, losses of the isovaleryl group (5 C-atoms) or of the acetyl group (2 C-atoms) were observed *in planta* (metabolites 1, 2, and 8). In agreement with that, Cole [9] reported that hydrolysation of xenobiotic esters is a common phase I metabolism route.

All metabolites showed $^{12}C/^{13}C$ signal intensity ratios of approximately 1/1, except for the T2 metabolites 15-acetyl-T2-tetraol-Glc (1) and dehydro-15-acetyl-T2-tetraol-Glc (2). These two substances were not recognised by the untargeted approach because of unexpected $^{12}C/^{13}C$ signal intensity ratios (2/1 and higher). Similar to the previous barley study [32], a minor impurity of non-labelled neosolaniol in the $^{12}C/^{13}C$-T2 treatment solution has probably led to higher $^{12}C/^{13}C$ signal intensity ratios of the mentioned metabolites due to a parallel metabolism process of neosolaniol in the treated plants.

Table 1. List of identified and annotated metabolites of HT-2 toxin in oats.

ID	Metabolite	Retention Time (min)	Elemental Composition [a]	Accurate Mass [b]	Adduct [b]	n_c [c]	Mass Error (ppm)
1	HT2 *	15.69	$C_{22}H_{32}O_8$	442.2446	$[M+NH_4]^+$	22	2.4
2	15-Acetyl-T2-tetraol-Glc **	5.45	$C_{23}H_{34}O_{12}$	520.2408	$[M+NH_4]^+$	17	3.7
3	Dehydro-15-acetyl-T2-tetraol-Glc [d],***	8.09	$C_{23}H_{32}O_{12}$	518.2250	$[M+NH_4]^+$	17	3.5
4	Hydroxy-HT2-HexGlc **	10.21	$C_{34}H_{52}O_{19}$	782.3468	$[M+NH_4]^+$	22	3.4
5	Hydroxy-HT2-HexGlc **	10.88	$C_{34}H_{52}O_{19}$	809.3100	$[M+HCOO]^-$	22	1.9
6	Hydroxy-HT2-Glc **	11.08	$C_{28}H_{42}O_{14}$	620.2927	$[M+NH_4]^+$	22	2.3
7	Hydroxy-HT2 **	11.71	$C_{22}H_{32}O_9$	458.2398	$[M+NH_4]^+$	22	2.9
8	Hydroxy-HT2-anhydro-HexGlc **	12.37	$C_{34}H_{50}O_{18}$	764.3355	$[M+NH_4]^+$	22	2.6
9	T2-triol-Glc **	13.81	$C_{26}H_{40}O_{12}$	589.2508	$[M+HCOO]^-$	20	1.1
10	Dehydro-HT2-Glc **	14.05	$C_{28}H_{40}O_{13}$	629.2458	$[M+HCOO]^-$	22	1.1
11	HT2-HexGlc **	14.23	$C_{34}H_{52}O_{18}$	766.3509	$[M+NH_4]^+$	22	2.2
12	HT2-HexGlc **	14.76	$C_{34}H_{52}O_{18}$	766.3511	$[M+NH_4]^+$	22	2.5
13	HT2-malylGlc **	14.85	$C_{32}H_{46}O_{17}$	720.3087	$[M+NH_4]^+$	22	1.9
14	HT2-3-O-β-D-Glc *	15.00	$C_{28}H_{42}O_{13}$	604.2973	$[M+NH_4]^+$	22	1.5
15	HT2-MalGlc [d],**	15.11	$C_{31}H_{44}O_{16}$	690.2994	$[M+NH_4]^+$	22	3.8
16	HT2-anhydro-HexGlc **	15.51	$C_{34}H_{50}O_{17}$	748.3404	$[M+NH_4]^+$	22	2.4
	T2 [d],*	16.87	$C_{24}H_{34}O_9$	484.2549	$[M+NH_4]^+$	22	1.6

HT2, HT-2 toxin; T2, T-2 toxin; Glc, glucoside; Glc, glucoside; MalGlc, malonylglucoside; HexGlc, hexosylglucoside; [a] Elemental composition of uncharged metabolite; [b] Accurate mass (mean m/z) of the most abundant adduct of MetExtract II derived features; [c] Number of C-atoms originating from parent toxin; [d] Detected by targeted search for barley-[32] and wheat-[33] derived metabolites; * Structure confirmation with standard by comparison of retention time, accurate mass and HRMS/MS-spectrum; ** Structure annotation with accurate mass, number of parent toxin-derived C-atoms, HRMS/MS-spectrum, and by assessment of retention time; *** Structure annotation with accurate mass, number of parent toxin-derived C-atoms, and by assessment of retention time.

Table 2. List of identified and annotated metabolites of T-2 toxin in oats.

ID	Metabolite	Retention Time (min)	Elemental Composition [a]	Accurate Mass [b]	Adduct [b]	n_c [c]	Mass Error (ppm)
	T2 *	16.83	$C_{24}H_{34}O_9$	484.2548	$[M+NH_4]^+$	24	1.4
1	15-Acetyl-T2-tetraol-Glc d,**	5.46	$C_{23}H_{34}O_{12}$	520.2407	$[M+NH_4]^+$	17	3.6
2	Dehydro-15-acetyl-T2-tetraol-Glc d,***	8.10	$C_{23}H_{32}O_{12}$	518.2248	$[M+NH_4]^+$	17	3.1
3	Hydroxy-HT2-HexGlc d,**	10.23	$C_{34}H_{52}O_{19}$	782.3462	$[M+NH_4]^+$	22	2.7
5	Hydroxy-HT2-Glc **	11.08	$C_{28}H_{42}O_{14}$	620.2926	$[M+NH_4]^+$	22	2.1
6	Hydroxy-HT2 **	11.71	$C_{22}H_{32}O_9$	458.2396	$[M+NH_4]^+$	22	2.5
7	Hydroxy-HT2-anhydro-HexGlc d,**	12.36	$C_{34}H_{50}O_{18}$	764.3348	$[M+NH_4]^+$	22	1.7
8	T2-triol-Glc **	13.81	$C_{26}H_{40}O_{12}$	589.2508	$[M+HCOO]^-$	20	1.1
9	Dehydro-HT2-Glc **	14.04	$C_{28}H_{40}O_{13}$	629.2458	$[M+HCOO]^-$	22	1.1
10	HT2-HexGlc **	14.23	$C_{34}H_{52}O_{18}$	766.3509	$[M+NH_4]^+$	22	2.2
11	HT2-HexGlc **	14.78	$C_{34}H_{52}O_{18}$	766.3511	$[M+NH_4]^+$	22	2.5
12	HT2-malylGlc **	14.84	$C_{32}H_{46}O_{17}$	720.3088	$[M+NH_4]^+$	22	2.0
13	HT2-3-O-β-D-Glc *	14.99	$C_{28}H_{42}O_{13}$	604.2975	$[M+NH_4]^+$	22	1.9
14	HT2-MalGlc d,**	15.11	$C_{31}H_{44}O_{16}$	690.2990	$[M+NH_4]^+$	22	3.2
15	HT2-anhydro-HexGlc **	15.50	$C_{34}H_{50}O_{17}$	729.2983	$[M–H]^-$	22	1.1
17	HT2 *	15.68	$C_{22}H_{32}O_8$	442.2446	$[M+NH_4]^+$	22	2.4
18	3-Acetyl-T2 *	18.07	$C_{26}H_{36}O_{10}$	526.2653	$[M+NH_4]^+$	24	1.2
19	Feruloyl-T2 d,**	19.17	$C_{34}H_{42}O_{12}$	660.3022	$[M+NH_4]^+$	24	1.1

HT2, HT-2 toxin; T2, T-2 toxin; Glc, glucoside; MalGlc, malonylglucoside; HexGlc, hexosylglucoside; [a] Elemental composition of uncharged metabolite; [b] Accurate mass (mean *m/z*) of the most abundant adduct of MetExtract II derived features; [c] Number of C-atoms originating from parent toxin; [d] Detected by targeted search for HT2 metabolites known to be produced in oats or for barley-[32] and wheat-[33] derived metabolites; * Structure confirmation with standard by comparison of retention time, accurate mass, and HRMS/MS-spectrum; ** Structure annotation with accurate mass, number of parent toxin-derived C-atoms, HRMS/MS-spectrum, and by assessment of retention time; *** Structure annotation with accurate mass, number of parent toxin-derived C-atoms, and by assessment of retention time.

As can be seen from the toxin metabolites in Tables 1 and 2 and Figure 1, hydroxylation and glucosylation metabolism processes dominated in oats. These findings are concordant with Cole [9] who described that xenobiotics are often hydroxylated by cytochrome P450-dependent hydroxylases and glucosylated by O-glucosyltransferases in plants. While the glucosylation of HT2, forming a β-linked D-glucose conjugate (confirmed by comparison with respective standard), was an important detoxification strategy in oats, no T2-Glc was detected; it was, however, found in barley [32], as well as in naturally contaminated wheat and oats [31]. Additionally, in contrast to previous xenobiotic metabolism studies [32,33] and [15,34,35] hardly any malonic acid conjugations were detected in oats, except the less abundant HT2-malonylglucoside (HT2-MalGlc (14)), which was close to the analytical limit of detection (LOD, signal-to-noise ratio of 3 in matrix) (see Figure 1). Thus, it is assumed that the quantity or activity of HT2/T2 metabolising O-malonyltransferases is lower in oats than in barley and wheat under the tested conditions.

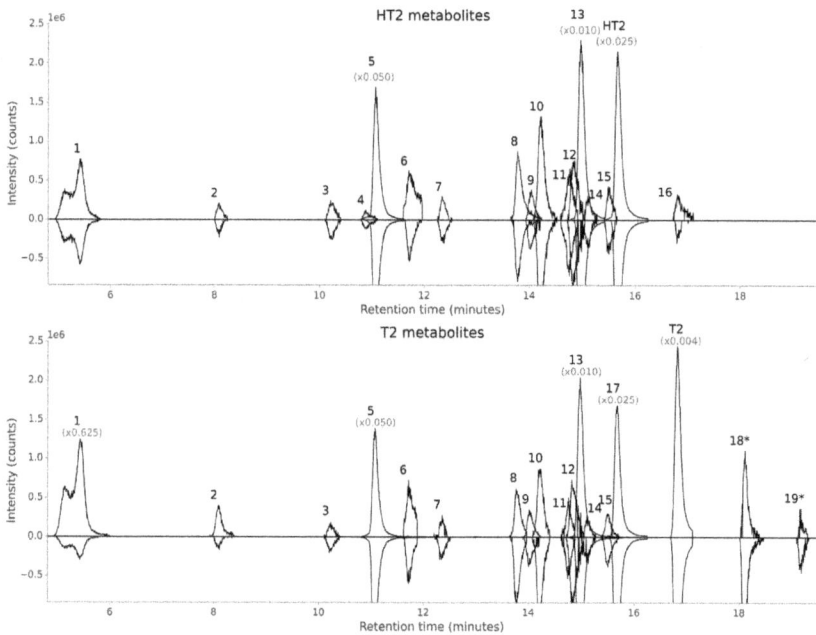

Figure 1. Overlay of extracted ion chromatograms (EICs) of HT2/T2 plant metabolites. (**a**) Overlaid EICs of HT2 metabolites based on HT2-treated oat sample (time point full-ripening) and Table 1; (**b**) overlaid EICs of T2 metabolites based on T2-treated oat samples (time point full-ripening and accumulated time points marked with an asterisk) and Table 2. Oat panicles were treated with a 1/1 mixture of non-labelled and uniformly ^{13}C-labelled toxin, extracted and analysed by LC-Orbitrap-MS in positive and negative ion mode and MetExtract II software. Non-labelled metabolite form is depicted with positive intensity (up) and corresponding ^{13}C-labelled metabolite form with negative intensity (down). HT2, HT-2 toxin; T2, T-2 toxin.

Interestingly, in the present study six novel HT2/T2 metabolites were detected for the first time in plants; these were putatively annotated as hydroxy-HT2-hexosylglucoside (hydroxy-HT2-HexGlc (two isomers 3 and 4)), hydroxy-HT2 (6), hydroxy-HT2-anhydro-HexGlc (7), HT2-malylGlc (12) and HT2-anhydro-HexGlc (15). While five of them seem to be closely related to already known HT2/T2 metabolites, HT2-malylGlc was presumably formed by the covalent binding of malic acid to a glucose moiety of the initially formed HT2-3-O-β-D-Glc. The conjugation with organic acids is widely known in plant metabolism. Abdel-Farid et al. [36], for example, described malate conjugates of cinnamic acid derivatives in *Brassica rapa* leaves and Fujisawa et al. [37] elucidated a malate conjugate of a pesticide in leaf cell suspension of *Brassica oleracea*.

Our data indicate that not only is there a deacetylation reaction from T2 into HT2, but also the reverse reaction from HT2 to T2 occurs in plants, although to a much lower extent (see tab:toxins-08-00364-t001 and Figure 1). While in wheat C-3 and C-4 acetylated HT2 (4-acetyl-HT2 is corresponding to T2) were probably detected [33], the present study revealed solely the putative C-4 acetylated HT2 with a $^{12}C/^{13}C$ mass shift of Δ22.074 u (theoretical value). It is worth mentioning that an impurity of ^{13}C-T2 in the $^{12}C/^{13}C$-HT2 treatment solution was also observed with the M' − 2 isotopolog having the same *m/z* value as the labelled metabolite form of acetylated HT2. However, a significant increase of the M' − 2 isotopolog of ^{13}C-T2 was recognised in the respective MS spectra and the metabolite $^{12}C/^{13}C$ signal intensity ratio was approximately 1/1, which confirmed that this toxin derivative was T2, formed by HT2.

2.2. Structure Elucidation by LC-HRMS/MS

As a starting point of structure annotation of detected HT2 and T2 metabolites, TracExtract output of LC-HRMS measurements provided the number of parent toxin-derived C-atoms and the assigned ion species of intact toxin derivatives (Tables 1 and 2). Further metabolite information was obtained by LC-HRMS/MS spectra acquired with a LC-Q-TOF-system.

All metabolites for which LC-HRMS/MS spectra could be acquired in the positive ion mode showed the same characteristic HT2 or T2 fragments as in [32]. Spectra were additionally checked for typical fragment ions of hexose (such as glucose) and malonylglucose moieties. The relative mass deviations between predicted, and measured precursor or fragment ions did not exceed 22 ppm.

2.2.1. Confirmation of Previously Found HT2 and T2 Metabolites

HT2, T2, HT2-3-*O*-β-D-Glc, and 3-acetyl-T2 were identified by the comparison of retention times (using additionally longer gradient method 3 from study [32]), accurate masses, and LC-HRMS/MS spectra with authentic standards. In the same way, the metabolites (1, 2, 5, 8–11, 14, 19) were confirmed by comparison with those known from previous barley [32] and wheat [33] studies.

As a result, the metabolites (1, 2, 5, 8–11, 13, 14, 16–19) were shown to be identical to those from the studies with barley and wheat. An LC-HRMS/MS spectrum in negative ion mode of T2-triol-Glc was additionally annotated and is illustrated in Supplementary Figure S1. The abundance of dehydro-15-acetyl-T2-tetraol-Glc (2) was too low to generate meaningful product ion spectra.

2.2.2. Elucidation of Novel HT2 and T2 Metabolites

Moreover, the structure of novel oat-derived HT2/T2 metabolites was elucidated by the comparative interpretation of LC-HRMS/MS spectra of native and corresponding ^{13}C-labelled metabolites, acquired in the same LC-HRMS/MS run. Within overlaid $^{12}C/^{13}C$ LC-HRMS/MS spectra, fragment ions containing a part of the parent toxins showed $^{12}C/^{13}C$ mass shifts proportional to the parent toxin-derived number of C-atoms. On the other hand, for fragment ions of conjugated moieties originating from the native plant, no $^{12}C/^{13}C$ mass shifts were observed. The product ion spectra were interpreted and calculated sum formulas as well as $^{12}C/^{13}C$ mass shifts partially confirmed by the help of the FragExtract module of the MetExtract II software. Additionally, the relative elution order of the putative toxin derivatives was checked for plausibility.

The six HT2/T2 metabolites (3, 4, 6, 7, 12, and 15), which are described here for the first time contained intact HT2 carbon skeleton with a $^{12}C/^{13}C$ mass shift of Δ22.074 u (theoretical value). LC-HRMS/MS spectra of these novel metabolites are illustrated in Supplementary Figures S2–S6. The positive and negative ion mode derived LC-HRMS/MS spectra of (3/4, 7, 12, and 15) suggest that at least one hexose (most probably glucose) molecule is conjugated to HT2 showing the typical fragment ions of a hexose moiety, namely *m/z* 145.0495, *m/z* 127.0390, and *m/z* 161.0455, corresponding to [hexose–2 H_2O+H]$^+$, [hexose–3 H_2O+H]$^+$ and [hexose–H_2O–H]$^-$, respectively.

Two putative hydroxy-HT2-HexGlc isomers (3) and (4) were detected with the same m/z values at different retention times. Since product ion spectra (Figure S2) were similar to those of hydroxy-HT2-Glc (5) and HT2-HexGlc (10, 11) respectively, it is assumed that (3) and (4) were formed by the hydroxylation of the two isomers of (10) and (11). The loss of the additional oxygen atom always coincided with the loss of isovaleric acid ($C_5H_{10}O_2$, $\Delta102.0681$ u) such as m/z 647.2540 [M–O–isoval acid+H]$^+$ and m/z 485.1987 [M–O–isoval acid–hexoside+H]$^+$, suggesting that the hydroxyl group is linked to the isovaleryl group. Moreover, after cleavage of two hexoside moieties corresponding to two times $\Delta162.053$ u (theoretical value) from the [M+H]$^+$ ion, m/z 441.2093 [hydroxy-HT2+H]$^+$ was left.

Hydroxy-HT2 (6) was annotated by the detection of an intact HT2 backbone and a mass shift of $\Delta15.994$ u between m/z 441.2105 [M+H]$^+$ and theoretical m/z 425.2170 [HT2+H]$^+$ (Figure S3). Two losses of water resulting in m/z 423.1968 [M–H_2O+H]$^+$ and m/z 405.1867 [M–2 H_2O+H]$^+$ were observed. Moreover, the product ion spectra of (6) showed the simultaneous losses of the hydroxy–oxygen atom together with isovaleric acid, indicating again that the location of the hydroxyl group was at the isovaleryl group of HT2.

For putative HT2-malylGlc (12) the intact HT2 backbone and fragment ions corresponding to a conjugated hexose (most probably glucose) molecule were confirmed (Figure S4). Subtracting the theoretical m/z 587.2698 [HT2-Glc+H]$^+$ from m/z 703.2830 [M+H]$^+$ gave $\Delta116.013$ u in accordance with malic acid minus water. The malylglucose moiety was easily cleaved off from the intact toxin derivative (12) by LC-HRMS/MS resulting in m/z 425.2157 [HT2+H]$^+$, m/z 407.2055 [HT2–H_2O+H]$^+$ and losses of water from malylglucose, namely m/z 279.0689 [malylglucose–H_2O+H]$^+$ and m/z 261.0604 [malylglucose–2 H_2O+H]$^+$. Typical fragments of malic acid reported in the PubChem database [38] were also detected such as m/z 135.0260 [malic acid+H]$^+$ and m/z 89.0234 [malic acid–formic acid+H]$^+$ in positive ion mode or m/z 115.0037 [malic acid–H_2O–H]$^-$ and m/z 71.0141 [malic acid–H_2O–CO_2–H]$^-$ (m/z 71.014 also detected in LC-HRMS/MS spectrum of HT2-3-O-β-D-Glc originating from glucose moiety) in negative ion mode (data not shown).

Two of the detected toxin derivatives are assumed to carry an anhydro-hexose moiety corresponding to hexose minus water. The tentative hydroxy-HT2-anhydro-HexGlc (7) and HT2-anhydro-HexGlc (15) were closely related, since the fragmentation patterns were very similar (Figures S5 and S6). Most fragment ions in positive ion mode LC-HRMS/MS spectra had identical m/z values. This can be explained by the simultaneous loss of the additional hydroxy–oxygen atom with each loss of isovaleric acid of (7) resulting in m/z 629.2390 [(7)–O–isoval acid+H]$^+$ and m/z 569.2193 [(7)–O–isoval acid–acetic acid+H]$^+$, for example, which were identical to m/z 629.2359 [(15)–isoval acid+H]$^+$ and m/z 569.2205 [(15)–isoval acid–acetic acid+H]$^+$. While for (15), m/z 425.2158 [HT2+H]$^+$ was observed, (7) has solely shown m/z 423.1992 [hydroxy-HT2–H_2O+H]$^+$. Typical fragment ions of hexose moieties were detected as m/z 145.0488/145.0490 [hexose–2 H_2O+H]$^+$, m/z 127.0375/127.0374 [hexose–3 H_2O+H]$^+$ and even m/z 163.0583/163.0590 [hexose–H_2O+H]$^+$. In contrast to all other HT2-Glc derivatives, the hexose-related fragment ions showed higher relative abundances in the LC-HRMS/MS spectra of (7) and (15). Moreover, product ions of putative dihexosides such as m/z 307.0993/307.1007 [dihexoside–2 H_2O+H]$^+$, m/z 289.0902/289.0890 [dihexoside–3 H_2O+H]$^+$, m/z 271.0803/271.0784 [dihexoside–4 H_2O+H]$^+$, m/z 253.0670/253.0696 [dihexoside–5 H_2O+H]$^+$, m/z 207.0645/207.0623 [dihexoside–6 H_2O–CO+H]$^+$, and m/z 177.0543/177.0551 [dihexoside–6 H_2O–CO–CH_2O+H]$^+$ were present with high relative intensities. Interestingly, some of these fragment ions were also found in positive ion mode product ion spectra of the two isomers HT2-HexGlc (10 and 11), however with very low abundances. This indicates that both isomers include a dihexoside, rather than two single sugar moieties that are linked to the C-3 and C-4 position of HT2. Hedin and Phillips [39] reported the structure elucidation of different sugars and related derivatives with chemical ionisation mass spectrometry. They described the fragmentation pattern of glucose (and other aldohexoses) to contain fragment ions at m/z 163, 145, and 127 as well as of dihexosides, showing, amongst others, m/z 307 and 289. Since in negative ion mode LC-HRMS/MS spectra (data not shown) of (15) m/z 585.2525 [HT2-Glc–H]$^-$ and of (7) the respective m/z 601.2484 [hydroxy-HT2-Glc–H]$^-$ were detected, our results suggest that the second conjugated hexose moiety was modified in the plant by the cleavage of one water molecule. This has probably led to more dominant further losses of water in positive ion mode LC-HRMS/MS spectra and thus higher relative intensities of respective fragment ions. Taken together, we conclude that (7) and (15) were most probably derived from one isomer of HT2-HexGlc and that (7) was presumably formed by the hydroxylation of the HT2 isovaleryl group of metabolite (15). Another plausible metabolic pathway is the formation of (7) by the cleavage of one hexose–water molecule from intermediate hydroxy-HT2-HexGlc (3/4).

2.3. Kinetics of Metabolite Formation and Distribution

Separate oat panicles were treated with 200 µg non-labelled HT2 (corresponding to 0.47 µmol) and T2 (corresponding to 0.43 µmol) (stated ^{12}C samples) in five biological replicates and harvested immediately (zero), after one, three, or seven days, or at the full-ripening stage (time course experiment). Plants were cut into three parts, namely treated spikelets, non-treated spikelets, and pooled stem plus side branches, each of which was analysed separately on a LC-Q-TOF-MS-system in full scan mode.

2.3.1. Absolute Quantification

Since standards were available, parent toxins plus HT2-3-O-β-D-Glc were quantified in treated spikelets. The absolute amounts were calculated in µmol/treated part and put in relation to the amounts of recovered HT2 and T2 at time point zero days. All measured concentration values were corrected by the respective standard purities and matrix effects, which were between 73% and 86% in 400-fold diluted oat extracts.

Figure 2 shows the kinetics of HT2/T2 metabolism. Since the harvesting procedure for each oat panicle took approximately 5–10 min until freezing in liquid nitrogen and T2 was rapidly converted to HT2, at time point zero 0.05 ± 0.01 µmol HT2, corresponding to approximately 11% of the added T2, were already found in T2-treated oats. Therefore, the initially formed HT2 amount was added to that of T2 at time point zero. The recoveries of parent toxins in oats immediately harvested after treatment were 90% (0.42 ± 0.02 µmol) for HT2 and 99% (0.43 ± 0.02 µmol) for T2 plus HT2.

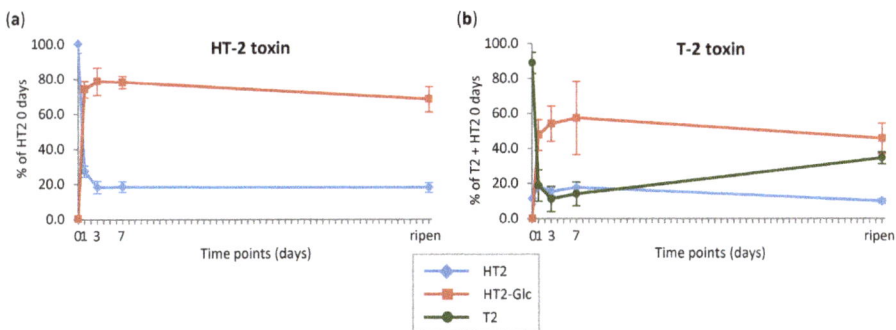

Figure 2. Time courses of native toxins and major plant-derived metabolites. (**a**) Time course of HT2 (HT-2 toxin) metabolism; (**b**) time course of T2 (T-2 toxin) metabolism. Oat panicles were treated with 200 µg non-labelled HT2 and T2 and harvested directly (zero), and one, three, or seven days after treatment, or at full-ripening stage (for each time point toxins were applied on separate panicles in five biological replicates). Quantification was performed on a LC-Q-TOF-system; absolute amounts were calculated in µmol/treated part and put in relation to the amounts of recovered HT2 and T2 at time point zero days (percent mean value ± standard deviation is illustrated). HT2-Glc, HT2-3-O-β-D-glucoside.

As depicted in Figure 2, parent toxins HT2 and T2 were metabolised very fast in oats. After one day, only 27% HT2 (0.12 ± 0.01 μmol in HT2-treated oats) and 19% T2 (0.08 ± 0.04 μmol in T2-treated oats) were left. However, in T2-treated oats an increase in parent toxin T2 to 35% (0.15 ± 0.01 μmol) was observed until the full-ripening stage. HT2-3-*O*-β-D-Glc was the main metabolite present with 74% (0.31 ± 0.02 μmol) in HT2-treated oats and 48% (0.20 ± 0.04 μmol) in T2-treated oats after one day. In comparison with the previous barley [32] and wheat [33] studies, oats was the plant species with the highest turnover from HT2 and T2 into HT2-3-*O*-β-D-Glc. Over time (time points three and seven days), the amounts of HT2-3-*O*-β-D-Glc showed a slight increase but dropped subsequently until the full-ripening point. The later decrease indicates that other HT2/T2 biotransformation products were formed by further metabolic reactions of HT2-3-*O*-β-D-Glc. As the sum of parent toxins plus these absolutely quantified metabolites made up between 86% and 101% in HT2-treated plant parts and between 80% and 100% in T2-treated plant parts, a maximum of 14% or 20% was left. This remaining proportion includes other HT2/T2 biotransformation products and/or not-recognised amounts of toxin derivatives due to transport in non-treated plant parts (for example non-treated spikelets, stem, side branches, leaves, and roots) or incorporation into the plant matrix.

2.3.2. Relative Quantification

Relative quantification of other HT2/T2 metabolites in treated plant parts is depicted in Figure 3. Only for three HT2 and four T2 metabolites were the abundances over time high enough to create meaningful time courses. The formation of T2 in ^{12}C-HT2-treated oat samples was not depicted, since no differentiation could be made between T2 contamination and T2 originating from parent toxin HT2.

Figure 3 shows that most HT2/T2 metabolites increased in concentration until the full-ripening point. The same applies to metabolites hydroxy-HT2-HexGlc (3, 4), hydroxy-HT2-anhydro-HexGlc (7), T2-triol-Glc (8), dehydro-HT2-Glc (9), HT2-HexGlc (10, 11, also in T2-treated plants) and HT2-MalGlc (14). Additionally, metabolites dehydro-15-acetyl-T2-tetraol-Glc (2), hydroxy-HT2 (6), HT2-malylGlc (12), and HT2-anhydro-HexGlc (15) were only obviously detected in ^{12}C/^{13}C samples harvested after ripening and measured with LC-Orbitrap-MS due to a lower injection volume of the ^{12}C samples applied on LC-Q-TOF-MS and different instrument sensitivities. These late maxima suggest again that they are derived from the rapidly formed main metabolic intermediate HT2-3-*O*-β-D-Glc. Figure 4 depicts the proposed metabolic fate of HT2 and T2 in oats based on our findings. For both parent toxins the central metabolic pathway was congruent, including the glucosylation of HT2 at the C-3 position, while for T2 the C-4 acetyl group was firstly cleaved to form HT2. Taking HT2-3-*O*-β-D-Glc as a starting point, hydrolytic cleavages of ester groups as well as further covalent linkages with different polar moieties were observed with highest amounts after ripening. 15-Acetyl-T2-tetraol-Glc (1), dehydro-15-acetyl-T2-tetraol-Glc (2), and T2-triol-Glc (8) resulted from the hydrolytic cleavages of the isovaleryl group at C-8 position or the acetyl group at C-15 position, respectively. After HT2-3-*O*-β-D-Glc formation, putative malonic acid (14) or malic acid (12) was conjugated to the glucoside. HT2-HexGlc (10, 11) was most probably formed by the extension of the glucoside with a second hexose molecule. The subsequent formation of tentative HT2-anhydro-HexGlc (15) resulted from the loss of one water molecule, presumably from one of the hexoside residues. Moreover, many metabolites were formed by hydroxylation of the C-8 isovaleryl group such as hydroxy-HT2-Glc (5), hydroxy-HT2-HexGlc (3, 4), hydroxy-HT2-anhydro-HexGlc (7), and hydroxy-HT2 (6). Putative dehydro-HT2-Glc (9) resulted most probably from the cleavage of the additional hydroxyl group of hydroxy-HT2-Glc (5), corresponding to a loss of one water molecule.

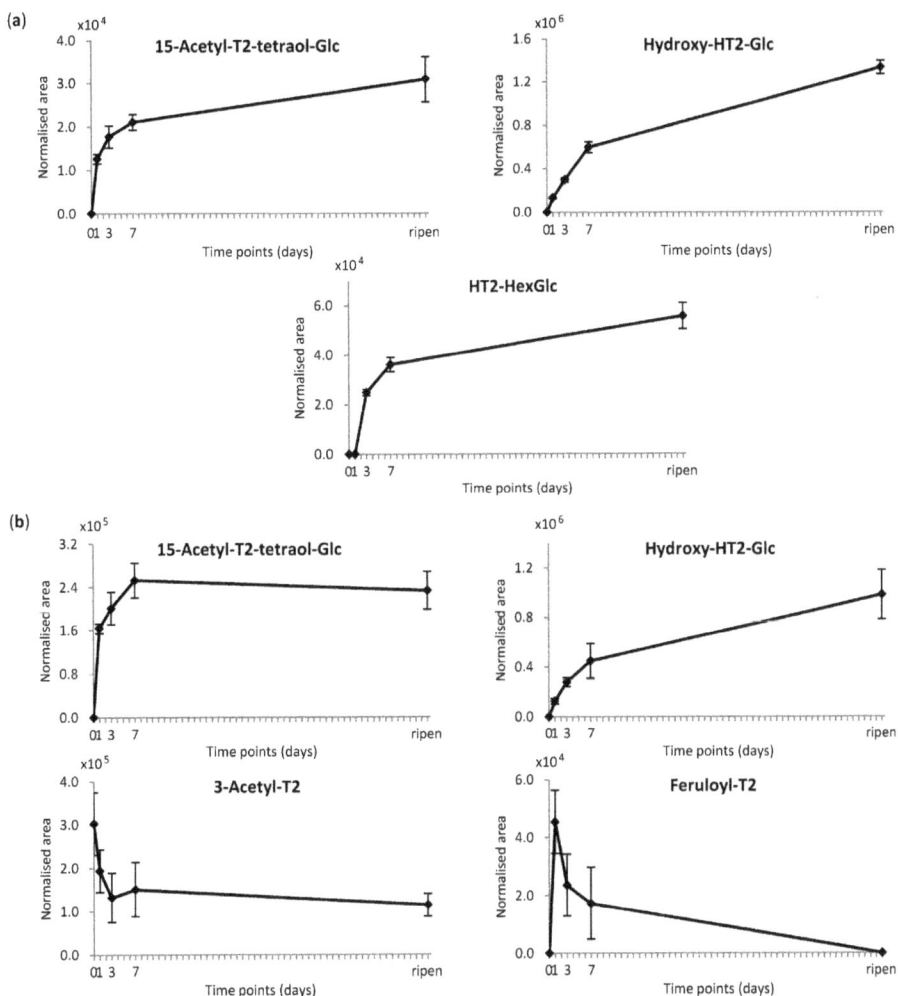

Figure 3. Relative time courses of plant-derived metabolites. (**a**) Relative time courses of HT2 (HT-2 toxin) metabolites; (**b**) relative time courses of T2 (T-2 toxin) metabolites. Oat panicles were treated with 200 µg non-labelled HT2 and T2 and harvested directly (zero), and one, three, or seven days after treatment, or at the full-ripening stage (for each time point toxins were applied on separate panicles in five biological replicates). Relative quantification was performed on a LC-Q-TOF-system; peak areas of ammonium adducts were normalised by respective weight of treated plant part (area mean value ± standard deviation is illustrated). Glc, glucoside; HexGlc, hexosylglucoside.

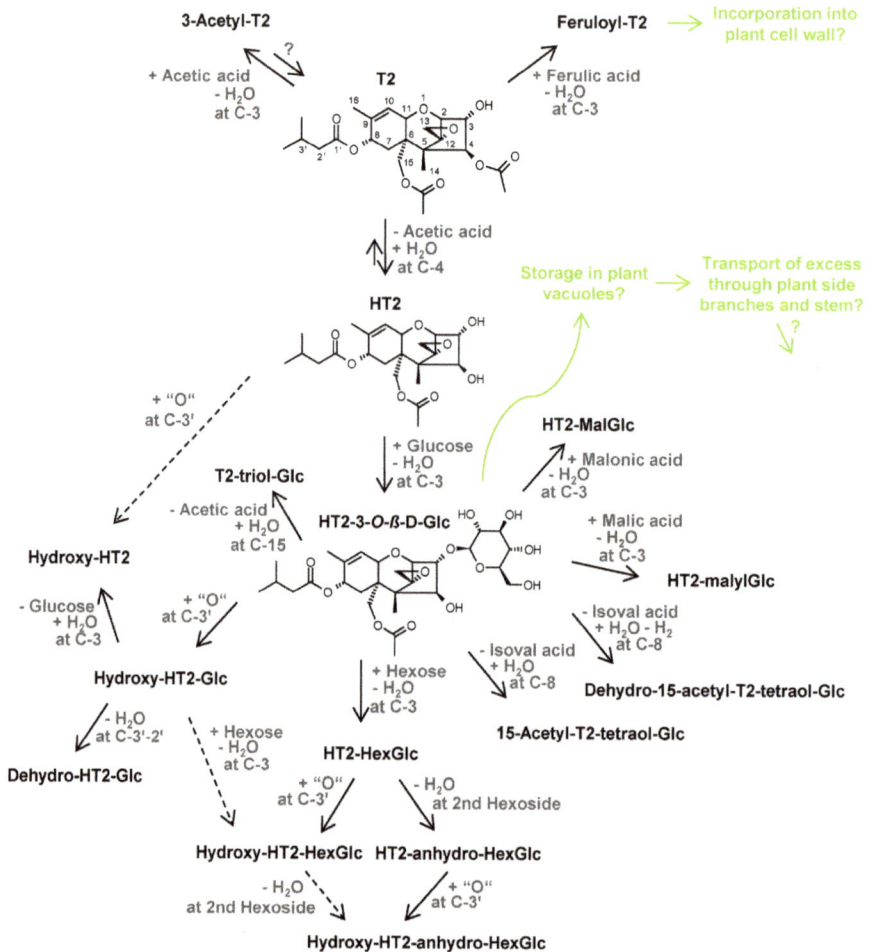

Figure 4. Proposed metabolic fate of HT2 (HT-2 toxin) and T2 (T-2 toxin) in oats. Analysis was performed by liquid chromatography–high-resolution mass spectrometry (LC-HRMS). Structure annotation was based on accurate masses, number of parent toxin-derived C-atoms, LC-HRMS/MS spectra, and on assessment of retention times. Structure identification was based on comparisons with available standards. Glc, glucoside; MalGlc, malonylglucoside; HexGlc, hexosylglucoside; isoval acid, isovaleric acid.

While it can be hypothesised that hydroxy-HT2 (6) followed another metabolic route, formed directly by the hydroxylation of unmodified HT2, its maximum abundance at the latest time point (full ripening) indicates that HT2-3-O-β-D-Glc was also transformed into hydroxy-HT2-Glc and then the Glc moiety was cleaved off (see Figure 4). This is in agreement with the observation that the hydroxylation reaction occurred more slowly and to a lesser extent than the glucosylation of the parent toxins, as can be seen in Figures 1–3 when comparing hydroxy-HT2-Glc (5) with HT2-3-O-β-D-Glc (13) and hydroxy-HT2-HexGlc (3, 4) with HT2-HexGlc (10, 11). Taken together, both typical phase I and II metabolic transformations were observed in oats, while phase II metabolism appeared to be faster and more dominant.

Not surprisingly, the time courses of 15-acetyl-T2-tetraol-Glc differed in HT2- and T2-treated oats (Figure 3). The abundances were approximately 10 times higher in T2-treated oats and the concentration maxima arose at different time points. This can be explained by the small impurity of non-labelled neosolaniol in the ^{12}C-T2 treatment solution, which might be additionally converted to 15-acetyl-T2-tetraol-Glc, but probably faster than T2.

Interestingly, metabolites including the intact T2 skeleton as 3-acetyl-T2 and feruloyl-T2 showed contrary kinetics of formation. The maximum abundance of 3-acetyl-T2 was already reached at zero days and quantification revealed an amount of <0.5% relative to the amounts of recovered T2 plus HT2. Feruloyl-T2 was at its highest concentration one day after treatment. Thus, similar to former barley [32] and wheat [33] studies, a second metabolic pathway was observed for T2 being modified by the covalent binding of acetic acid and putative ferulic acid before the C-4 acetyl group has been hydrolysed to form HT2 (see Figure 4). Moreover, Figure 3 shows that with increasing time the amounts of 3-acetyl-T2 and feruloyl-T2 decreased. Since the hydrolysis of acetyl groups is a common metabolic process monitored in this study, it is assumed that the reverse reaction from 3-acetyl-T2 to T2 could be responsible for that decrease. This hypothesis additionally correlates well with the observed increase of T2 between seven days after treatment and the full-ripening time point (Figure 2). Although the maximum of the T2 formation from parent toxin HT2 (2.1) could not be determined, both reverse reactions from 3-acetyl-T2 and HT2 into T2 have probably led to a rise in the amount of T2. For the decrease of feruloyl-T2 after one day, we presume continuous incorporation into the plant cell wall. This is in agreement with Iiyama et al. [40], who have reported that phenolic acids such as ferulic acid are involved in cell wall strengthening of plants, being a significant strategy to reduce the access of pathogens. Additionally, McKeehen et al. [41] have described the involvement of ferulic acid in resistance mechanisms of grains against *Fusarium* species.

2.3.3. Mobility of Parent Toxins and HT2-3-*O*-β-D-Glc in Oats

Semi-quantitative analysis of HT2, T2, and HT2-3-*O*-β-D-Glc in non-treated plant parts (non-treated spikelets and pooled stem plus side branches) was performed and amounts were related to the amounts of recovered HT2 and T2 in treated plant parts at zero days.

While HT2 was below the LOD in any of the HT2- and T2-treated oat samples, T2 was found to be <0.2% in non-treated plant parts (T2-treated oats). This difference may well be due to a higher analytical sensitivity for T2 compared with HT2 rather than a different mobility behaviour of the two toxins. As Tables 3 and 4 show, solely HT2-3-*O*-β-D-Glc was detected, with considerable amounts higher than 1%.

Table 3. Mobility of HT2-3-*O*-β-D-Glc in HT2-treated oat panicles.

Time Points (Days)	% of HT2 at 0 Days [a] Stem plus Side Branches	% of HT2 at 0 Days Non-Treated Spikelets
0	0.0 ± 0.0	<LOD
1	4.4 ± 2.6	<LOD
3	3.0 ± 2.6	<LOD
7	2.4 ± 2.2	<LOD
full-ripening	1.2 ± 1.2	<LOD

HT2, HT-2 toxin; Glc, glucoside; LOD, limit of detection was estimated to correspond to the signal-to-noise ratio of three in matrix; Oat panicles were treated with 200 µg non-labelled HT2 (for each time point toxins were applied on separate panicles in five biological replicates), harvested directly (zero), and one, three, or seven days after treatment or at the full-ripening stage. For non-treated plant parts (non-treated spikelets and pooled stem plus side branches), semi-quantitative analysis was performed ([a] percent mean value ± standard deviation in percentage points is stated).

Table 4. Mobility of HT2-3-*O*-β-D-Glc in T2-treated oat panicles.

Time Points (Days)	% of T2 + HT2 at 0 Days [a] Stem plus Side Branches	% of T2 + HT2 at 0 Days Non-Treated Spikelets
0	0.0 ± 0.0	<LOD
1	1.3 ± 0.9	<LOD
3	0.7 ± 0.3	<LOD
7	1.0 ± 0.8	<LOD
full-ripening	0.3 ± 0.3	<LOD

HT2, HT-2 toxin; T2, T-2 toxin; Glc, glucoside; LOD, limit of detection was estimated to correspond to the signal-to-noise ratio of three in matrix; Oat panicles were treated with 200 µg non-labelled T2 (for each time point toxins were applied on separate panicles in five biological replicates), harvested directly (zero), and one, three, or seven days after treatment or at the full-ripening stage. For non-treated plant parts (non-treated spikelets and pooled stem plus side branches), semi-quantitative analysis was performed ([a] percent mean value ± standard deviation in percentage points is stated).

Our results demonstrate that the main metabolite HT2-3-*O*-β-D-Glc was slightly mobile in oat plants, while the less polar parent toxins were hardly transported. As can be seen in Tables 3 and 4, low proportions of HT2-3-*O*-β-D-Glc moved through the stem and side branches but did not end up in the adjacent non-treated spikelets. The maximum was reached at one day, with approximately 4% in HT2-treated and 1% in T2-treated oats, and a decrease was observed over time until the full-ripening stage.

According to Coleman et al. [10], it can be hypothesised that the xenobiotics HT2 and T2 passively diffused into the plant cells of treated spikelets, landing in cytosol, where many enzymes such as *O*-glucosyltransferases were responsible for the metabolic reactions. As the main metabolic process, forming HT2-3-*O*-β-D-Glc, occurred very fast and with high conversion rates in treated spikelets, it is suggested that these parts of oat underwent a saturation process with HT2-3-*O*-β-D-Glc. Although the storage of conjugates in cell vacuoles is an important mechanism for xenobiotics disposal [9], we assume that the local capacity for HT2-3-*O*-β-D-Glc storage was temporarily exhausted and the plants consequently transported the excess through side branches and stem towards roots (see Figure 4). This movement of HT2-3-*O*-β-D-Glc was probably conducted via the phloem, which is known to act as a transport system for information and nutrients from source to the sink tissue [42]. As in general the uptake of sugars into the phloem is mediated by carrier proteins, we assume that our main toxin derivative was either recognised by such a carrier or moved passively (diffusion) through the cell walls and membranes. Moreover, the observed decrease of HT2-3-*O*-β-D-Glc in the pooled stem plus side branches over time was probably due to the further transport to lower, not investigated plant parts, further metabolic transformation, the incorporation into the plant matrix or the further storage in plant vacuoles, which are distributed over the whole oat plant.

Adding the estimated values of HT2-3-*O*-β-D-Glc in non-treated plant parts to the total amounts of quantified HT2, T2, and HT2-3-*O*-β-D-Glc (2.3.1) in treated plant parts over time led to a maximum remaining quantity of approximately 13% in HT2-treated and 19% in T2-treated oats.

3. Conclusions

The application of an untargeted analytical approach based on isotopic labelling, LC-HRMS, and LC-HRMS/MS measurements revealed 16 HT2 and 17 T2 metabolites in oats. Many related derivatives were formed by typical phase I and II metabolic processes, including hydrolysis of ester groups, glycosylation, and hydroxylation reactions (Figure 4). Six of the detected biotransformation products are described for the first time in plants, namely two isomers of putative hydroxy-HT2-HexGlc as well as tentative hydroxy-HT2, hydroxy-HT2-anhydro-HexGlc, HT2-malylGlc, and HT2-anhydro-HexGlc. In contrast to HT2/T2 metabolism in barley [32] and wheat [33], malonylation played only a minor role in oats.

After time course experiments, HT2-3-*O*-β-D-Glc has been found to be the main metabolic intermediate of both parent toxins HT2 and T2, which was rapidly formed and further metabolised. Under the tested conditions, the sum of HT2, T2, and HT2-3-*O*-β-D-Glc amounted to between 88% and 106% in HT2-treated oats and between 81% and 100% in T2-treated oats, leading to a maximum remaining proportion of approximately 13% and 19%, respectively, for all other derivatives including those that have not been captured here. Relative time courses of other metabolites showed that most of them were derived from HT2-3-*O*-β-D-Glc, except 3-acetyl-T2 and feruloyl-T2, which followed a separate metabolic pathway. Interestingly, semi-quantitative analysis of HT2, T2, and HT2-3-*O*-β-D-Glc of non-treated plant parts revealed that low fractions of HT2-3-*O*-β-D-Glc were mobile in oats and probably transported by the phloem, while the movement of parent toxins was negligible. The elucidation of HT2/T2 metabolism in plants might be an important contribution to the risk assessment of contaminated cereals. However, further studies are needed to determine the toxicity of the detected metabolites, including the health hazards associated with potential reactivation by their hydrolysis in the intestinal tract of mammals.

4. Materials and Methods

4.1. Chemicals and Reagents

Methanol and acetonitrile (HPLC grade) were obtained from VWR (Vienna, Austria), ammonium formate solution (5 M) was provided by Agilent Technologies (Waldbronn, Germany), and formic acid (LC-MS grade) was purchased from Sigma-Aldrich (Vienna, Austria). Water was successively purified by reverse osmosis and an ELGA Purelab Ultra Mk2 Analytic system from Veolia (Vienna, Austria). For plant treatment, crystalline standards of non-labelled as well as uniformly ^{13}C-labelled (degree of enrichment between 99.3 atom% and 99.6 atom% [13]C) HT2 and T2 (purity between 85% and 99%), were purchased from Romer Labs (Tulln, Austria). Oat treatment solutions of HT2 and T2 for qualitative screening of plant-derived toxin derivatives consisted of a 1/1 (*v*/*v*) mixture of non-labelled and labelled toxin (1000 mg/L per toxin), whilst for time course experiments, solutions of non-labelled toxin with 2000 mg/L were prepared. Besides low amounts of non-labelled neosolaniol, none of the known type A trichothecenes were present in treatment solutions as impurities over 1%. The solvent for all treatment solutions was acetonitrile/water 1/1 (*v*/*v*), which additionally served as a blank treatment solution (mock). HT2 and T2 standards for quantification purposes were obtained from Romer Labs as stock solutions of 100 mg/L and 101 mg/L, respectively, in acetonitrile. The analytical standards HT2-3-*O*-β-D-Glc and 3-acetyl-T2 (purity ≥ 95%) were enzymatically or chemically synthesised, characterised by nuclear magnetic resonance measurements (unpublished data), and dissolved in acetonitrile to obtain concentrations of 1000 and 5000 mg/L, respectively.

4.2. Plant Cultivation

The oat (*Avena sativa* L.) variety "Eneko", originating from a cross between "Triton" × "Flaemingsprofi", was selected for the experiments. This is a commercial spring oat variety bred by Saatzucht Edelhof in Austria [43]. Seeds of "Eneko" were germinated in pots filled with a substrate as described by Meng-Reiterer et al. [32]. In each pot five seedlings were planted. During the whole experiment the pots were watered when required. Plants were germinated in the greenhouse and after germination transferred to a growth chamber. At the end of tillering, 2 g of a mineral fertiliser (COMPO Blaukorn ENTEC, Muenster, Germany; N/P/K/Mg: 14/7/17/2) was applied per pot. Settings for light, temperature, and relative air humidity in the growth chamber were computer-controlled. Light intensity was 560 μmol·s^{-1}·m^{-2} at 1 m above the soil and relative air humidity was set between 60% and 70% during plant growth. Temperature (day/night) and duration of illumination (hours) varied according to the development stage of the plants (for details, see [32]). From the start of flowering until the end of the experiments, including application of the treatment solutions and sampling, settings were 20 °C/18 °C (day/night), 16 h light and 60%–70% relative air humidity.

4.3. Treatment and Sampling of Oat Plants

Oat panicles at flowering stage (spikelets in the middle of panicle flowered) were randomly selected for the three treatment groups HT2, T2 and mock. For each treatment variant, 20 spikelets in the upper area of the panicle were labelled with wool threads and each spikelet was treated with 5 μL of the respective treatment solutions with an electronic pipette, starting with the highest ones. Thus, an amount of 100 μg of non-labelled and 100 μg of labelled toxin per panicle was applied for qualitative screening experiment or 200 μg of non-labelled toxin per panicle for time course experiments. After each treatment step, small transparent plastic bags were sprayed with purified water, put over the panicles to prevent drying and promote diffusion of the treatment solutions into the plant cells, and removed 24 h ($\pm$2 h) later. On the day of harvest, treated oat panicles were cut into three parts, which were the 20 treated spikelets, the remaining non-treated spikelets, and pooled stem plus side branches. After weighing, the plant parts were rapidly frozen in liquid nitrogen and stored at -80 °C until further analysis.

4.3.1. Qualitative Screening Experiment

For qualitative screening, the treatment time points 7, 5, 3, 2, and 1 day(s) before harvest were accumulated on each single panicle by injecting 5 μL of the ^{12}C/^{13}C treatment solutions into each of four spikelets per time point, starting with the highest ones and continuing with the next four spikelets below or close beside it. All treatment variants (HT2, T2, and mock) were applied in biological triplicate. Additionally, time point full-ripening was produced as single treatment per condition by treatment of 20 spikelets at flowering stage and harvesting approximately eight weeks later. Within the scope of this experiment, only treated spikelets were further processed and analysed.

4.3.2. Time Course Experiment

Treatment for time course experiment included five time points on separate oat panicles with five biological replicates per treatment and time point. Plants were treated with non-labelled treatment solutions at flowering stage (20 spikelets at once) and harvested immediately (0), 1, 3, or 7 days later and at full-ripening stage (approximately eight weeks after treatment). Treated spikelets, non-treated spikelets, and pooled stem plus side branches were investigated separately to study the formation of toxin derivatives over time and to get an insight into the mobility of both toxins and their major metabolite within the plants.

4.4. Sample Preparation

Frozen plant material was milled and extracted with a mixture of acetonitrile/water/formic acid 79/20.9/0.1 (*v*/*v*/*v*) according to [32] prior to the LC-HRMS and LC-HRMS/MS measurements.

4.4.1. Qualitative Screening Experiment

For qualitative screening, 200 μL portions of the sample extracts (^{12}C/^{13}C-toxin-treated and mock-treated plants) in solvent acetonitrile/water/formic acid 79/20.9/0.1 (*v*/*v*/*v*) were dried under vacuum at room temperature with CentriVap Refrigerated Concentrator from Labconco (Kansas City, MO, USA) and re-dissolved in 200 μL acetonitrile/water/formic acid 20/80/0.1 (*v*/*v*/*v*).

4.4.2. Time Course Experiment

Non-labelled extracts for quantitative analysis were measured undiluted or after 400-fold dilution with solvent acetonitrile/water 1/1 (*v*/*v*). Extracts of non-treated plant parts from ^{12}C samples (non-treated spikelets and pooled stem plus side branches) were measured undiluted or after 10-fold dilution with acetonitrile/water 1/1 (*v*/*v*).

4.5. Analysis by LC-HRMS and LC-HRMS/MS

4.5.1. Qualitative Screening Experiment

Measurements of ^{12}C/^{13}C-toxin- and mock-treated samples were performed with an UltiMate 3000 HPLC system combined with an Exactive Plus Orbitrap mass spectrometer (Thermo Fisher Scientific, Bremen, Germany). Chromatographic settings were as follows: column Kinetex C18 (150 × 2.1 mm, 2.6 µm; Phenomenex, Aschaffenburg, Germany); column temperature 25 °C; eluents 0.1% formic acid and 5 mM ammonium formate in water (eluent A) and in methanol (eluent B); flow rate 250 µL/min. Injection volume was set to 10 µL and Gradient method 1 (30 min gradient, from 10% to 100% B plus re-equilibration) was used as described in [32]. Orbitrap measurements were performed in positive and negative electrospray ionisation mode separately with a scan range from *m/z* 130 to 1300. All other mass spectrometric settings were concordant with those of Kluger et al. [25]. Data were acquired and evaluated with Thermo Xcalibur 4.0.27.10 and 2.2 software (both Thermo Fisher Scientific), respectively.

4.5.2. Structure Annotation and Quantification (Time Course Experiment)

For all other measurements a 1290 Infinity UHPLC system combined with a 6550 iFunnel Q-TOF-MS (Agilent Technologies) was used. The chromatographic separation was carried out as recently published for barley [32]. In brief, a Zorbax SB-C18 Rapid Resolution HD column (150 × 2.1 mm, 1.8 µm; Agilent Technologies), a flow rate of 250 µL/min, a column temperature of 30 °C and the same eluents as for Orbitrap measurements (4.5.1) were used. For structure annotation of detected HT2/T2 metabolites, LC-HRMS/MS spectra were acquired using Gradient method 2 (25 min gradient) and undiluted extracts of ^{12}C as well as ^{12}C/^{13}C samples. Time course experiments were carried out with short Gradient method 4 (10 min gradient) and in MS full scan mode. The mass spectrometric settings were similar to Meng-Reiterer et al. [32] with some modifications: Capillary voltage was set to 3000 V, drying gas flow to 16 L/min and sheath gas flow to 11 L/min. Data acquisition and evaluation were made by MassHunter Acquisition software B.06.01 and MassHunter Qualitative Analysis B.07.00 and Quantitative Analysis B.07.01 (Agilent Technologies), respectively.

4.6. Metabolite Recognition by MetExtract II (Module TracExtract)

LC-HRMS raw data from qualitative screening experiment were converted to the centroid mode and the mzXML format with ProteoWizard [44]. The subsequent processing with the TracExtract module of the MetExtract II software (in-house programme) [25] enabled the selective recognition of HT2- and T2-derived biotransformation products. To this end, MetExtract II software searched for pairs of corresponding native (M or ^{12}C) and partially ^{13}C-labelled (M' or ^{13}C) ions of the same metabolite in each MS scan. An intensity ratio of approximately 1/1 (±0.4) and a minimum intensity abundance of 50,000 counts were required for two signals M and M' to be considered. The *m/z* difference between two corresponding M and M' signals was proportional to the number of toxin-derived ^{13}C-atoms (number of ^{13}C-atoms × 1.00335 u). M, M', and their carbon isotopologs (M + 1 and M' − 1) had to be present within a mass tolerance of ±4 ppm and a maximum relative intensity abundance error of ±20%. For all corresponding M and M' ions, EICs were generated (mass tolerance of accurate mass ±5 ppm). Chromatographic peaks present in both EICs were detected with the algorithm of Du et al. [45] as well as examined for coelution (tolerance window of ±10 scans) and similar peak shape (minimum Pearson correlation of 0.85). As the final step, recognised biotransformation product ions were automatically convoluted into feature groups, each of which represents a toxin-derived plant metabolite (minimum Pearson correlation of 0.85).

4.7. Annotation of Unknown Metabolites by MetExtract II (Module FragExtract)

FragExtract, developed by Neumann and Lehner et al. [46], was used to confirm the manual interpretation of overlaid LC-HRMS/MS spectra of coeluting M and M′ ions by comparing sum formulas and/or ^{12}C/^{13}C mass shifts. This module of MetExtract II software automatically detected and annotated fragment ions of successive M and M′ LC-HRMS/MS spectra. Any fragment ion signal present in the LC-HRMS/MS spectra of M was tested for a corresponding fragment ion signal in the LC-HRMS/MS spectra of M′ indicating that these two signals were derived from the analysed biotransformation product. For this, both LC-HRMS/MS spectra were scaled relative to the intensity of their respective precursor signal intensity and only those signals with a minimum scaled relative abundance of 1% were considered for further investigation. The *m/z* difference between two corresponding fragment ion signals was proportional to the number of toxin-derived ^{13}C-atoms (number of ^{13}C-atoms × 1.00335 u). Zero ^{13}C-atoms, which led to no mass shifts of fragment ion signals in M and M′ LC-HRMS/MS spectra, were allowed to account for fragments originating from native conjugated molecules carrying no ^{13}C-atoms. If two corresponding fragment ion signals showed a maximum mass tolerance of ±30 ppm and a maximum relative intensity abundance error of ±40%, they were annotated with the number of carbon atoms originating from the labelled tracer compound. Other signals from both LC-HRMS/MS spectra not successfully matching were discarded as noise. For matched fragment signal pairs, putative sum formulas were generated with the Seven Golden Rules [47]. These sum formulas had to have at least as many carbon atoms as their fragment ion signals were annotated with.

4.8. Quantification Experiments

LC-Q-TOF (4.5.2) with Gradient method 4 and full scan mode was used for all quantitative measurements. For absolute quantification of HT2, T2, and HT2-3-*O*-β-D-Glc in extracts of ^{12}C-HT2- and ^{12}C-T2-treated oat spikelets, samples were diluted 400-fold with acetonitrile/water 1/1 (*v/v*). Standard mixtures in the same solvent with concentrations at five levels between 3–300 µg/L were prepared (mostly, four of them were used) to perform external linear calibration (1/x weighted). Sodium adducts of the respective analytes were extracted (exact mass ±20 ppm) and integrated when EICs were above the limit of quantification (LOQ, signal-to-noise ratio of 10 in matrix). Amounts of HT2, T2, and HT2-3-*O*-β-D-Glc were calculated in µmol/treated part and corrected by matrix effects obtained for the respective metabolite. The matrix effects were determined with mock samples harvested at 1 day and full-ripening, separately, in biological triplicate. To this end, 50 µL of a 200 µg/L (per analyte) stock solution including HT2, T2, and HT2-3-*O*-β-D-Glc were evaporated and dissolved in 200 µL of 400-fold diluted matrix to obtain 50 µg/L (per analyte). The EIC peak areas of matrix-affected sodium adducts were then divided by the respective areas of a 50 µg/L standard mix and multiplied by 100.

All other metabolites were measured in undiluted extracts of ^{12}C-HT2- and ^{12}C-T2-treated oat spikelets by generating EICs of *m/z* traces corresponding to ammonium adducts (exact mass ±20 ppm). Metabolite peak areas were normalised by the respective weight of the treated plant part.

Semi-quantitative estimation of HT2, T2, and HT2-3-*O*-β-D-Glc in non-treated plant parts (non-treated spikelets and pooled stem plus side branches) of the ^{12}C samples was based on undiluted or 10-fold diluted extracts, linear calibration (1/x weighted) in the range of 10–1000 µg/L, and EICs of ammonium adducts (exact mass ±25 ppm), which are less prone to matrix effects compared to the corresponding sodium adducts. Since the measured concentrations of HT2-3-*O*-β-D-Glc were partially out of the calibration range (up to five times higher than the highest calibrant), linear calibration was extrapolated. Due to this procedure and no corrections of matrix effects, the indicated amounts in µmol/non-treated part represent semi-quantitative estimations.

Supplementary Materials: The following are available online at www.mdpi.com/2072-6651/8/12/364/s1, Figure S1: LC-HRMS/MS-spectrum and putative structure formula of the HT2/T2 plant metabolite T2-triol-Glc (8). Formate adduct (marked with a diamond) was fragmented with a collision energy of 20 eV on a LC-Q-TOF-system. Characteristic fragments are highlighted. HT2, HT-2 toxin; T2, T-2 toxin; Glc, glucoside; isoval acid, isovaleric acid, Figure S2: LC-HRMS/MS-spectrum and putative structure formula of the HT2/T2 plant metabolite Hydroxy-HT2-HexGlc (3). Ammonium adduct (marked with a diamond) was fragmented with a collision energy of 12 eV on a LC-Q-TOF-system. Characteristic fragments are highlighted. Additional hydroxyl group was confirmed to be located at isovaleric acid moiety. * Sum formula and/or $^{12}C/^{13}C$ mass shift confirmed by FragExtract module of the MetExtract II software. HT2, HT-2 toxin; T2, T-2 toxin; HexGlc, hexosylglucoside; isoval acid, isovaleric acid, Figure S3: LC-HRMS/MS-spectrum and putative structure formula of the HT2/T2 plant metabolite Hydroxy-HT2 (6). Ammonium adduct (marked with a diamond) was fragmented with a collision energy of 8 eV on a LC-Q-TOF-system. Characteristic fragments are highlighted. Additional hydroxyl group was confirmed to be located at isovaleric acid moiety. HT2, HT-2 toxin; T2, T-2 toxin; isoval acid, isovaleric acid. Figure S4: LC-HRMS/MS-spectrum and putative structure formula of the HT2/T2 plant metabolite HT2-malylGlc (12). Ammonium adduct (marked with a diamond) was fragmented with a collision energy of 13 eV on a LC-Q-TOF-system. Characteristic fragments are highlighted. * Sum formula and/or $^{12}C/^{13}C$ mass shift confirmed by FragExtract module of the MetExtract II software. Glc, glucoside; isoval acid, isovaleric acid, Figure S5: LC-HRMS/MS-spectrum of the HT2/T2 plant metabolite Hydroxy-HT2-anhydro-HexGlc (7). Ammonium adduct (marked with a diamond) was fragmented with a collision energy of 15 eV on a LC-Q-TOF-system. Characteristic fragments are highlighted. Additional hydroxyl group was confirmed to be located at isovaleric acid moiety. * Sum formula and/or $^{12}C/^{13}C$ mass shift confirmed by FragExtract module of the MetExtract II software. HT2, HT-2 toxin; T2, T-2 toxin; HexGlc, hexosylglucoside; isoval acid, isovaleric acid, Figure S6: LC-HRMS/MS-spectrum of the HT2/T2 plant metabolite HT2-anhydro-HexGlc (15). Ammonium adduct (marked with a diamond) was fragmented with a collision energy of 12 eV on a LC-Q-TOF-system. Characteristic fragments are highlighted. * Sum formula and/or $^{12}C/^{13}C$ mass shift confirmed by FragExtract module of the MetExtract II software. HT2, HT-2 toxin; T2, T-2 toxin; HexGlc, hexosylglucoside; isoval acid, isovaleric acid.

Acknowledgments: We thank Herbert Michlmayr and Alexandra Malachová for the synthesis of HT2-3-*O*-β-D-Glc within the course of a Vienna Science and Technology Fund project (WWTF LS12-012). Philipp Fruhmann and Gerhard Adam are acknowledged for providing 3-acetyl-T2, which was synthesised within the FWF project SFB F37. The development of MetExtract II software was also enabled by SFB F37 project. We furthermore thank Imer Maloku for the assistance with plant experiments and Elisabeth Varga for the help in LC-HRMS method development. This work was primarily funded by the Austrian Science Fund (FWF) (Project P26213). Furthermore, the authors would like to thank the Federal Ministry of Science, Research and Economy, the National Foundation for Research, Technology and Development, and BIOMIN Holding GmbH for funding the Christian Doppler Laboratory for Mycotoxin Metabolism, which provided the LC-Q-TOF-system for analysis.

Author Contributions: Rainer Schuhmacher, Marc Lemmens, Franz Berthiller, and Jacqueline Meng-Reiterer conceived and designed the experiments; Jacqueline Meng-Reiterer performed the experiments, analysed the samples, and evaluated all data; Justyna Rechthaler provided the LC-Orbitrap-MS instrument and developed/optimised full scan MS methods; Christoph Bueschl developed the data processing software MetExtract II and optimised parameters; Rainer Schuhmacher, Marc Lemmens, and Franz Berthiller supervised the experimental work and data analysis. Jacqueline Meng-Reiterer drafted the manuscript and all authors contributed to the paper.

Conflicts of Interest: The authors declare no conflict of interest. The founding sponsors had no role in the design of the study; in the collection, analyses, or interpretation of data; in the writing of the manuscript, and in the decision to publish the results.

References

1. Edwards, S.G.; Barrier-Guillot, B.; Clasen, P.-E.; Hietaniemi, V.; Pettersson, H. Emerging issues of HT-2 and T-2 toxins in European cereal production. *World Mycotoxin J.* **2009**, *2*, 173–179. [CrossRef]
2. Lemmens, M.; Steiner, B.; Sulyok, M.; Nicholson, P.; Mesterhazy, A.; Buerstmayr, H. Masked mycotoxins: Does breeding for enhanced *Fusarium* head blight resistance result in more deoxynivalenol-3-glucoside in new wheat varieties? *World Mycotoxin J.* **2016**. [CrossRef]
3. Gottschalk, C.; Barthel, J.; Engelhardt, G.; Bauer, J.; Meyer, K. Simultaneous determination of type A, B and D trichothecenes and their occurrence in cereals and cereal products. *Food Addit. Contam.* **2009**, *26*, 1273–1289. [CrossRef]
4. Kirinčič, S.; Škrjanc, B.; Kos, N.; Kozolc, B.; Pirnat, N.; Tavčar-Kalcher, G. Mycotoxins in cereals and cereal products in Slovenia—Official control of foods in the years 2008–2012. *Food Control* **2015**, *50*, 157–165. [CrossRef]

5.   Marin, S.; Ramos, A.J.; Cano-Sancho, G.; Sanchis, V. Mycotoxins: Occurrence, toxicology, and exposure assessment. *Food Chem. Toxicol.* **2013**, *60*, 218–237. [CrossRef] [PubMed]
6.   Krska, R.; Malachova, A.; Berthiller, F.; van Egmond, H.P. Determination of T-2 and HT-2 toxins in food and feed: An update. *World Mycotoxin J.* **2014**, *7*, 131–142. [CrossRef]
7.   European Food Safety Authority (EFSA) Panel on Contaminants in the Food Chain (CONTAM). Scientific Opinion on the risks for animal and public health related to the presence of T-2 and HT-2 toxin in food and feed. *EFSA J.* **2011**, *9*. [CrossRef]
8.   Commission Recommendation 2013/165/EU of 27 March 2013 on the presence of T-2 and HT-2 toxin in cereals and cereal products. *Off. J. Eur. Commun.* **2013**, *L91*, 12–15.
9.   Cole, D.J. Detoxification and Activation of Agrochemicals in Plants. *Pestic. Sci.* **1994**, *42*, 209–222. [CrossRef]
10. Coleman, J.O.D.; Blake-Kalff, M.M.A.; Davies, T.G.E. Detoxification of xenobiotics by plants: Chemical modification and vacuolar compartmentation. *Trends Plant Sci.* **1997**, *2*, 144–151. [CrossRef]
11. Gougler, J.A.; Geiger, D.R. Uptake and Distribution of N-Phosphonomethylglycine in Sugar Beet Plants. *Plant Physiol.* **1981**, *68*, 668–672. [CrossRef] [PubMed]
12. Schröder, P.; Scheer, C.E.; Diekmann, F.; Stampfl, A. How Plants Cope with Foreign Compounds. Translocation of xenobiotic glutathione conjugates in roots of barley (*Hordeum vulgare*). *Environ. Sci. Pollut. Res.* **2007**, *14*, 114–122.
13. Wu, Q.; Dohnal, V.; Kuča, K.; Yuan, Z. Trichothecenes: Structure-Toxic Activity Relationships. *Curr. Drug Metab.* **2013**, *14*, 641–660. [CrossRef] [PubMed]
14. Berthiller, F.; Werner, U.; Sulyok, M.; Krska, R.; Hauser, M.-T.; Schuhmacher, R. Liquid chromatography coupled to tandem mass spectrometry (LC-MS/MS) determination of phase II metabolites of the mycotoxin zearalenone in the model plant *Arabidopsis thaliana*. *Food Addit. Contam.* **2006**, *23*, 1194–1200. [CrossRef] [PubMed]
15. Kluger, B.; Bueschl, C.; Lemmens, M.; Michlmayr, H.; Malachova, A.; Koutnik, A.; Maloku, I.; Berthiller, F.; Adam, G.; Krska, R.; et al. Biotransformation of the Mycotoxin Deoxynivalenol in *Fusarium* Resistant and Susceptible Near Isogenic Wheat Lines. *PLoS ONE* **2015**, *10*. [CrossRef]
16. Pierron, A.; Mimoun, S.; Murate, L.S.; Loiseau, N.; Lippi, Y.; Bracarense, A.-P.F.L.; Liaubet, L.; Schatzmayr, G.; Berthiller, F.; Moll, W.-D.; et al. Intestinal toxicity of the masked mycotoxin deoxynivalenol-3-β-D-glucoside. *Arch. Toxicol.* **2016**, *90*, 2037–2046. [CrossRef] [PubMed]
17. Gareis, M.; Bauer, J.; Thiem, J.; Plank, G.; Grabley, S.; Gedek, B. Cleavage of zearalenone glycoside, a 'masked' mycotoxin during digestion in swine. *J. Vet. Med. B* **1990**, *37*, 236–240. [CrossRef]
18. Berthiller, F.; Brera, C.; Crews, C.; Iha, M.H.; Krska, R.; Lattanzio, V.M.T.; MacDonald, S.; Malone, R.J.; Maragos, C.; Solfrizzo, M.; et al. Developments in mycotoxin analysis: An update for 2014–2015. *World Mycotoxin J.* **2016**, *9*, 5–29. [CrossRef]
19. Berthiller, F.; Crews, C.; Dall'Asta, C.; De Saeger, S.; Haesaert, G.; Karlovsky, P.; Oswald, I.P.; Seefelder, W.; Speijers, G.; Stroka, J. Masked mycotoxins: A review. *Mol. Nutr. Food Res.* **2013**, *57*, 165–186. [CrossRef] [PubMed]
20. Sulyok, M.; Krska, R.; Schuhmacher, R. Application of an LC-MS/MS based multi-mycotoxin method for the semi-quantitative determination of mycotoxins occurring in different types of food infected by moulds. *Food Chem.* **2010**, *119*, 408–416. [CrossRef]
21. Sulyok, M.; Beed, F.; Boni, S.; Abass, A.; Mukunzi, A.; Krska, R. Quantitation of multiple mycotoxins and cyanogenic glucosides in cassava samples from Tanzania and Rwanda by an LC-MS/MS-based multi-toxin method. *Food Addit. Contam. Part A* **2015**, *32*, 488–502. [CrossRef] [PubMed]
22. Lattanzio, V.M.T.; Ciasca, B.; Powers, S.; Visconti, A. Improved method for the simultaneous determination of aflatoxins, ochratoxin A and *Fusarium* toxins in cereals and derived products by liquid chromatography-tandem mass spectrometry after multi-toxin immunoaffinity clean up. *J. Chromatogr. A* **2014**, *1354*, 139–143. [CrossRef] [PubMed]
23. Bueschl, C.; Krska, R.; Kluger, B.; Schuhmacher, R. Isotopic labeling-assisted metabolomics using LC-MS. *Anal. Bioanal. Chem.* **2013**, *405*, 27–33. [CrossRef] [PubMed]
24. Kluger, B.; Bueschl, C.; Lemmens, M.; Berthiller, F.; Häubl, G.; Jaunecker, G.; Adam, G.; Krska, R.; Schuhmacher, R. Stable isotopic labelling-assisted untargeted metabolic profiling reveals novel conjugates of the mycotoxin deoxynivalenol in wheat. *Anal. Bioanal. Chem.* **2013**, *405*, 5031–5036. [CrossRef] [PubMed]

25. Kluger, B.; Bueschl, C.; Neumann, N.; Stückler, R.; Doppler, M.; Chassy, A.W.; Waterhouse, A.L.; Rechthaler, J.; Kampleitner, N.; Thallinger, G.G.; et al. Untargeted Profiling of Tracer-Derived Metabolites Using Stable Isotopic Labeling and Fast Polarity-Switching LC–ESI-HRMS. *Anal. Chem.* **2014**, *86*, 11533–11537. [CrossRef] [PubMed]

26. Lattanzio, V.M.T.; Visconti, A.; Haidukowski, M.; Pascale, M. Identification and characterization of new *Fusarium* masked mycotoxins, T2 and HT2 glycosyl derivatives, in naturally contaminated wheat and oats by liquid chromatography-high-resolution mass spectrometry. *J. Mass Spectrom.* **2012**, *47*, 466–475. [CrossRef] [PubMed]

27. Busman, M.; Poling, S.M.; Maragos, C.M. Observation of T-2 Toxin and HT-2 Toxin Glucosides from *Fusarium sporotrichioides* by Liquid Chromatography Coupled to Tandem Mass Spectrometry (LC-MS/MS). *Toxins* **2011**, *3*, 1554–1568. [CrossRef] [PubMed]

28. De Angelis, E.; Monaci, L.; Pascale, M.; Visconti, A. Fate of deoxynivalenol, T-2 and HT-2 toxins and their glucoside conjugates from flour to bread: An investigation by high-performance liquid chromatography high-resolution mass spectrometry. *Food Addit. Contam. Part A* **2013**, *30*, 345–355. [CrossRef] [PubMed]

29. Nakagawa, H.; Sakamoto, S.; Sago, Y.; Nagashima, H. Detection of Type A Trichothecene Di-Glucosides Produced in Corn by High-Resolution Liquid Chromatography-Orbitrap Mass Spectrometry. *Toxins* **2013**, *5*, 590–604. [CrossRef] [PubMed]

30. Veprikova, Z.; Vaclavikova, M.; Lacina, O.; Dzuman, Z.; Zachariasova, M.; Hajslova, J. Occurrence of mono- and di-glycosylated conjugates of T-2 and HT-2 toxins in naturally contaminated cereals. *World Mycotoxin J.* **2012**, *5*, 231–240. [CrossRef]

31. McCormick, S.P.; Kato, T.; Maragos, C.M.; Busman, M.; Lattanzio, V.M.T.; Galaverna, G.; Dall-Asta, C.; Crich, D.; Price, N.P.J.; Kurtzman, C.P. Anomericity of T-2 Toxin-glucoside: Masked Mycotoxin in Cereal Crops. *J. Agric. Food Chem.* **2015**, *63*, 731–738. [CrossRef] [PubMed]

32. Meng-Reiterer, J.; Varga, E.; Nathanail, A.V.; Bueschl, C.; Rechthaler, J.; McCormick, S.P.; Michlmayr, H.; Malachová, A.; Fruhmann, P.; Adam, G.; Berthiller, F.; Lemmens, M.; Schuhmacher, R. Tracing the metabolism of HT-2 toxin and T-2 toxin in barley by isotope-assisted untargeted screening and quantitative LC-HRMS analysis. *Anal. Bioanal. Chem.* **2015**, *407*, 8019–8033. [CrossRef] [PubMed]

33. Nathanail, A.V.; Varga, E.; Meng-Reiterer, J.; Bueschl, C.; Michlmayr, H.; Malachova, A.; Fruhmann, P.; Jestoi, M.N.; Peltonen, K.; Adam, G.; Lemmens, M.; Schuhmacher, R.; Berthiller, F. Metabolism of the Fusarium Mycotoxins T-2 Toxin and HT-2 Toxin in Wheat. *J. Agric. Food Chem.* **2015**, *63*, 7862–7872. [CrossRef] [PubMed]

34. Taguchi, G.; Ubukata, T.; Nozue, H.; Kobayashi, Y.; Takahi, M.; Yamamoto, H.; Hayashida, N. Malonylation is a key reaction in the metabolism of xenobiotic phenolic glucosides in Arabidopsis and tobacco. *Plant J.* **2010**, *63*, 1031–1041. [CrossRef] [PubMed]

35. Bockers, M.; Rivero, C.; Thiede, B.; Jankowski, T.; Schmidt, B. Uptake, Translocation, and Metabolism of 3,4-Dichloroaniline in Soybean and Wheat Plants. *Z. Naturforsch.* **1994**, *49c*, 719–726.

36. Abdel-Farid, I.B.; Kim, H.K.; Choi, Y.H.; Verpoorte, R. Metabolic Characterization of *Brassica rapa* Leaves by NMR Spectroscopy. *J. Agric. Food Chem.* **2007**, *55*, 7936–7943. [CrossRef] [PubMed]

37. Fujisawa, T.; Matoba, Y.; Katagi, T. Application of Separated Leaf Cell Suspension to Xenobiotic Metabolism in Plant. *J. Agric. Food Chem.* **2009**, *57*, 6982–6989. [CrossRef] [PubMed]

38. U.S. National Library of Medicine. National Center for Biotechnology Information. PUBCHEM Database. Compound Summary for CID 525. MS-MS-Spectra of NIST Number 1118542 and 1052306. Available online: https://pubchem.ncbi.nlm.nih.gov/compound/malic_acid#section=MS-MS (accessed on 16 August 2016).

39. Hedin, P.A.; Phillips, V.A. Chemical Ionization (Methane) Mass Spectrometry of Sugars and Their Derivatives. *J. Agric. Food Chem.* **1991**, *39*, 1106–1109. [CrossRef]

40. Iiyama, K.; Lam, T.B.-T.; Stone, B.A. Covalent Cross-Links in the Cell Wall. *Plant Physiol.* **1994**, *104*, 315–320. [CrossRef] [PubMed]

41. McKeehen, J.D.; Busch, R.H.; Fulcher, R.G. Evaluation of Wheat (*Triticum aestivum* L.) Phenolic Acids during Grain Development and Their Contribution to *Fusarium* Resistance. *J. Agric. Food Chem.* **1999**, *47*, 1476–1482. [CrossRef] [PubMed]

42. Carpaneto, A.; Geiger, D.; Bamberg, E.; Sauer, N.; Fromm, J.; Hedrich, R. Phloem-localized, proton-coupled sucrose carrier ZmSUT1 mediates sucrose efflux under the control of the sucrose gradient and the proton motive force. *J. Biol. Chem.* **2005**, *280*, 21437–21443. [CrossRef] [PubMed]

43. Saatzucht Edelhof. Available online: http://www.saatzucht.edelhof.at/en/ (accessed on 28 November 2016).

44. Kessner, D.; Chambers, M.; Burke, R.; Agusand, D.; Mallick, P. ProteoWizard: Open source software for rapid proteomics tools development. *Bioinformatics* **2008**, *24*, 2534–2536. [CrossRef] [PubMed]

45. Du, P.; Kibbe, W.A.; Lin, S.M. Improved peak detection in mass spectrum by incorporating continuous wavelet transform-based pattern matching. *Bioinformatics* **2006**, *22*, 2059–2065. [CrossRef] [PubMed]

46. Neumann, N.K.; Lehner, S.M.; Kluger, B.; Bueschl, C.; Sedelmaier, K.; Lemmens, M.; Krska, R.; Schuhmacher, R. Automated LC-HRMS(/MS) approach for the annotation of fragment ions derived from stable isotope labeling-assisted untargeted metabolomics. *Anal. Chem.* **2014**, *86*, 7320–7327. [CrossRef] [PubMed]

47. Kind, T.; Fiehn, O. Seven Golden Rules for heuristic filtering of molecular formulas obtained by accurate mass spectrometry. *BMC Bioinform.* **2007**, *8*, 105–124. [CrossRef] [PubMed]

© 2016 by the authors. Licensee MDPI, Basel, Switzerland. This article is an open access article distributed under the terms and conditions of the Creative Commons Attribution (CC BY) license (http://creativecommons.org/licenses/by/4.0/).

toxins

MDPI

Article

Co-Occurrence of Regulated, Masked and Emerging Mycotoxins and Secondary Metabolites in Finished Feed and Maize—An Extensive Survey

Paula Kovalsky [1,†], Gregor Kos [2,†], Karin Nährer [1], Christina Schwab [1], Timothy Jenkins [1], Gerd Schatzmayr [1], Michael Sulyok [3,*] and Rudolf Krska [3]

[1] BIOMIN Research Center, Tulln 3430, Austria; paula.kovalsky@biomin.net (P.K.); karin.naehrer@biomin.net (K.N.); christina.schwab@biomin.net (C.S.); timothy.jenkins@biomin.net (T.J.); gerd.schatzmayr@biomin.net (G.S.)

[2] Department of Atmospheric and Oceanic Sciences, McGill University, Montreal QC H3A 0B9, Canada; greg@meteo.mcgill.ca

[3] Department of Agrobiotechnology (IFA-Tulln), University of Natural Resources and Life Sciences, Vienna 1180, Austria; rudolf.krska@boku.ac.at

* Correspondence: michael.sulyok@boku.ac.at; Tel.: +43-2272-66280

† These authors contributed equally to this work.

Academic Editor: Aldo Laganà
Received: 10 August 2016; Accepted: 21 November 2016; Published: 6 December 2016

Abstract: Global trade of agricultural commodities (e.g., animal feed) requires monitoring for fungal toxins. Also, little is known about masked and emerging toxins and metabolites. 1926 samples from 52 countries were analysed for toxins and metabolites. Of 162 compounds detected, up to 68 metabolites were found in a single sample. A subset of 1113 finished feed, maize and maize silage samples containing 57 compounds from 2012 to 2015 from 44 countries was investigated using liquid chromatography and mass spectrometry. Deoxynivalenol (DON), zearalenone (ZEN) and fumonisins showed large increases of annual medians in Europe. Within a region, distinct trends were observed, suggesting importance of local meteorology and cultivars. In 2015, median DON concentrations increased to 1400 $\mu g \cdot kg^{-1}$ in Austria, but were stable in Germany at 350 $\mu g \cdot kg^{-1}$. In 2014, enniatins occurred at median concentrations of 250 $\mu g \cdot kg^{-1}$ in Europe, at levels similar to DON and ZEN. The latter were frequently correlated with DON-3-glucoside and ZEN-14-sulfate. Co-occurrence of regulated toxins was frequent with e.g., enniatins, and moniliformin. Correlation was observed between DON and DON-3-glucoside and with beauvericin. Results indicate that considerably more than 25% of agricultural commodities could be contaminated with mycotoxins as suggested by FAO, although this is at least partly due to the lower limits of detection in the current survey. Observed contamination percentages ranged from 7.1 to 79% for B trichothecenes and 88% for ZEN.

Keywords: mycotoxin; secondary metabolites; survey; global; masked mycotoxins; emerging mycotoxins; concentration data

1. Introduction

The contamination of agricultural commodities with mycotoxins (secondary fungal metabolites) is of global concern, due to their toxicity and impacts on animal health [1,2]. The most important genera producing mycotoxins and fungal secondary metabolites are *Aspergillus*, *Fusarium*, and *Penicillium* [3]. Contamination with mycotoxins by fungal growth occurs in the field, during storage and transportation. A highly concentrated, localised "hot-spot" (i.e., a highly inhomogeneous distribution of a toxin) can spoil an entire batch [4].

The movement of animal feed products across the globe facilitated by international trade agreements requires constant monitoring of mycotoxin levels by authorities and traders; see Figure 1 for an example of the global maize trade. In 2014 compound feed production was 153 million tonnes in the European Union (EU–28), with global production having reached 964 million tonnes [5]. In 2014 compound feed accounted for 80% of all of the purchased feedstuffs in the EU [6].

■ Maize exports (tons) ■ Maize imports (tons)

Figure 1. Simplified global maize trade for 2015. Map shows the largest importer (red) and exporter (blue) on each continent with the five largest countries of origin and destination, respectively [7].

1.1. Regulated Toxins and Mycotoxins with Guidance Levels

Legislation has been put in place for compounds commonly called "regulated toxins" and "mycotoxins with guidance levels", which are comprised of aflatoxins (AFLA), some type A and B trichothecenes, zearalenone (ZEN), fumonisins (FUM) and ochratoxin A (OTA), because of their acute and chronic toxic effects. Values specified in feed are either maximum allowed levels or guidance values.

Animal consumption of commodities contaminated with AFLA causes extensive functional and structural damage to the liver, including links to liver cancer [8]. Livestock animals (e.g., poultry, swine) show sensitivity to AFLA [9]. While aflatoxin B_1 is the most potent toxin and carcinogen, its congeners, nevertheless, also show high toxicity [10]. Among agricultural commodities, peanuts, maize and rice, mostly originating from subtropical and tropical regions are affected by elevated AFLA concentrations [11].

Trichothecenes are sesquiterpenes with an epoxy ring produced mainly, but not exclusively, by *Fusarium* species. Depending on the chemical structure more toxic, but less prevalent type A trichothecenes (e.g., T-2 and HT-2 toxins) and widely occurring type B trichothecenes (e.g., DON, nivalenol) are defined. Contamination typically occurs before harvest in maize and cereals. Toxicity mostly affects the gastrointestinal tract causing vomiting, diarrhea due to inhibition of protein synthesis [12]. Lower feed conversion ratios [13] and reduced feed intake have also been reported, especially at low contamination and chronic dose conditions.

The type B trichothecene DON, the most prevalent trichothecene in temperate climates is a stable molecule, thus moving through feed production intact. Formation of DON is commonly observed with concentrations up to the lower mg·kg^{-1}. The main symptoms are vomiting (hence its common name "vomitoxin"), feed refusal, skin damage and hemorrhage [14,15] notably in swine. For poultry, feed contaminated with <5 mg·kg^{-1} DON results in a decreased immune response and increase of infectious diseases [16].

The type A trichothecenes T-2 and HT-2 toxins occur less often and at lower concentrations in feed than e.g., DON, but are more toxic. Young pigs exposed to T-2 toxin showed decreases in red blood cell and leucocyte count [17]. Low-concentration dosage increased immunotoxic effects [18], decreased body weight gain and possible tissue damage in poultry [19]. For a comprehensive summary see Grenier and Oswald [20]. HT-2 toxin is the main metabolite of T-2 toxin and of similar toxicity. Data for the two mycotoxins are usually presented together [21].

Like DON, ZEN is also produced by *Fusarium* species and found in the same group of crops under cool and wet conditions. It is, together with some of its metabolites, known to be the cause of hyperestrogenism, causing breeding problems notably in swine and poultry. Concentrations in cereals were observed in the µg·kg^{-1} range.

FUM are a group of *Fusarium* produced toxins, frequently co-occurring with DON and ZEN. Symptoms observed with contaminated feed are low appetite and activity, as well as pulmonary oedema in pigs [22]. Due to their stability, FUM are readily detected in processed feed. In poultry, effects on broiler chicks showed morphological changes such as reduced villus height [23].

Produced by *Aspergillus* and *Penicillium* species, OTA is a mycotoxin commonly produced in storage facilities, as opposed to previously discussed toxins, originating mainly from the field. It has nephrotoxic effects in mammals, notably swine, and is suspected to be a carcinogen [24]. In animals, especially poultry, ochratoxicosis is characterised by poor weight gain and decreased egg production [25]. Observed concentrations are in the lower µg·kg^{-1} range. OTA is readily transferred into finished feed products [26].

In general, some of the drastic acute symptoms described above were due to ingestion of large toxin quantities, which were linked to the consumption of heavily mould-infested feed. However, continuous low concentration exposure is also of relevance. For AFLA, chronic effects include the development of hepatitis and jaundice [9]. In general, chronic toxicity has been less investigated as were synergistic and additive effects due to co-occurrence of multiple compounds [20]. Available data is still scarce [27,28].

1.2. Masked Mycotoxins

The study of "masked mycotoxins", i.e., plant metabolites of mycotoxins, or following Rychlik et al. [29]'s systematic definition "biologically modified" mycotoxins, is especially challenging since the chemical modifications introduced by the plant's metabolism potentially has effects on both, toxicity (which could be increased or decreased compared to the parent toxin molecule) and analytical detectability. For the latter, masked toxins are either bound to carbohydrates or proteins and, therefore, not extractable with existing protocols aimed at the extraction of the toxin, or they are not detectable using established chromatography routines; hence their name "masked" mycotoxins; see [30]. Furthermore, because of structural similarities, some masked compounds, which sometimes differ in toxicity, are co-detected with the toxin itself by e.g., immunoassays [31]. Because of these analytical challenges and subsequent lack of established methodologies for routine testing, data are still scarce [32,33]. Among the group of masked mycotoxins, ZEN-14-sulfate and DON-3-glucoside are most commonly observed in feed. Their toxicological properties are currently being investigated, including the conversion of DON-3-glucoside to DON and ZEN-14-sulfate to ZEN by microbiota of the intestinal tract in an effort to assess the risks with exposure to masked mycotoxins [34]. ZEN-14-sulfate is a natural *Fusarium* metabolite [35]. Since it is readily hydrolysed, ZEN is produced upon ingestion by animals, triggering an estrogenic response typical for ZEN in e.g., swine. In contrast, the glucosides resist hydrolysis and are, therefore, not active.

1.3. Emerging Toxins

"Emerging toxins" are a group of chemically diverse mycotoxins for which to date no regulations exist. Ongoing studies employing advanced LC–MS/MS (liquid chromatography tandem mass spectrometry) for structure elucidation provide a steady stream of insights about newly discovered metabolites as do plant breeding efforts adapting to a changing climate [36]. Risk assessment studies are currently underway in preparation for legislation, if deemed necessary [37]. Commonly mentioned in this group are aflatoxin precursors, ergot alkaloids, enniatins, beauvericin (BEA) and moniliformin (MON). For a detailed list of individual substances grouped by these terms, see Table 1. Jestoi [38] has published an extensive review regarding this diverse set of compounds, summarising available data, analytical methods and toxicity studies available.

Table 1. Summary statistics of investigated concentrations of 57 regulated mycotoxins and mycotoxins with guidance levels, masked and emerging metabolites in 3 matrices (1113 samples from 46 countries). Abbreviations: <LOD—below limit of detection (replaced with zero for the median calculation), *n*—number of samples above threshold *t*, %—percentage of samples above threshold *t*. Indented compound concentrations were summed up as part of a group (group name in italics). A threshold (*t*) of relevant concentrations was established to be $>1.0\ \mu g \cdot kg^{-1}$ or >LOD, whichever was higher. For sums of compound concentrations, the highest LOD within a group was employed as *t*.

Metabolite (Group)	Median Concentration ($\mu g \cdot kg^{-1}$)	75th Percentile ($\mu g \cdot kg^{-1}$)	95th Percentile ($\mu g \cdot kg^{-1}$)	Maximum Concentration ($\mu g \cdot kg^{-1}$)	*n* >*t*	% >*t*
Regulated toxins and toxins with guidance levels						
AFLA, t > 1.5 $\mu g \cdot kg^{-1}$						
Aflatoxin B1	<LOD	<LOD	1.4	1077	49	4.9
Aflatoxin B2	<LOD	<LOD	<LOD	112	14	1.4
Aflatoxin G1	<LOD	<LOD	<LOD	95	19	1.9
Aflatoxin G2	<LOD	<LOD	<LOD	12	8	0.80
ZEN, *t* > 1 $\mu g \cdot kg^{-1}$	20	77	596	11,192	884	88
DON, *t* > 1.5 $\mu g \cdot kg^{-1}$	193	546	2278	13,488	799	79
T-2 and HT-2 toxins, t > 10 $\mu g \cdot kg^{-1}$						
T-2 toxin	< LOD	3	22	852	105	10
HT-2 toxin	< LOD	0.0	51	2328	189	19
FUM, t > 4.0 $\mu g \cdot kg^{-1}$						
Fumonisin B1	42	248	1842	31,784	678	67
Fumonisin B2	14	84	696	12,968	580	58
Fumonisin B3	< LOD	34	284	3345	400	40
Fumonisin B4	< LOD	10	192	4341	284	28
Fumonisin B6	< LOD	< LOD	< LOD	30	1	0.10
OTA, *t* > 1.5 $\mu g \cdot kg^{-1}$	< LOD	< LOD	1.0	67	45	4.5
Masked toxins						
DON-3-glucoside, *t* > 1 $\mu g \cdot kg^{-1}$	12	44	424	3159	701	70
ZEN-14-sulfate, *t* > 2 $\mu g \cdot kg^{-1}$	1.3	17	132	4318	471	47

Table 1. *Cont.*

Metabolite (Group)	Median Concentration ($\mu g \cdot kg^{-1}$)	75th Percentile ($\mu g \cdot kg^{-1}$)	95th Percentile ($\mu g \cdot kg^{-1}$)	Maximum Concentration ($\mu g \cdot kg^{-1}$)	n $>t$	% $>t$
Emerging toxins						
BEA, $t > 1.0\ \mu g \cdot kg^{-1}$	8.5	25	114	1610	831	83
MON, $t > 2.0\ \mu g \cdot kg^{-1}$	16	61	236	1367	793	79
Ergot alkaloids, $t > 1.5\ \mu g \cdot kg^{-1}$						
Agroclavine	<LOD	<LOD	<LOD	108	37	3.7
Chanoclavine	<LOD	0.053	0.76	21	19	1.9
Dihydrolysergol	<LOD	<LOD	<LOD	5.2	2	0.20
Elymoclavine	<LOD	<LOD	<LOD	0.24	0	0
Ergine	<LOD	<LOD	<LOD	0.40	0	0
Ergocornine	<LOD	<LOD	1.9	48	56	6
Ergocorninine	<LOD	<LOD	1.8	21	57	5.7
Ergocristine	<LOD	<LOD	13	449	114	11
Ergocristinine	<LOD	<LOD	4.3	118	84	8.3
Ergocryptine	<LOD	<LOD	7.9	65	101	10
Ergocryptinine	<LOD	<LOD	2.0	20	66	6.6
Ergometrine	<LOD	<LOD	12	405	145	14
Ergometrinine	<LOD	<LOD	1.2	53	41	4.1
Ergosine	<LOD	<LOD	6.5	560	102	10
Ergosinine	<LOD	<LOD	1.4	102	49	4.9
Ergotamine	<LOD	<LOD	8.8	334	89	8.8
Ergotaminine	<LOD	<LOD	1.1	65	48	4.8
Festuclavine	<LOD	<LOD	<LOD	22	7	0.70
Enniatins, $t > 1.0\ \mu g \cdot kg^{-1}$						
Enniatin A	0.22	1.5	8.1	92	319	32
Enniatin A1	2.0	10	57	481	596	59
Enniatin B	5.9	29	137	1514	711	71
Enniatin B1	5.4	29	145	1846	693	69
Enniatin B2	<LOD	0.90	4.3	98	233	23
Enniatin B3	<LOD	0.010	0.070	138	30	3.0
Aflatoxin precursors, $t > 4.0\ \mu g \cdot kg^{-1}$						
Norsolorinic acid	<LOD	<LOD	<LOD	24	3	0.30
Averufin	<LOD	<LOD	2.2	139	30	3.0
Averufanin	<LOD	<LOD	<LOD	13	2	0.20
Versicolorin A	<LOD	<LOD	0.12	15	3	0.30
Versicolorin C	<LOD	<LOD	6.1	906	55	5.5
Averantin	<LOD	<LOD	<LOD	9.1	2	0.20
Sterigmatocystin	<LOD	<LOD	1.9	6296	23	2.3
Trichothecenes						
Type A trichothecenes, $t > 15\ \mu g \cdot kg^{-1}$ (incl. T-2, HT-2 toxins)						
Diacetoxyscirpenol	<LOD	<LOD	<LOD	41	2	0.20
15-Monoacetoxyscirpenol	<LOD	<LOD	<LOD	94	7	0.70
Neosolaniol	<LOD	<LOD	2.2	125	35	3.5
T2-Tetraol	<LOD	<LOD	<LOD	290	13	1.3
T2-Triol	<LOD	<LOD	<LOD	93	1	0.10
Type B trichothecenes, $t > 15\ \mu g \cdot kg^{-1}$ (incl. DON, DON-3-glucoside)						
15-Acetyldeoxynivalenol	<LOD	<LOD	178	2177	128	13
3-Acetyldeoxynivalenol	<LOD	<LOD	24	527	71	7.1
Nivalenol	4.7	18	127	11,232	286	28

1.4. Co-Occurrence

Serrano et al. [39] investigated samples from the Mediterranean region and found contamination with multiple toxins, notably nivalenol and BEA in a number of cereal products with contamination rates between 33% and 95%, highlighting the prevalence of multi-mycotoxin contamination. Similar observations were made during other comprehensive studies focusing on animal feed samples, with 38%–75% of samples being contaminated with more than one toxin [40,41]. Among masked toxins DON-3-glucoside has been reported to frequently co-occur with DON in commodities such as wheat, maize and barley and is also formed during processing [42]. It is formed by the plant following DON production by *Fusarium* fungi [43].

Among *Fusarium*–produced toxins, co-occurrence is frequently observed for compounds such as ZEN, DON and FUM and synergistic effects of *Fusarium* were reported in the past [44]. These included reduced weight gain in pigs (DON and fumonisin B1) [45] and adverse effects on broiler chicks [19]. Sensitivity to DON, even at low dosages, and overall toxicity is determined by co-occurring compounds, such as FUM, present in feed [46]. Synergistic effects at concentrations close to EU guidance levels have been reported to impact antioxidant activity of cells [47]. Co-ocurrence of DON, ZEN and nivalenol was reported in cereal samples by Tanaka et al. [48]. Other, similar findings, on a global scale were summarised by Placinta et al. [49]. For ZEN and ZEN-14-sulfate, co-occurrence with DON and nivalenol is well established [50].

1.5. Regulations

In the EU, maximum allowable concentrations vary with commodity, degree of processing (e.g., maize, finished feed) and intended consumers (e.g., animal feed). Lowest levels are in place for AFLA with a maximum limit of 20 $\mu g \cdot kg^{-1}$, 5 $\mu g \cdot kg^{-1}$ for compound feed for dairy cattle and calves, sheep, piglets and young poultry animals [51]. In the EU, 1000 $mg \cdot kg^{-1}$ are the set maximum allowable level for ergot sclerotia. Recommended tolerance levels for ergot alkaloids in feed in Canada for swine are 4–6 $mg \cdot kg^{-1}$ and chicks at 6–9 $mg \cdot kg^{-1}$ [52].

For other compounds guidance levels have been put in place. FUM in animal feed has a set guidance level of 60 $mg \cdot kg^{-1}$, but with lower values for pigs (5 $mg \cdot kg^{-1}$) and poultry and young animals (20 $mg \cdot kg^{-1}$). Guidance values for DON (0.9–12 $mg \cdot kg^{-1}$), ZEN (0.1–2 $mg \cdot kg^{-1}$) and OTA (0.1–0.25 $\mu g \cdot kg^{-1}$) are significantly lower. The upper level provides a general value with exceptions applying for animals showing high sensitivity. Lowest values are typically established for swine, poultry and young animals [51]. Recommended values for the sum of T-2 and HT-2 toxins are lowest with 250 $\mu g \cdot kg^{-1}$ for compound feed, 500 $\mu g \cdot kg^{-1}$ for cereal products and 2000 $\mu g \cdot kg^{-1}$ for oat milling products [53].

1.6. Global Surveys

The analysis of global mycotoxin occurrence data is of particular interest, because it helps with the identification of geographical areas, which are highly contaminated and, thus, might affect global trade of agricultural commodities. It also supports the study of emerging toxins due to changing climatic conditions or specific meteorological events, such as exceptionally dry or wet growing seasons [54,55].

Previous survey reports focused mostly on regulated toxins and toxins with guidance levels and their co-occurrence among regulated mycotoxins, e.g., Streit et al. [40], Tanaka et al. [48], Murugesan et al. [56], Streit et al. [57], Rodrigues and Naehrer [58]. Sharman et al. [59] monitored MON concentrations on a global scale. While a limited number of global survey papers exist, other reports cover smaller geographic regions, e.g., Romania, Argentina and The Netherlands [48,60,61]. While these studies contribute to the availability of global data, different analytical methodologies employed make a comparison of concentrations challenging.

Recently increased awareness has led to the study of masked and emerging toxins. The need to investigate co-occurrence and possible correlation of regulated toxins, toxins with guidance levels and masked and emerging toxin concentrations has been stressed by Schatzmayr and Streit [62], Streit et al. [41] and Jestoi [38]. A summary of global survey data published during the past 5 years is provided in Table 2. During the last 5 years only a small number of true global surveys was carried out. Overwhelmingly, the focus of the reported work was on regulated toxins and toxins with guidance levels, using a variety of analytical methods, including relatively simple thin layer chromatography. This complicates direct comparisons due to differing limits of detection and analytical performance parameters, leading to potential underreporting due to different limits of detection (LOD). Van Der Fels-Klerx et al. [63] explicitly acknowledged this issue, and as a result the LOD for data coming from different sources in their study (i.e., national monitoring programmes) was fixed at the highest LOD most frequently reported for a specific toxin.

Table 2. Summary of global survey data for regulated toxins and mycotoxins with guidance levels, masked and emerging toxins. Abbreviations: SBM—soybean meal, DDGS—dried distillers grain with solubles, REG—Toxins and secondary metabolites (regulated or with guidance levels), AFB1—aflatoxin B1, FUS—toxins produced by *Fusarium* spp., NIV—nivalenol, 3-Ac-DON—3-acetyl-DON, DAS—diacetoxyscirpenol, A & B— type A & B trichothecenes, HPLC—high performance liquid chromatography, FLD—fluorescence detector, MS—mass spectrometry, Elisa—enzyme linked immunosorbent assay, TLC—thin layer chromatography, NA—Not available.

Region, Country	Matrix	Year(s)	Toxin	Method	Reference
Global	Feed and ingredients	2004–2013	REG	HPLC, Elisa	[56]
Europe, Asia	Feed and ingredients	2004–2011	REG	HPLC, Elisa	[57]
Global	Feed and ingredients	2004–2012	REG	HPLC, Elisa	[62]
Americas, Europe, Asia	Corn, wheat, SBM, DDGS	2009–2011	REG	HPLC, Elisa	[58]
Middle East, Africa	Feed and ingredients	2009	REG, A&B	HPLC–FLD, LC–MS	[64]
Finland, Sweden, Norway, The Netherlands	Cereal grains	1989–2009	REG, NIV, 3-Ac-DON	NA	[63]
Belgium	Oats, pig/poultry feed	2012	FUS	LC–MS/MS	[65]
China	Dairy cow feed	2010	AFLA	HPLC–FLD	[66]
The Netherlands	Maize	2010	REG, FUS	LC–MS/MS	[67]
Pakistan	Poultry feed	2009–2010	AFB1	TLC	[68]
USA	DDGS	2009–2010	REG	HPLC–FLD, TLC	[69]
Portugal	Pig and poultry feed	2009–2010	OTA	HPLC–FLD	[70]
Argentina	Poultry feed	2008–2009	REG, DAS	LC–MS/MS	[71]
South Africa	Compound feeds	2010	REG	LC–MS/MS	[72]
Romania	Cereals	2008–2010	REG	Elisa	[73]
Serbia	Wheat	2007	FUS	LC–MS/MS	[74]
China	Feed and ingredients	2008–2009	REG	LC–MS	[75]

The resulting requirements call for an analytical methodology that is able to provide consistent high-quality and comparable results for a wide range of metabolites as co-occurrence and correlation of compounds are increasingly studied. The lack of information on masked and emerging toxins, including co-occurrence with regulated toxins and toxins with guidance levels calls for expanded survey data. In this way, new trends and emerging research questions that go hand-in-hand with increasing frequency of extreme weather events and a changing climate can be efficiently addressed [54,55,76].

This study presents global survey data for the years 2012–2015 for regulated toxins, toxins with guidance levels and selected masked and emerging toxins and metabolites. Toxin concentrations were summed up creating groups of toxins and metabolites that were studied as a sum (e.g., AFLA were reported as the sum of concentrations of aflatoxins B1, B2, G1 and G2; for details regarding the other groups, see Table 1). Concentrations were determined with a single liquid chromatography mass spectrometric method. Mycotoxin and metabolite concentrations of 57 compounds in finished feed, maize and maize silage were investigated in 1113 samples from 46 countries. Representative results from regions and countries across the globe were compiled. Specifically, the co-occurrence and correlation of regulated mycotoxins and toxins with guidance levels with selected masked and emerging toxins and metabolites was investigated.

2. Results

Results of regulated toxins, toxins with guidance levels, masked and emerging toxin concentrations from 1113 samples of global origin from 3 commodities (finished feed, maize and maize silage) collected from 2012 to 2015 are presented. Tables 1 and 5 provide a description and basic statistical information of the data set and the subsets investigated. Samples contained on average 16 and up to 35 out of 57 compounds at the same time. The number of metabolites found in a single sample was 35 in finished feed, 29 in maize and 28 in maize silage. The lower part of Table 5 provides subset information with regard to number of samples available per matrix.

Regarding the reported trends for the years which observations were reported for, it has to be noted that because the data are limited to only four consecutive years and it is well known that especially for mycotoxin contamination there are large year–to–year variations, no long–term trends can be deduced at this point.

Results from subsets with sample numbers greater than 40 were reported in detail; for details regarding the number of samples in each subset, see Table 5. If, for completeness, subsets with lower sample number were chosen, this was clearly indicated. Summary data in Table 1 highlights global relevance of ZEN and DON in the investigated matrices with 88% and 79% of 1113 samples contaminated with concentrations above the threshold. Prevalence of the related masked toxins DON-3-glucoside (70%) and ZEN-14-sulfate (47%) was high. Among emerging toxins, MON and BEA showed high occurrence in 79 and 83% of samples, respectively. Slightly lower occurrence was observed for enniatins with enniatin A, A1, B and B1 occurring in 32%, 59%, 71% and 69% of samples.

Detailed regional results were most useful to study trends and concentrations for the purpose of regulation and monitoring. Figure 2 provides a summary of concentrations of regulated toxins and toxins with guidance levels in Central Europe. Maximum observed concentrations from Table 1 and data available from Figure 2 show that observed concentrations are relevant for regulatory purposes and animal health.

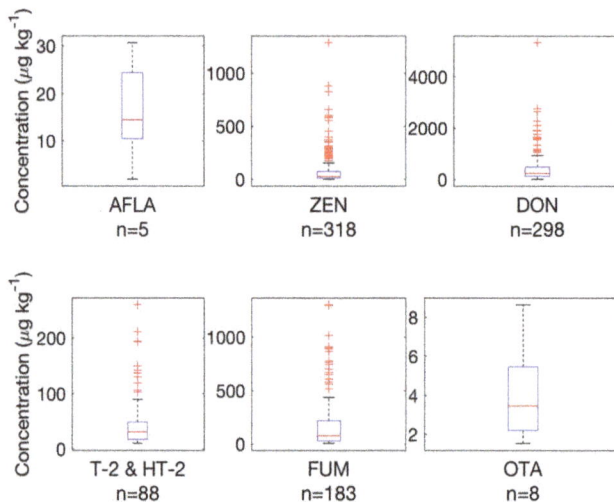

Figure 2. Survey results for regulated toxins and toxins with guidance levels in 335 finished feed samples in Central Europe above defined thresholds listed in Table 1. *n* provides number of samples. Boxplots follow definition by McGill et al. [77].

This occurrence pattern is further elaborated on in Figure 3. The percentage of samples with concentrations above the threshold was low (<5%) for AFLA and OTA, but the percentage of above-threshold concentrations in samples for the other four regulated mycotoxins and mycotoxins with guidance levels was greater than 50%, and high for DON and ZEN (with >80% of investigated finished feed samples showing concentrations above threshold (see Figure 3a). Regarding yearly median concentrations 2014 and 2015 showed marked increases for DON and ZEN and FUM with doubled median concentrations compared to 2012 and 2013 (Figure 3b,c). Significance codes indicate significant differences between yearly medians as a result of Kruskal–Wallis tests, while error bars provide Wilcoxon confidence intervals. Different letters indicate a significant difference between the groups. The size is due to a large concentration range and few available samples (e.g., for 2012). Concentrations for other regulated toxins and toxins with guidance levels remained stable.

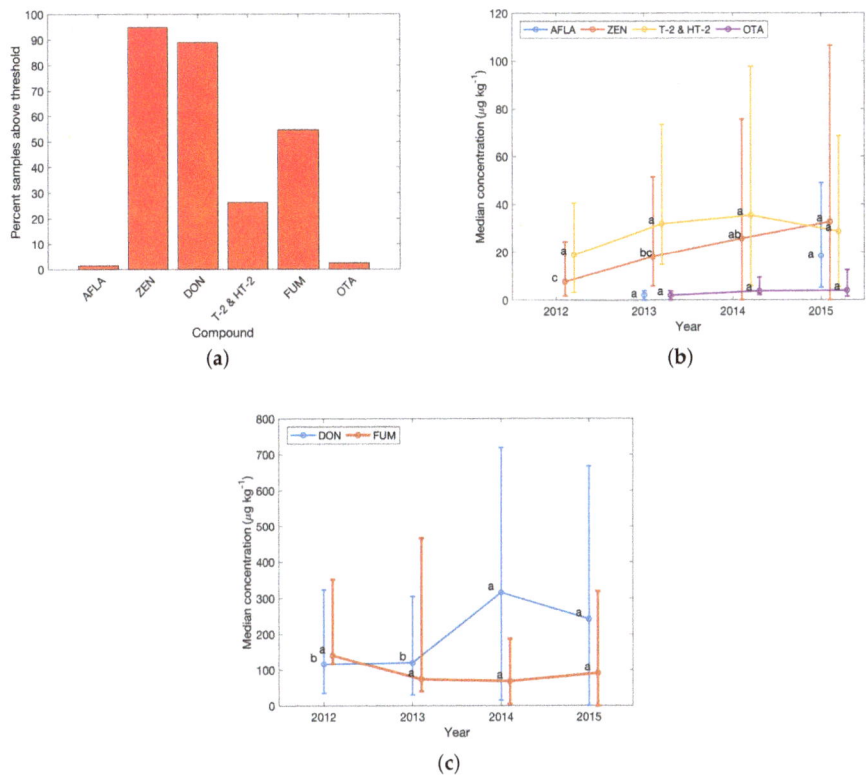

(a)

(b)

(c)

Figure 3. Survey results for regulated toxins in finished feed samples in Central Europe. (a) Percentage of samples with concentrations above thresholds; see Table 1 for details; (b,c) Yearly median concentrations from 2012 to 2015 (missing point indicates that no data were available). Error bars reflect the Wilcoxon confidence interval (CI). Lower error were replaced with the median, if the Wilcoxon CI would have resulted in negative concentrations. Significance codes show differences between yearly medians from a Kruskal–Wallis test result. Different letters indicate a significant difference between the groups. Data points were offset on the x-axis for clarity. Sample numbers for calculation of the median of each year are availale in Table 3.

The situation regarding masked and emerging toxins and secondary metabolites is illustrated in Figure 4. Compared to regulated toxins and toxins with guidance levels, masked toxin concentrations were in a range of up to 800 µg·kg^{-1} with a larger number of samples at concentrations around <300 µg·kg^{-1} in finished feed samples.

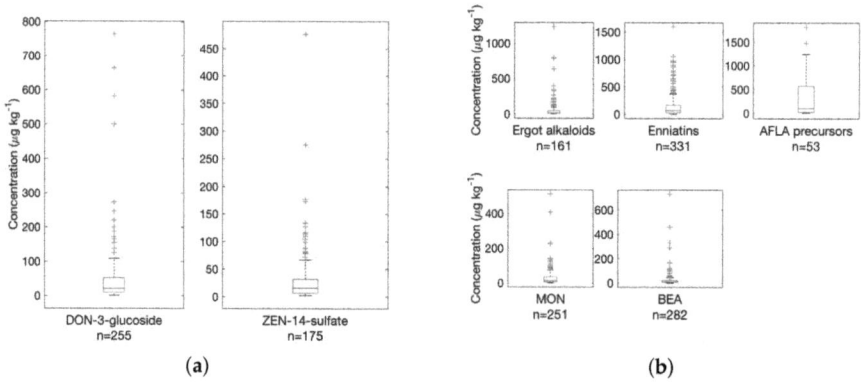

(a)   (b)

Figure 4. Survey results for (**a**) masked and (**b**) emerging toxins in finished feed samples from Central Europe (335 samples) above threshold concentrations; see Table 1 for details.

Table 3. Number of samples at concentrations >*t* used for the calculation of median plots and statistical analysis. NA—sample number was <1 and no median could be calculated.

Region/Country	Compound	2012	2013	2014	2015
Central Europe					
	AFLA	NA	NA	NA	4
	ZEN	21	20	124	153
	T-2 & HT-2	2	7	57	22
	DON	21	25	123	129
	FUM	6	11	72	94
	OTA	NA	NA	2	5
	Ergots	3	21	84	53
	Enniatins	21	25	128	157
	PreAflas	NA	NA	13	32
	MON	21	19	105	114
	BEA	17	17	98	102
	DON-3-glucoside	20	14	100	121
	ZEN-14-sulfate	2	5	76	92
Austria					
	AFLA	NA	NA	NA	NA
	ZEN	NA	NA	34	24
	T-2 & HT-2	NA	4	24	3
	DON	NA	4	32	22
	FUM	NA	4	24	12
	OTA	NA	NA	NA	NA

Table 3. *Cont.*

Region/Country	Compound	2012	2013	2014	2015
Germany					
	AFLA	NA	NA	NA	NA
	ZEN	NA	15	35	33
	T-2 & HT-2	NA	NA	8	9
	DON	NA	15	36	32
	FUM	NA	4	17	14
	OTA	NA	NA	NA	NA
Italy					
	AFLA	NA	NA	NA	NA
	ZEN	2	5	13	34
	T-2 & HT-2	NA	NA	4	4
	DON	2	4	12	32
	FUM	2	4	13	30
	OTA	NA	NA	NA	NA
The Netherlands					
	AFLA	NA	NA	NA	4
	ZEN	NA	2	21	15
	T-2 & HT-2	NA	NA	7	NA
	DON	NA	2	20	13
	FUM	NA	NA	17	12
	OTA	NA	NA	NA	NA

Results for masked and emerging toxins are complemented in a similar fashion as shown in Figure 3 to highlight the percentage of samples in finished feed above the defined threshold; see Figure 5a. Enniatins were ubiquitous in finished feed samples and MON and BEA were found in more than 70% and 80% of samples, respectively, i.e., very similar to DON and ZEN regarding occurrence, underlining the importance to study emerging toxins and secondary metabolites.

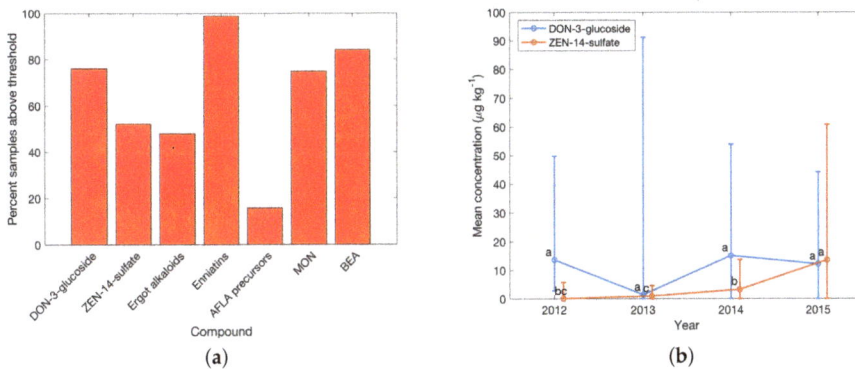

(a)

(b)

Figure 5. *Cont.*

(c)

(d)

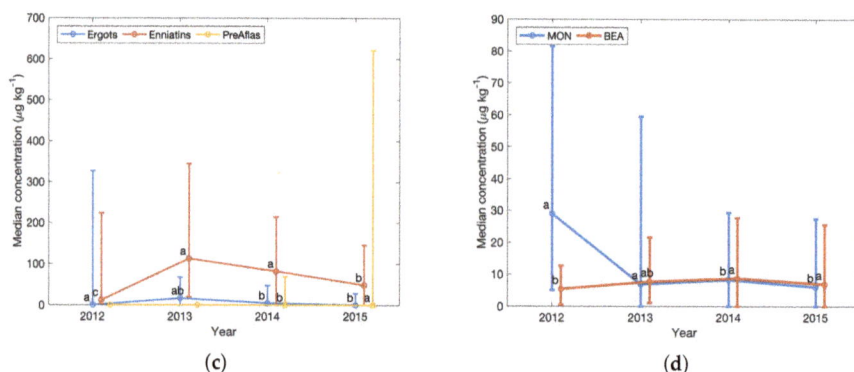

Figure 5. (**a**) Survey results for masked (DON-3-glucoside and ZEN-14-sulfate) and emerging toxins in finished feed samples from Central Europe above threshold levels; see Table 1 for details. Subfigures (**b**) masked, (**c**) and (**d**) show yearly median data for emerging toxins for the years 2012–2015 in Central Europe. Error bars reflect the Wilcoxon confidence interval (CI). Lower error were replaced with the median, if the Wilcoxon CI would have resulted in negative concentrations. Significance codes show differences between yearly medians from a Kruskal–Wallis test result. Different letters indicate a significant difference between the groups. Data points were offset on the *x*-axis for clarity. Sample numbers for calculation of the median of each year are available in Table 3.

Correlation analysis was conducted to investigate how concentrations varied between compounds and if correlation was observed. In accordance with Van Der Fels-Klerx et al. [63] a correlation coefficient >0.5 was considered high enough to be reported. In finished feed samples from Central Europe a correlation coefficient of 0.6 between DON and DON-3-glucoside was observed (Figure 6a). For clarity, both axes were plotted in logarithmic form. Additional correlation data of regulated toxins with emerging compounds in Eastern Europe are found in Figure 6.

Country-specific data provide information about differences within a geographic region as demonstrated in Figure 7. The subplots show clear differences in observed yearly medians, e.g., for DON in Austria and Germany (1500 vs. 400 μg·kg^{-1} in 2014). Notable are also differences between Italy and Austria regarding yearly median FUM concentrations (1500 vs. <100 μg·kg^{-1}) in 2015.

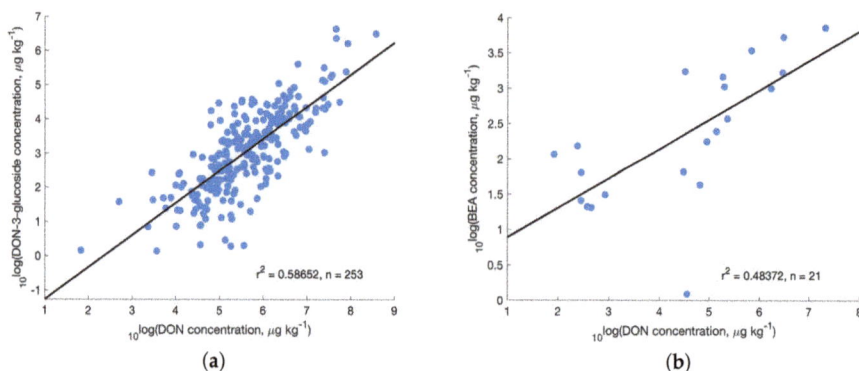

(a)

(b)

Figure 6. *Cont.*

(c)　　　　　　　　　　　　　　　　　　　　　**(d)**

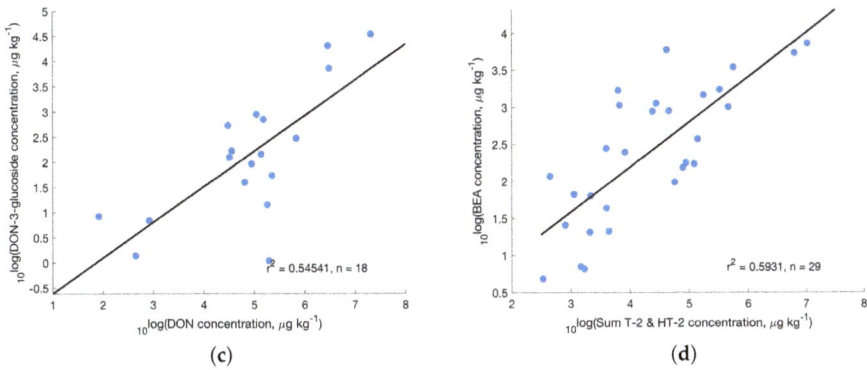

Figure 6. Correlation of (**a**) DON and DON-3-glucoside in Central Europe; (**b**) DON with BEA (Eastern Europe); (**c**) DON with DON-3-glucoside (Eastern Europe) and (**d**) sum of T-2 and HT-2 toxins with BEA in finished feed samples (Eastern Europe).

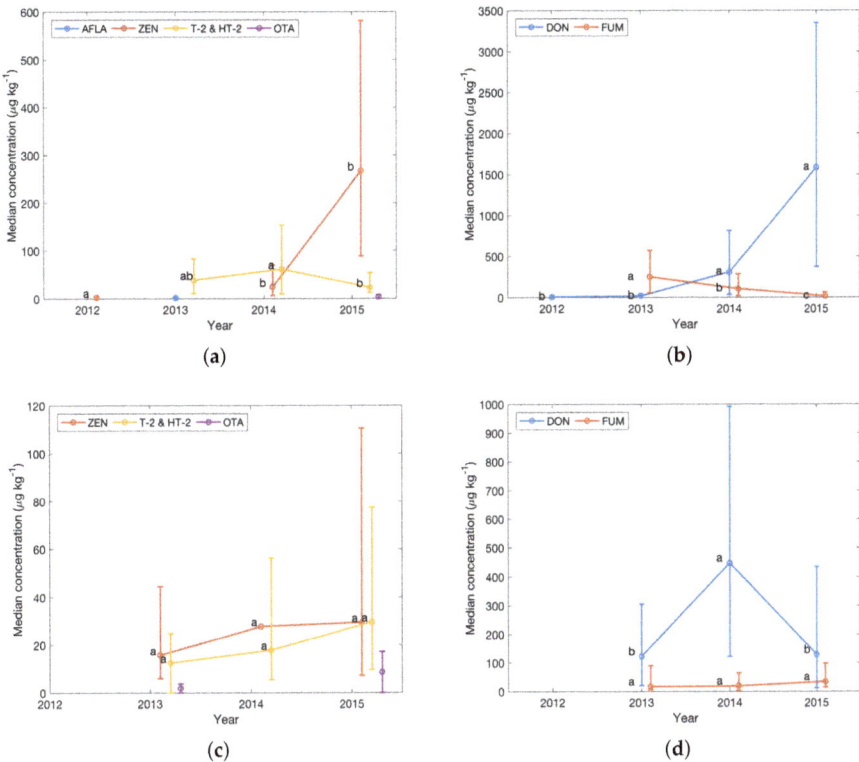

(a)　　　　　　　　　　　　　　　　　　　　　**(b)**

(c)　　　　　　　　　　　　　　　　　　　　　**(d)**

Figure 7. *Cont.*

127

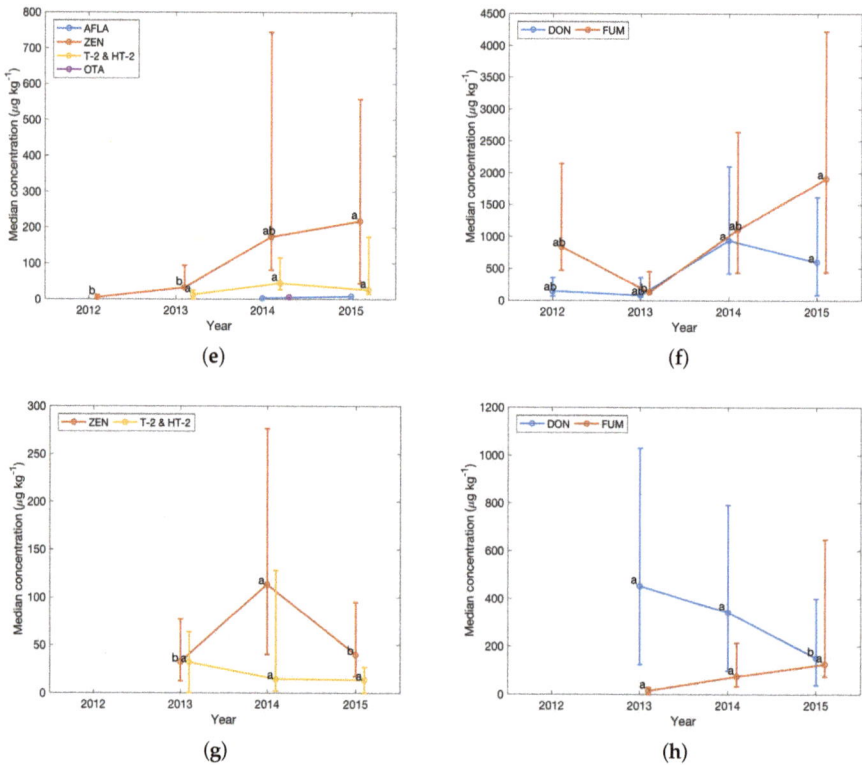

Figure 7. Yearly median concentrations of regulated toxins and compounds with guidance levels in finished feed from (**a**) and (**b**) Austria, (**c**) and (**d**) Germany from 2012 to 2015. Error bars reflect the Wilcoxon confidence interval (CI). Lower error were replaced with the median, if the Wilcoxon CI would have resulted in negative concentrations. Significance codes show differences between yearly medians from a Kruskal–Wallis test result. Different letters indicate a significant difference between the groups. Data points were offset on the *x*-axis for clarity. Sample numbers for calculation of the median of each year are availale in Table 3. Yearly median concentrations of regulated toxins and compounds with guidance levels in finished feed from (**e**) and (**f**) Italy, and (**g**) and (**h**) The Netherlands from 2012 to 2015. Error bars reflect the Wilcoxon confidence interval (CI). Lower error were replaced with the median, if the Wilcoxon CI would have resulted in negative concentrations. Significance codes show differences between yearly medians from a Kruskal–Wallis test result. Different letters indicate a significant difference between the groups. Data points were offset on the *x*-axis for clarity. Sample numbers for calculation of the median of each year are available in Table 3.

3. Discussion

3.1. Finished Feed in Central Europe

Box plots in Figure 2 illustrate occurrence and concentrations of regulated toxins and toxins with guidance levels in samples from Central Europe [77]. A large number of samples showed detectable concentrations for all compounds, although the majority of samples showed low concentrations, especially when compared to the maximum levels and guidance values. While overall the number of infected samples, i.e., with concentrations above the detection limit is quite high, this is comparable with other recent studies, e.g., reports by Streit et al. [57] for feed samples and Van Asselt et al. [67] for maize samples, where authors have chosen similar reporting limits for LC–MS/MS data, thus making comparisons feasible. The report by Van Asselt et al. [67] also revealed a high percentage of contaminated samples (e.g., >80% samples were contaminated with DON in 2006 and 2007. Information provided in this study (e.g., in Table 1) shows additional detail on concentration distribution, similar to data by Van Der Fels-Klerx et al. [63] for selected type A and B trichothecenes in cereal samples, restricted to Northern Europe. Figure 2 highlights that DON and ZEN were found at above-threshold concentrations in finished feed. Samples contained a sum of T-2 and HT-2 toxin concentrations above-threshold (up to 250 $\mu g \cdot kg^{-1}$). FUM with was found at maximum concentrations of 1200 $\mu g \cdot kg^{-1}$. Overall, legal limits and guidance values for regulated toxins and toxins with guidance levels are not reached for the vast majority of samples. Single samples exceed regulations (e.g., 5 samples with AFLA concentrations between 10 and 30 $\mu g \cdot kg^{-1}$), but given the overall samples analysed for Central Europe (n = 335) the percentage (1.5%) remains low. For DON, 27 samples (8.1%) showed concentrations above 0.9 ppm, the lowest guidance level established for piglets and calves. For all other regulated mycotoxins the established guideline values were not exceeded. The general occurrence and level of contamination with very high concentrations for few samples confirms data from previous studies [67]. Care has to be taken for reports stating the percentage of contaminated samples to ensure that comparable lower reporting limits for positive results were chosen, but even in older studies high percentages of positives were reported, e.g., 58% for ZEN in corn from a survey conducted in 19 countries [78].

As indicated in Table 4 there is a high degree of co-occurrence for both investigated masked toxins of samples containing both compounds. Typically, between 50% and (in some cases) almost all samples contained 3 or more (out of 6) regulated toxins or toxins with guidance levels, both masked toxins and 3 or more (out of 5) emerging toxins. This is found for all matrices investigated, where enough samples were available for a detailed assessment. Concentrations above 1 $\mu g \cdot kg^{-1}$ were observed in 30%–40% of finished feed samples. A previous report investigating co-occurrence of DON and the masked DON-3-glucoside [79] in samples from Central European countries also found a very high degree of co-occurrence with all 77 field samples containing both compounds.

With the exception of AFLA precursors and ergot alkaloids, all other investigated masked and emerging compounds occur in between 79% and 98% of samples, thus showing a high degree of co-occurrence with the most prevalent regulated toxins and toxins with guidance levels (i.e., DON, ZEN and FUM). Overall, a high degree of contamination was observed, which is due to low detection limits, but in line with previous studies on smaller sample sets, e.g., for enniatins among the emerging compounds and toxins with guidance levels, e.g., DON [41]. Maximum concentrations were in the 500–1500 $\mu g \cdot kg^{-1}$ range for 25 (7.6%) samples containing a summarised enniatin concentration >500 $\mu g \cdot kg^{-1}$. Yearly medians were in the 30 and 100 $\mu g \cdot kg^{-1}$ range for masked and emerging compounds, respectively and quite variable between 2012 and 2015. The high occurrence of masked and emerging compounds in finished feed samples from Central Europe is further illustrated in Figure 5a. With the exception of AFLA precursors, close to 50% or more samples showed concentrations >t for all investigated compounds. European samples containing AFLA, either the aflatoxins themselves or the precursors, were generally rare showing the lowest percentage of positives of all regulated toxins or toxins with guidance levels, which was comparable to previous reports Rodrigues et al. [64]. A fairly large number of samples also showed high to very high concentrations between 100 and 1500 $\mu g \cdot kg^{-1}$ for enniatins, ergot alkaloids and AFLA precursors. Occurrence of MON and BEA is similar with about two thirds of samples contaminated, but at lower concentrations up to 700 $\mu g \cdot kg^{-1}$.

Table 4. Summary of co-occurrence and correlation data for investigated samples. For each matrix (finished feed, FF; maize, M; maize silage, MSI) and toxin group (regulated, masked, emerging) the percentage of samples with 3 or more (2 for masked compounds) co-occurring toxins is provided. Correlation between 2 specific species with Pearson coefficients of correlation > 0.5 is listed in the last column. empty cell – not calculated, because of too small subset. (H)T-2 – Sum of T-2 and HT-2 toxins.

Region/Toxins	Co-Occurrence FF (%)	M (%)	MSI (%)	Correlation Compounds (Matrix, r^2)
Africa				
Regulated	100			AFLA & AFLA precursors (FF, 0.93) ZEN & Enniatins (FF, 0.54) DON & DON-3-glucoside (FF, 0.51)
Masked	92			
Emerging	100			
South Africa				
Regulated	90	59		AFLA & AFLA precursors (M, 0.69) AFLA & BEA (M, 0.77) DON & DON-3-glucoside (M, 0.76) ZEN & ZEN-14-sulfate (M, 0.65)
Masked	50	28		
Emerging	90	36		
Central Europe				
Regulated	73	56	54	DON & DON-3-glucoside (FF, 0.57) DON & DON-3-glucoside (M, 0.80) ZEN & ZEN-14-sulfate (M, 0.56) DON & DON-3-glucoside (MSI, 0.77) ZEN & ZEN-14-sulfate (MSI, 0.79)
Masked	57	72	44	
Emerging	93	83	83	
Eastern Europe				
Regulated	74			(H)T-2 & DON-3-glucoside (FF, 0.50) DON & DON-3-glucoside (FF, 0.81) ZEN & ZEN-14-sulfate (FF, 0.76) (H)T-2 & BEA (FF, 0.57)
Masked	22			
Emerging	91			
Northern Europe				
Regulated	45			DON & DON-3-glucoside (FF, 0.58) ZEN & ZEN-14-sulfate (FF, 0.65)
Masked	47			
Emerging	82			
Southern Europe				
Regulated	89			DON & DON-3-glucoside (FF, 0.63)
Masked	73			
Emerging	96			

Table 4. *Cont.*

Region/Toxins	Co-Occurrence FF (%)	M (%)	MSI (%)	Correlation Compounds (Matrix, r^2)
Middle East				
Regulated	93			DON & DON-3-glucoside (FF, 0.50) FUM & MON (FF, 0.51)
Masked	91			
Emerging	95			
North America				
Regulated	88	63		DON & DON-3-glucoside (FF, 0.50) ZEN & ZEN-14-sulfate (FF, 0.81)
Masked	52	40		
Emerging	100	77		
South America				
Regulated		54		AFLA & AFLA precursors (M, 0.98) DON & DON-3-glucoside (M, 0.76) ZEN & ZEN-14-sulfate (M, 0.93) DON & MON (M, 0.56) ZEN & MON (M, 0.64)
Masked		21		
Emerging		71		

The analysis for other regions and selected countries from Table 5 was conducted in an identical fashion; the following subsection provides a summary of results obtained from corresponding plots (data are shown in the supplementary information for the regions discussed).

3.2. Finished Feed in Global Regions and Countries

The situation for regulated toxins in other European regions was in general similar to that in Central Europe, with some notable differences. Contamination with DON and ZEN were still of concern (as these toxins are of global relevance), but was less prevalent in Eastern Europe at lower concentrations, i.e., less than 400 and up to 1500 $\mu g \cdot kg^{-1}$. Contamination with FUM was in a similar concentration range (up to 1500 $\mu g \cdot kg^{-1}$). A similar situation was also observed for Southern Europe.

Generally, very few samples were contaminated with AFLA in Central and Eastern Europe. The maximum concentration found in a single sample was almost three times higher in Eastern compared to Central Europe with 90 and 30 $\mu g \cdot kg^{-1}$, respectively. Northern and Southern Europe showed markedly lower maximum concentrations at 3 and 7 $\mu g \cdot kg^{-1}$, respectively. However, for these single samples established limits for young animals such as piglets and calves were exceeded. Therefore, a continued monitoring of AFLA levels is advisable, especially as finished feed varies in composition [80].

Samples from Central and Southern Europe showed the maximum ZEN concentration >1000 $\mu g \cdot kg^{-1}$, especially samples originating from Germany, Austria and Italy, which indicates high prevalence in Central Europe. In a report by Streit et al. [40] ZEN in feed materials in Europe was found to be at maximum concentration of 1045 $\mu g \cdot kg^{-1}$, so very much comparable with the data presented here. A previous report summarising occurrence of ZEN in different matrices found similar concentrations, e.g., max. 950 $\mu g \cdot kg^{-1}$ in feeds and grains [28]. Samples from North America and Northern Europe, at 800 and 400 $\mu g \cdot kg^{-1}$, followed. Just like DON, almost all samples contained ZEN at concentration levels >t, regardless the region or country of origin, but generally not at levels exceeding regulations. The least contamination was observed for OTA with low occurrence (typically around 10%), but individual samples with concentrations of up to 3000 $\mu g \cdot kg^{-1}$. The sum of T-2 and HT-2 toxins was in the 100–200 $\mu g \cdot kg^{-1}$ range.

In other non-European regions, OTA concentrations were comparatively elevated such as in samples from the Middle East ($n = 23$) and Africa ($n = 24$), with the caveat that sample numbers were lower. In samples from South America, South Africa and the Middle East, FUM was more prevalent than DON with a larger percentage containing FUM than DON above the threshold, e.g., 95% of samples containing FUM and 80% containing DON for South America. These results are mirrored in a report, where 94% of 224 samples tested positive for FUM and only 13% of 130 samples contained DON [58]. Contamination of finished feed with AFLA in African samples was of concern with >35% of samples at concentrations >t and a maximum concentration of 60 $\mu g \cdot kg^{-1}$, even though the number of samples was low ($n = 24$). Therefore, special precautions are required, when importing feed from Africa to ensure that established limits are met [81]. Samples from South Africa were not included here, as only 5% of samples showed notable concentrations of AFLA with a much larger number of samples ($n = 74$).

Regarding masked and emerging toxins, samples from Eastern Europe showed similar concentrations compared to Central Europe for masked mycotoxins in finished feed. Most concentrations were <50 $\mu g \cdot kg^{-1}$ with maxima of 90 $\mu g \cdot kg^{-1}$ (DON-3-glucoside) and 250 $\mu g \cdot kg^{-1}$ (ZEN-14-sulfate). For emerging toxins, the situation was quite similar in Eastern Europe compared to Central Europe. Concentrations of enniatins (with exception of a single highly contaminated sample) were generally lower, at around 500 $\mu g \cdot kg^{-1}$ for the higher contaminated samples. In Northern Europe a single, highly contaminated sample of AFLA precursors at 12,000 $\mu g \cdot kg^{-1}$ was found, with 15 samples containing AFLA precursors. Occurrence of high levels of AFLA precursors without elevated AFLA concentrations indicated microbial producers other than aflatoxigenic *Aspergillus* species. Sterigmatocystin is produced by approximately 15 different fungal species the most important one being *A. versicolor* [82]. It was observed that MON and BEA occurred at higher concentrations and more often in samples from Northern Europe, reaching concentration levels similar to DON, thus making all three compounds relevant for e.g., monitoring [83]. Low concentrations of <150 $\mu g \cdot kg^{-1}$ were found for emerging toxins in South Africa with MON showing highest concentrations at about 100 $\mu g \cdot kg^{-1}$. The situation was similar in Southern Europe, where MON showed highest concentrations at <100 $\mu g \cdot kg^{-1}$.

The number of contaminated samples generally suggest a high degree of co-occurrence, e.g., 298 of 335 samples in Central Europe showed concentrations >t for DON and 318 for ZEN; see Figure 2 and Table 4. The situation in other global regions is quite similar (e.g., for enniatins, BEA and MON in South Africa with 51, 64 and 71, respectively out of 74 samples). However, correlation of concentrations between regulated toxins, toxins with guidance levels and emerging toxins was less often observed as Table 4 demonstrates. DON is usually well correlated with DON-3-glucoside and so is ZEN with ZEN-14-sulfate. No correlation for ZEN with ZEN-14-sulfate was observed in samples from South Africa. Regarding AFLA precursors and AFLA, however, generally no correlation was observed, except for African finished feed data (for a relatively low sample number of $n = 24$) and maize samples in South Africa and South America. A larger number of notable correlations with r^2 around 0.5 was found in samples from Eastern Europe for DON and the sum of T-2 and HT-2 toxins with the masked compound DON-3-glucoside and the emerging toxin BEA; see Figure 6 and Table 4 for details, where also single cases of MON correlated with regulated toxins are listed (e.g., for DON and ZEN, respectively in maize from South America).

3.3. Yearly Median Concentrations 2012–2015

For finished feed samples yearly median concentrations show some marked changes between 2012 and 2015. Figure 4c shows marked increases for DON in Central Europe for the years 2014 and 2015, which are significantly different compared to 2012 and 2013. In Eastern Europe, the median yearly FUM concentration was high (up to 600 $\mu g \cdot kg^{-1}$) in 2013, followed by an increase in DON for 2014.

Samples from Northern Europe also showed median yearly DON increases similar to Central Europe. FUM concentrations were higher than DON for Southern Europe. A notable contrast was seen for neighbouring Germany and Austria, where DON concentrations decreased in 2015 for the former, but strongly increased for the latter (see Figure 7) at much higher concentrations. Significant differences are also seen for Italy, neighbouring Austria and The Netherlands, neighbouring Germany. This suggests that regional climate and cultivars might play an important role for infection rates. Therefore, it is advisable to be careful, when pooling data ensuring that data subsets are fit for purpose, resulting in different subsets for e.g., regulatory or environmental impact studies.

Overall, DON and FUM were the dominating compounds on a regional as well as country scale. Other regulated toxins and toxins with guidance levels only played a secondary role regarding yearly medians and the overall toxin load. However, there were countries with low yearly medians, e.g., Russia for DON in the 50–100 $\mu g \cdot kg^{-1}$ range or South Africa for DON with 50–300 $\mu g \cdot kg^{-1}$ yearly median concentrations.

Yearly median concentrations of masked and emerging compounds were quite variable, lacking some of the distinctive trends, e.g., observed for Central Europe for the years 2014 and 2015, where large increases in DON and FUM were observed (see comparison of Figures 5b–d and 7 further supporting the hypothesis that co-occurrence does not imply correlation. Median concentrations were usually by a factor of 10 (emerging toxins) and 100 (masked mycotoxins) lower than e.g., regulated compounds for the same observation period. Another notable observation was a 5-fold increase to an median 250 $\mu g \cdot kg^{-1}$ of enniatin concentrations in Eastern European samples.

A large source of variability could be in the types of finished feed sampled and variation in the composition of each of the feeds. However, the extent of signal suppression was not significantly higher e.g., in feed compared to the other matrices. It is mostly relative matrix effects that are typically a source of concern, but these were considered to be sufficiently low for the purpose of this manuscript. These concerns add to to the variability in the data, generally increasing the difficulty of achieving statistical significance and introducing some potential for unintended bias in results. The finished feed types include poultry, swine and ruminant feed. These compound feeds can indeed vary in their contents (between and within feed types). A major source of variability between years can be the number of samples per country which is a large potential source of unintended bias. This is the reason for showing the individual countries in Figure 7 for regulated toxins and mycotoxins with guidance levels. Overall, the chosen thresholds are realistic, as they reflect the concentration that may be reliably determined leaving a safety margin for variation of the instrument's performance and the complexity of the sample.

Due to the already discussed year-to-year variations, long-term trends could not be identified. Furthermore, the country of origin for the raw ingredients of finished feed samples is not necessarily also the country of origin, which makes the evaluation of contamination levels for specific geographical regions challenging.

3.4. Maize and Maize Silage

The observations of finished feed concentrations for the years 2012–2015 for regulated, masked and emerging toxins set a general trend that is largely observed for maize and maize silage, though with some key differences.

For regulated toxins and toxins with guidance levels in maize and maize silage from Central Europe a dozen samples out of 78 showed very high concentrations of ZEN up to 10,000 $\mu g \cdot kg^{-1}$. Concentrations were lower in maize silage (up to 3000 $\mu g \cdot kg^{-1}$ for ZEN), but otherwise similar trends were observed. The situation was analogous for DON. High yearly medians of 2500 $\mu g \cdot kg^{-1}$ for DON in maize silage were observed in 2014, which was also the case for maize, but at a higher concentration level of 4000 $\mu g \cdot kg^{-1}$. Maize samples in South Africa showed regulated toxins and toxins with guidance levels in a similar concentration range as in Central Europe with lower maxima (e.g., 8000 $\mu g \cdot kg^{-1}$ for a single DON sample). It has been reported previously that in maize high concentrations of AFLA were observed [61]. The presented data show high concentrations for individual samples only, for both, AFLA and AFLA precursors for all investigated regions. Even in maize samples from South America, only 11 out of 77 samples contained concentrations >1 µg·kg^{-1} with 3 samples featuring very high concentrations around 200 (for 2 samples) and a single sample at 1300 µg·kg^{-1}.

Low concentrations were observed (<150 µg·kg^{-1}) for emerging toxins in maize in South Africa with MON at highest concentrations at about 400 µg·kg^{-1} for the 2012–2015 years. The situation was similar in Southern Europe, where MON showed highest concentrations between 2012 and 2015, but generally at <100 µg·kg^{-1}. In both cases only very few samples showed these elevated concentrations. Generally, a higher incidence of MON and BEA was observed in South African maize, whereas the occurrence of AFLA precursors, enniatins and ergot alkaloids was generally low. Overall, concentrations of emerging toxins in South American maize samples were lowest.

For masked toxins, maize silage showed the lowest concentrations of the three matrices with a range of 300–600 µg·kg^{-1} for the highest concentrations. Otherwise observations are similar to maize. Maximum concentrations in South African maize were higher for few individual samples with 1000 and 300 µg·kg^{-1} for DON-3-glucoside and ZEN-14-sulfate, respectively. In South America, maximum concentrations were even higher with 900 and 1200 µg·kg^{-1} for DON-3-glucoside and ZEN-14-sulfate, although yearly medians were well <100 µg·kg^{-1} since only very few samples were affected. 2014 and 2015 showed medians of 30 µg·kg^{-1} with virtually no masked toxins detected in 2013.

3.5. Type A and B Trichothecenes

The overall occurrence of type A and B trichothecenes was assessed by summation of compounds as described in Table 1. Trichothecenes occur together in about 40% of samples, with concentrations of type B trichothecenes being higher (9 samples showing concentrations between 4500 and up to 16,000 µg·kg^{-1}). Maximum concentrations for type A trichothecenes on the other hand were 30 times lower in the 500 µg·kg^{-1} range. 55% of samples were contaminated with type A trichothecene concentrations >t and 80% samples contained type B trichothecenes. Guidance values for the sum of trichothecenes currently do not exist, so a comparison was made with the guidance value of DON alone. The sum of type A and, the sum of type B trichothecenes exceeded the guidance value for DON in 100+ samples; thus a detailed investigation of the impact of multi-toxin contamination is advisable due to the high degree of co–occurrence. For DON alone, concentrations around 3500 µg·kg^{-1}, a level, which has previously shown adverse effects in animal studies (e.g., see [16]) were observed in all matrices investigated, e.g., in maize samples from South Africa and South America.

Type A and B trichothecenes were typically both found in most investigated finished feed samples with South Africa being a notable exception. Figure 8 compares occurrence (as sum of the respective trichothecene class indicated in Table 1. The absence of type A trichothecenes is clearly illustrated, since these were found in only in 3 samples, whereas type B trichothecenes are found in 69 out of 74 samples. While occurrence of T-2 toxin has been reported (e.g., by Placinta et al. [49], little quantitative data exists. T-2 and HT-2 toxins (and also DON) were determined in 92 commercial feed samples in a recent study by Njobeh et al. [72] and none of the samples contained these toxins, with the caveat that stated limits of detection were comparable with data reported here at 1 and 2.5 µg·kg^{-1}, respectively. Overall data on A trichothecenes in South Africa remains scarce.

Trichothecene concentrations found in maize samples were similar to those finished feed. The situation for maize silage generally resembled maize with regard to concentration ranges and medians. Specifically, samples from Brazil showed similarly high occurrence for type A and B trichothecenes with 31 and 30, respectively (out of 37) having a concentration of >t, so both types were found in samples at the same time (which is in stark contrast to South African samples, highlighting the importance of regional assessments). For South America as a whole, the situation was similar, although type A trichothecenes were not as frequently observed; only 40% of samples ($n = 77$) contained both types and more than 70% of samples had an type A trichothecene load of <t. South African maize on the other hand, was not contaminated with type A trichothecenes (only 1 sample with a toxin load of 1.6 µg·kg^{-1} was found to contain type A trichothecenes). T-2 toxin was also absent in a study by Sydenham et al. [84], which were collected Eastern Cape province. On the other hand, 44 out of 53 samples contained type B trichothecenes at concentrations $>t$ mirroring results from finished feed.

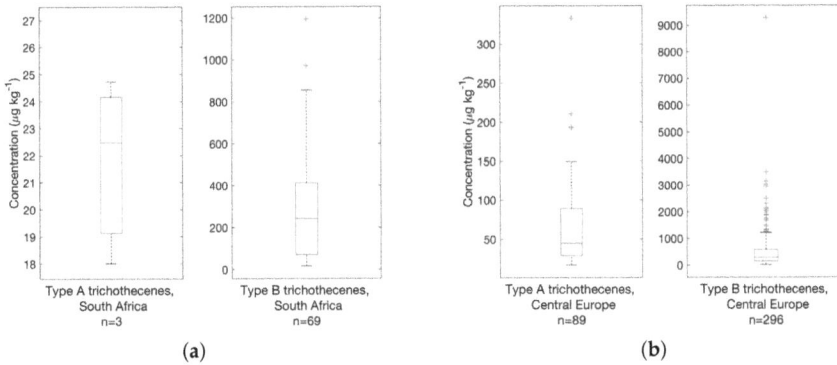

(a)   (b)

Figure 8. Survey results for type A and type B trichotheces from (**a**) South Africa and (**b**) Central Europe in finished feed samples above threshold concentrations; see Table 1 for details.

4. Conclusions

The data presented focused on the analysis of 57 mycotoxins and secondary metabolites from regulated, masked and emerging compounds in 1113 samples from three different matrices (i.e., finished feed, maize and maize silage) for the years 2012–2015 obtained as part of a global survey. The single LC-MS/MS analysis method used makes the data well comparable, while providing similar sensitivity for the compounds investigated.

The majority of samples showed low concentrations (i.e., generally below established guidelines for animal feed), however, there were usually a number of samples with high to very high concentrations for all regulated toxins and toxins with guidance levels considered. This highlights the importance of global surveys to maintain concentration levels of commodities below regulatory limits and guidance levels. Observed concentrations are also highly relevant for animal health. For the regulated AFLA, a few individual samples from Africa and Europe showed concentrations exceeding the 20 µg·kg^{-1} limit. Among the toxins with guidance levels DON, ZEN and to a certain extent FUM remain of global concern with highest concentrations observed, e.g., ZEN and DON concentrations in Central Europe were $>t$ in 80% of investigated finished feed samples, i.e., very few uncontaminated samples were observed.

This strongly indicates, that—similar to recently published reports, e.g., by [57]—considerably more than the FAO suggested figure of 25% of global agricultural commodities are contaminated with mycotoxins. In fact $>80\%$ of the agricultural commodities could be affected [57,85,86]; in the presented study contamination varied between e.g., 7.1 to 79% for B Trichothecenes and 88% for ZEN. The data of the present study serves as a starting point for a more detailed investigation of contamination rates on a global scale, by also including data and observations from other surveys (see e.g., Table 2). There are multiple issues to consider in an attempt to compare contamination rates. Any increases are in part due to improved analytical methodology and lower limits of detection. Furthermore, the number of samples, their geographical distribution, overall crop yields play a role in the calculation of an updated figure. On the other hand, concentrations at chronic exposure levels having adverse effects on the target population have also changed as new data became available.

Other concentration levels, including most of those observed here, e.g., for DON were well below the maximum allowable concentration guidelines.

However, highly sensitive methods provide critical tools to study the adverse impact of low level concentrations of mycotoxins in feed. Results are, therefore, relevant for animal health studies, especially, when synergistic and antagonistic effects of regulated, masked and emerging toxins and secondary metabolites are investigated.

The years 2014 and 2015 showed large increases of yearly median concentrations in Europe. In some regions such as South America, South Africa and the Middle East, FUM played a more significant role with regard to median concentrations than in other regions. Concentrations of OTA play a secondary role with much lower median and maximum concentrations. It has to be noted that even within a fairly small geographic area, e.g., Austria and Italy or Germany and The Netherlands within the Central European region, quite distinct trends and concentration ranges were observed for regulated toxins and toxins with guidance levels. Other parameters such as local meteorological conditions and varieties used need to be considered in order to explain these differences.

The absence of type A trichothecenes from South African samples has to be noted, while the occurrence of type B trichothecenes was quite similar to European samples regarding occurrence and concentrations observed. In general, type A trichothecene concentrations, also exemplified by the sum of T-2 and HT-2 toxin concentrations, were an order of magnitude lower than type B trichothecene concentrations in the very same samples.

Emerging toxins showed high occurrence for enniatins, MON and BEA (e.g., in South Africa), whereas AFLA precursors and ergot alkaloids were much less prevalent with the large majority of samples (up to 90%) showing concentrations of <1 µg·kg^{-1}, if any, i.e., staying below the LOD. In 2014, enniatin concentrations in finished feed were exceptionally high in European samples with an median of 250 µg·kg^{-1}, thus reaching median concentrations in the order of magnitude of DON and ZEN.

While co-occurrence of regulated toxins and toxins with guidance levels with other investigated compounds was frequent and wide-spread for e.g., enniatins, BEA and MON, correlation was limited to relatively few cases, e.g., DON and ZEN with MON in maize samples from South America. An exception was correlation of DON and ZEN with their masked metabolites DON-3-glucoside and ZEN-14-sulfate, which was observed frequently, but not in all cases. AFLA and AFLA precursors were well correlated in samples from Africa.

5. Materials and Methods

A total of 1926 samples from 46 countries with concentrations of 380 toxins and metabolites were collected for the years 2012 to 2015. Samples were provided by the Biomin Mycotoxin Survey and analysis was carried out employing a LC–MS/MS multi-mycotoxin analysis method for determination of toxins and metabolites. Of the toxins and metabolites investigated, 162 compounds were detected and quantified. For the presented study data for three different matrices (finished feed, FF; maize, M and maize silage, MSI) and 57 toxins and metabolites were chosen for detailed analysis, resulting in a subset of 1113 samples; see Table 1. These included regulated toxins and compounds with guidance levels, frequently reported in previous surveys, e.g., [57]. Concentrations of masked and emerging toxins and secondary metabolites using a single method of analysis for all mycotoxins and secondary metabolites that were not previously available in a global data set were reported with a focus on co-occurrence with regulated toxins and toxins with guidance levels.

A threshold (*t*) of relevant concentrations was established to be >1.0 µg·kg^{-1} or the LOD, whichever was higher. For compound groups (e.g., AFLA) the highest LOD in the group was employed [63]. Threshold levels are listed in Table 1. Extracted ion chromatograms of maize samples spiked near the t-value are shown in the supporting information. All plots show the fraction above the threshold unless otherwise indicated.

Figure 9 and Table 5 provide information about sample origin and frequency for each region and country and number of samples analysed in each subset, respectively together with some initial concentration information.

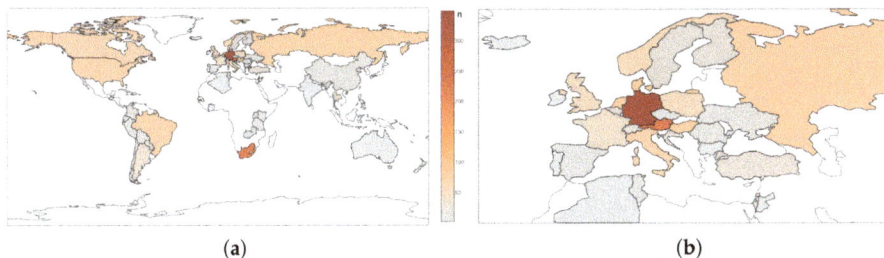

Figure 9. Number of samples n in the investigated data set from a (**a**) Global and (**b**) European perspective.

Table 5. (a) Maximum (Max.) and average (Av.) number (no.) of compounds found in samples and defined subsets; (b) Number of samples in region and country subsets. Unless explicitly noted, only subsets with 40 or more samples (in bold in the table) were employed for detailed analysis in order to ensure representativity. Statistical data available in Table 1; (c) Countries assigned to regions used for analysis of data. Data for regions in italics were not reported due to a too small sample size (with the exception of Africa).

(a) All 1926 Samples,	Number of Metabolites All Matrices		
380 compounds measured, 162 quantified			
Max. no. of compounds per sample	68		
Compounds in samples, conc. >1 $\mu g \cdot kg^{-1}$	59		
Av. no. of compounds in all samples	28		
Av. no. of compounds, conc. >1 $\mu g \cdot kg^{-1}$	24		

Subset 1113 samples,	Number of metabolites		
	Finished feed	**Maize**	**Maize silage**
57 metabolites quantified			
Max. no. of compounds per sample	35	29	28
Compounds in samples, conc. >1 $\mu g \cdot kg^{-1}$	31	26	20
Av. no. of compounds in all samples	16	12	11
Av. no. of compounds, conc. >1 $\mu g \cdot kg^{-1}$	13	10	9

(b)	Number of Samples		
	Finished Feed	**Maize**	**Maize Silage**
Region subsets			
All samples	708	267	138
Africa	24	7	1
South Africa	**74**	**53**	**28**
Central Europe	**335**	**76**	**78**
Eastern Europe	**45**	9	1
Northern Europe	**68**	4	12
Southern Europe	**90**	11	2
Middle East	23	0	0
North America	27	30	15
South America	22	**77**	1
Country subsets			
Austria	**64**	18	26
Germany	**89**	32	38
Hungary	**67**	2	3
Italy	**53**	2	2
The Netherlands	**40**	7	9

Table 5. *Cont.*

(c) Countries in region (total of 46)
Africa: Algeria, Ivory Coast, Kenya, Senegal, Tunisia, Tanzania, Uganda, Zambia
South Africa: South Africa
Central Europe: Austria, Belgium, Czech Republic, France, Germany, Hungary, The Netherlands, Poland, Romania, Switzerland
Eastern Europe: Bulgaria, Russia, Ukraine
Northern Europe: Denmark, Finland, Iceland, Ireland, Norway, Sweden, United Kingdom
Southern Europe: Croatia, Italy, Spain, Portugal, Turkey
Middle East: Israel, Jordan
North America: United States, Canada
South America: Argentina, Brazil, Bolivia, Chile, Colombia, Ecuador, Paraguay, Peru

Samples were taken by or under the instruction of trained staff with a protocol for the taking of stratified subsamples, homogenizing and then a minimum 500 g sample being submitted to laboratory. In the laboratory, samples were finely milled and homogenized immediately after reception at the Department of Agrobiotechnology (IFA-Tulln) at the University of Natural Resources and Life Sciences Vienna (BOKU) in Tulln, Austria. Analysis was carried out right after milling and homogenisation. Finished feed (complete feed) for poultry, swine and ruminants were sampled. Composition would be variable both between and within the different feed types.

Samples were analysed using a single "dilute and shoot" LC–MS/MS multi-mycotoxin method previously reported by Malachová et al. [87]. In brief, samples were ground for homogenisation and extracted for 90 min with a mixture of acetonitrile, water and acetic acid (79:20:1, per volume) on a rotary shaker. After centrifugation, the supernatant was transferred into a glass vial, following dilution with solvent (as above, but with a 20:79:1 volume ratio). The extract was injected into a LC–MS/MS system (electrospray ionisation and mass spectrometric detection employing a quadrupole/ion-trap combination mass filter). All samples were analysed using this instrument. Quantification was performed based on external calibration using serial dilutions of a multi-analyte stock solution. Results were corrected for apparent recoveries, which have been determined by spiking experiments using 9 different types of feed (unpublished data). For maize, the respective values obtained in [Malachova2014] were used. Regarding identification and quantitation, the acquisition of two MS/MS transition yields 4.0 identification points according to Commission Decision 2002/657/EC (with the exception of moniliformin and 3-nitropropionic acid, which exhibit only one fragment ion). In addition, the intensity ratios as well as retention times were not to deviate from the respective standards within certain limits, as stated in the same document. However, a fixed ion ratio criteria of 30% and a very strict retention time criteria of ±0.03 min has been applied following a recent suggestion in the field of veterinary drug analysis [88].

The method follows the guidelines established by the Directorate General for Health and Consumer affairs of the European Commission, published in document No 12495/2011 [89]. Apparent recoveries for feed have been determined by spiking experiments using 9 different types of feed (unpublished data). The external verification of the method accuracy by participation in proficiency testing schemes resulted in 104 acceptable, 6 questionable and 2 unsatisfactory results, respectively, for samples of animal feed (results until March 2016 included). While the method has continually evolved over time with more and more species being added, all analytes investigated here, were analysed in the full four year period being considered.

The use of this single unique multi-mycotoxin method provides comparable data for a large number of compounds quantified in samples of global origin. Thus some of the issues raised with previous surveys were overcome that typically employed data from several analytical methods and different analysis protocols (e.g., from several national monitoring agencies, see Table 2).

All concentration data were collected in a single file and sample information such as sampling year and month, country and region of origin and sample matrix were added for subsetting. The complete data set was then imported into Matlab (version R2015b) for subsetting and further statistical analysis.

Wilcoxon and Kruskal–Wallis tests of data were completed in R (version 3.3.1) using the "agricolae" package (version 1.2–4). For the latter, a confidence level (CL) of 95% was chosen; if it cold not be attained, a CL of 90% was set instead.

Author Contributions: P.K., K.N. and C.S. conceived and designed the study; G.S. and R.K. were responsible for the measurement programme implementation; M.S. performed the LC–MS/MS experiments; G.K. analyzed the data with input from T.J.; G.K. wrote the paper.

Conflicts of Interest: The authors declare no conflict of interest. The founding sponsors had no role in the design of the study; in the collection, analyses, or interpretation of data; in the writing of the manuscript, and in the decision to publish the results.

Abbreviations

The following abbreviations are used in this manuscript:

3-Ac-DON	3-acetyl-deoxynivalenol
A & B	type A & B trichothecenes
AF	aflatoxin
AFB1	aflatoxin B1
AFLA	aflatoxins
BEA	beauvericin
CL	confidence level
DAS	diacetoxyscirpenol
DDGS	dried distillers grain with solubles
DON	deoxynivalenol
Elisa	enzyme linked immunosorbent assay
EU	European Union
FAO	Food and Agriculture Organization of the United Nations
FF	finished feed
FLD	fluorescence detector
FUM	fumonisins
FUS	toxins produced by *Fusarium* spp.
HPLC	high performance liquid chromatography
LC–MS/MS	liquid chromatography tandem mass spectrometry
LOD	limit of detection
M	maize
MON	moniliformine
MS	mass spectrometry
MSI	maize silage
NIV	nivalenol
OTA	ochratoxin A
REG	toxins and secondary metabolites (regulated or with guidance levels)
SBM	soybean meal
TLC	thin layer chromatography
ZEN	zearalenone

References

1. Smith, J.E.; Solomons, G.; Lewis, C.; Anderson, J.G. Role of mycotoxins in human and animal nutrition and health. *Nat. Toxins* **1995**, *3*, 187–192.
2. Wild, C.P.; Gong, Y.Y. Mycotoxins and human disease: A largely ignored global health issue. *Carcinogenesis* **2010**, *31*, 71–82.
3. Sweeney, M.J.; Dobson, A.D. Mycotoxin production by *Aspergillus*, *Fusarium* and *Penicillium* species. *Int. J. Food Microbiol.* **1998**, *43*, 141–158.
4. Rivas Casado, M.; Parsons, D.J.; Weightman, R.M.; Magan, N.; Origgi, S. Modelling a two-dimensional spatial distribution of mycotoxin concentration in bulk commodities to design effective and efficient sample selection strategies. *Food Addit. Contam.* **2009**, *26*, 1298–1305.
5. European Feed Manufacturers' Federation (FEFAC). *FEFAC Annual Report 2014–2015*; FEFAC: Brussels, Belgium, 2015.

6. Pinotti, L.; Ottoboni, M.; Giromini, C.; DellOrto, V.; Cheli, F. Mycotoxin contamination in the EU feed supply chain: A focus on cereal byproducts. *Toxins* **2016**, *8*, 45.

7. Centre, I.T. Trade Map—Trade Statistics for International Business Development. Available online: http://www.trademap.org/ (accessed on 28 November 2016).

8. Hussein, H.S.; Brasel, J.M. Toxicity, metabolism, and impact of mycotoxins on humans and animals. *Toxicology* **2001**, *167*, 101–134.

9. Mathison, G.W. *The Toxicology of Aflatoxins: Human Health, Veterinary, and Agricultural Significance*; Academic Press, Inc.: New York, NY, USA, 2013; Volume 68, pp. 368–369.

10. Carnaghan, R.B.A.; Hartley, R.D.; O'Kelly, J. Toxicity and fluorescence properties of the aflatoxins. *Nature* **1963**, *4911*, 1101.

11. Kumar, V.; Basu, M.S.; Rajendran, T.P. Mycotoxin research and mycoflora in some commercially important agricultural commodities. *Crop Prot.* **2008**, *27*, 891–905.

12. Langseth, W.; Rundberget, T. The occurrence of HT-2 toxin and other trichothecenes in Norwegian cereals. *Mycopathologia* **1999**, *147*, 157–165.

13. Eriksen, G.S.; Pettersson, H. Toxicological evaluation of trichothecenes in animal feed. *Anim. Feed Sci. Technol.* **2004**, *114*, 205–239.

14. Pittet, A. Natural occurrence of mycotoxins in foods and feeds: An update review. *Revue Med. Vet.* **1998**, *6*, 479–492

15. Hochsteiner, W.; Schuh, M. Occurrence of the fusariotoxins deoxynivalenol and zearalenone in Austrian feedstuff in the period from 1995 to 1999. *DTW Deutsche Tierärztliche Wochenschrift* **2001**, *108*, 19–23.

16. Awad, W.; Ghareeb, K.; Böhm, J.; Zentek, J. The toxicological impacts of the Fusarium mycotoxin, deoxynivalenol, in poultry flocks with special reference to immunotoxicity. *Toxins* **2013**, *5*, 912–925.

17. Rafai, P.; Tuboly, S.; Bata, A.; Tilly, P.; Vanyi, A.; Papp, Z.; Jakab, L.; Tury, E. Effect of various levels of T-2 toxin in the immune system of growing pigs. *Vet. Rec.* **1995**, *136*, 511–514.

18. Meissonnier, G.M.; Laffitte, J.; Raymond, I.; Benoit, E.; Cossalter, A.M.; Pinton, P.; Bertin, G.; Oswald, I.P.; Galtier, P. Subclinical doses of T-2 toxin impair acquired immune response and liver cytochrome P450 in pigs. *Toxicology* **2008**, *247*, 46–54.

19. Kubena, L.F.; Edrington, T.S.; Harvey, R.B.; Buckley, S.A.; Phillips, T.D.; Rottinghaus, G.E.; Casper, H.H. Individual and combined effects of fumonisin B1 present in Fusarium moniliforme culture material and T-2 toxin or deoxynivalenol in broiler chicks. *Poult. Sci.* **1997**, *76*, 1239–1247.

20. Grenier, B.; Oswald, I. Mycotoxin co-contamination of food and feed: Meta-analysis of publications describing toxicological interactions. *World Mycotoxin J.* **2011**, *4*, 285–313.

21. Van der Fels-Klerx, H.; Stratakou, I. T-2 toxin and HT-2 toxin in grain and grain-based commodities in Europe: Occurrence, factors affecting occurrence, co-occurrence and toxicological effects. *World Mycotoxin J.* **2010**, *3*, 349–367.

22. Colvin, B.M.; Harrison, L.R. Fumonisin-induced pulmonary edema and hydrothorax in swine. *Mycopathologia* **1992**, *117*, 79–82.

23. Antonissen, G.; Van Immerseel, F.; Pasmans, F.; Ducatelle, R.; Janssens, G.P.J.; De Baere, S.; Mountzouris, K.C.; Su, S.; Wong, E.A.; De Meulenaer, B.; et al. Mycotoxins deoxynivalenol and fumonisins alter the extrinsic component of intestinal barrier in broiler chickens. *J. Agric. Food Chem.* **2015**, *63*, 10846–10855.

24. Pfohl-Leszkowicz, A.; Manderville, R.A. Ochratoxin A: An overview on toxicity and carcinogenicity in animals and humans. *Mol. Nutr. Food Res.* **2007**, *51*, 61–99.

25. Page, R.K.; Stewart, G.; Wyatt, R.; Bush, P.; Fletcher, O.J.; Brown, J. Influence of low levels of ochratoxin A on egg production, egg-shell stains, and serum uric-acid levels in Leghorn-type hens. *Avian Dis.* **1980**, *24*, 777–780.

26. Duarte, S.C.; Lino, C.M.; Pena, A. Ochratoxin A in feed of food-producing animals: An undesirable mycotoxin with health and performance effects. *Vet. Microbiol.* **2011**, *154*, 1–13.

27. Casteel, S.W.; Turk, J.R.; Cowart, R.P.; Rottinghaus, G.E. Chronic toxicity of fumonisin in weanling pigs. *J. Vet. Diagn. Investig.* **1993**, *5*, 413–417.

28. Zinedine, A.; Soriano, J.M.; Moltó, J.C.; Mañes, J. Review on the toxicity, occurrence, metabolism, detoxification, regulations and intake of zearalenone: An oestrogenic mycotoxin. *Food Chem. Toxicol.* **2007**, *45*, 1–18.

29. Rychlik, M.; Humpf, H.U.; Marko, D.; Dänicke, S.; Mally, A.; Berthiller, F.; Klaffke, H.; Lorenz, N. Proposal of a comprehensive definition of modified and other forms of mycotoxins including masked mycotoxins. *Mycotoxin Res.* **2014**, *30*, 197–205.

30. Gareis, M.; Bauer, J.; Thiem, J.; Plank, G.; Grabley, S.; Gedek, B. Cleavage of zearalenone-glycoside, a masked mycotoxin, during digestion in swine. *J. Vet. Med. Ser. B* **1990**, *37*, 236–240.

31. Berthiller, F.; Crews, C.; Dall'Asta, C.; Saeger, S.D.; Haesaert, G.; Karlovsky, P.; Oswald, I.P.; Seefelder, W.; Speijers, G.; Stroka, J. Masked mycotoxins: A review. *Mol. Nutr. Food Res.* **2013**, *57*, 165–186.

32. Berthiller, F.; Dall'Asta, C.; Schuhmacher, R.; Lemmens, M.; Adam, G.; Krska, R. Masked mycotoxins: Determination of a deoxynivalenol glucoside in artificially and naturally contaminated wheat by liquid chromatography-tandem mass spectrometry. *J. Agric. Food Chem.* **2005**, *53*, 3421–3425.

33. Lancova, K.; Hajslova, J.; Poustka, J.; Krplova, A.; Zachariasova, M.; Dostálek, P.; Sachambula, L. Transfer of Fusarium mycotoxins and masked deoxynivalenol (deoxynivalenol-3-glucoside) from field barley through malt to beer. *Food Addit. Contam.* **2008**, *25*, 732–744.

34. Dall'Erta, A.; Cirlini, M.; Dall'Asta, M.; Del Rio, D.; Galaverna, G.; Dall'Asta, C. Masked mycotoxins are efficiently hydrolyzed by human colonic microbiota releasing their aglycones. *Chem. Res. Toxicol.* **2013**, *26*, 305–312.

35. Plasencia, J.; Mirocha, C.J. Isolation and characterization of zearalenone sulfate produced by *Fusarium* spp. *Appl. Environ. Microbiol.* **1991**, *57*, 146–150.

36. Krska, R. How does climate change impact on the occurrence and the determination of natural toxins. In Proceedings of the 7th International Symposium on Recent Advances in Food Analysis, Prague, Czech, 3–6 November 2015.

37. Vaclavikova, M.; Malachova, A.; Veprikova, Z.; Dzuman, Z.; Zachariasova, M.; Hajslova, J. 'Emerging' mycotoxins in cereals processing chains: Changes of enniatins during beer and bread making. *Food Chem.* **2013**, *136*, 750–757.

38. Jestoi, M. Emerging Fusarium-mycotoxins fusaproliferin, beauvericin, enniatins, and moniliformin—A review. *Crit. Rev. Food Sci. Nutr.* **2008**, *48*, 21–49.

39. Serrano, A.; Font, G.; Ruiz, M.; Ferrer, E. Co-occurrence and risk assessment of mycotoxins in food and diet from Mediterranean area. *Food Chem.* **2012**, *135*, 423–429.

40. Streit, E.; Schatzmayr, G.; Tassis, P.; Tzika, E.; Marin, D.; Taranu, I.; Tabuc, C.; Nicolau, A.; Aprodu, I.; Puel, O. Current situation of mycotoxin contamination and co-occurrence in animal feed—Focus on Europe. *Toxins* **2012**, *4*, 788–809.

41. Streit, E.; Schwab, C.; Sulyok, M.; Naehrer, K.; Krska, R.; Schatzmayr, G. Multi-mycotoxin screening reveals the occurrence of 139 different secondary metabolites in feed and feed ingredients. *Toxins* **2013**, *5*, 504–523.

42. Malachova, A.; Dzuman, Z.; Veprikova, Z.; Vaclavikova, M.; Zachariasova, M.; Hajslova, J. Deoxynivalenol, deoxynivalenol-3-glucoside, and enniatins: The major mycotoxins found in cereal-based products on the Czech market. *J. Agric. Food Chem.* **2011**, *59*, 12990–12997.

43. Berthiller, F.; Schuhmacher, R.; Adam, G.; Krska, R. Formation, determination and significance of masked and other conjugated mycotoxins. *Anal. Bioanal. Chem.* **2009**, *395*, 1243–1252.

44. D'Mello, J.; Placinta, C.; Macdonald, A. Fusarium mycotoxins: a review of global implications for animal health, welfare and productivity. *Anim. Feed Sci. Technol.* **1999**, *80*, 183–205.

45. Harvey, R.; Edrington, T.; Kubena, L.; Elissalde, M.; Casper, H.; Rottinghaus, G.; Turk, J. Effects of dietary fumonisin B1-containing culture material, deoxynivalenol-contaminated wheat, or their combination on growing barrows. *Am. J. Vet. Res.* **1996**, *57*, 1790–1794.

46. Grenier, B.; Bracarense, A.P.F.; Schwartz, H.E.; Lucioli, J.; Cossalter, A.M.; Moll, W.D.; Schatzmayr, G.; Oswald, I.P. Biotransformation approaches to alleviate the effects induced by fusarium mycotoxins in swine. *J. Agric. Food Chem.* **2013**, *61*, 6711–6719.

47. Antonissen, G.; Croubels, S.; Pasmans, F.; Ducatelle, R.; Eeckhaut, V.; Devreese, M.; Verlinden, M.; Haesebrouck, F.; Eeckhout, M.; De Saeger, S.; et al. Fumonisins affect the intestinal microbial homeostasis in broiler chickens, predisposing to necrotic enteritis. *Vet. Res.* **2015**, *46*, 98.

48. Tanaka, T.; Yamamoto, S.; Hasegawa, A.; Aoki, N.; Besling, J.R.; Sugiura, Y.; Ueno, Y. A survey of the natural occurrence of *Fusarium* mycotoxins, deoxynivalenol, nivalenol and zearalenone, in cereals harvested in the Netherlands. *Mycopathologia* **1990**, *110*, 19–22.

49. Placinta, C.; D'Mello, J.; Macdonald, A. A review of worldwide contamination of cereal grains and animal feed with *Fusarium* mycotoxins. *Anim. Feed Sci. Technol.* **1999**, *78*, 21–37.

50. Ali, N.; Sardjono; Yamashita, A.; Yoshizawa, T. Natural co-occurrence of aflatoxins and Fusarium mycotoxins (fumonisins, deoxynivalenol, nivalenol and zearalenone) in corn from Indonesia. *Food Addit. Contam.* **1998**, *15*, 377–384.

51. Commission, E. Commission Regulation (EU) No 574/2011 of 16 June 2011 amending Annex I to Directive 2002/32/EC of the European Parliament and of the Council as regards maximum levels for nitrite, melamine, *Ambrosia* spp. and carry-over of certain coccidiostats and histomonostats and consolidating Annexes I and II thereto. *Off. J. Eur. Union* **2011**, *L159*, 7–24.

52. Food and Agriculture Organization (FAO). *Worldwide Regulations for Mycotoxins*; Food and Agriculture Organization of the United Nations Food and Nutrition Paper; FAO: Rome, Italy, 1997.

53. Commission, E. Commission Recommendation of 17 August 2006 on the presence of deoxynivalenol, zearalenone, ochratoxin A, T-2 and HT-2 and fumonisins in products intended for animal feeding. *Off. J. Eur. Union* **2006**, *L229*, 7–9.

54. Paterson, R.R.M.; Lima, N. How will climate change affect mycotoxins in food? *Food Res. Int.* **2010**, *43*, 1902–1914.

55. Magan, N.; Medina, A.; Aldred, D. Possible climate-change effects on mycotoxin contamination of food crops pre- and postharvest. *Plant Pathol.* **2011**, *60*, 150–163.

56. Murugesan, G.R.; Ledoux, D.R.; Naehrer, K.; Berthiller, F.; Applegate, T.J.; Grenier, B.; Phillips, T.D.; Schatzmayr, G. Prevalence and effects of mycotoxins on poultry health and performance, and recent development in mycotoxin counteracting strategies. *Poult. Sci.* **2015**, *94*, 1298–1315.

57. Streit, E.; Naehrer, K.; Rodrigues, I.; Schatzmayr, G. Mycotoxin occurrence in feed and feed raw materials worldwide: Long-term analysis with special focus on Europe and Asia. *J. Sci. Food Agric.* **2013**, *93*, 2892–2899.

58. Rodrigues, I.; Naehrer, K. A three-year survey on the worldwide occurrence of mycotoxins in feedstuffs and feed. *Toxins* **2012**, *4*, 663–675.

59. Sharman, M.; Gilbert, J.; Chelkowski, J. A survey of the occurrence of the mycotoxin moniliformin in cereal samples from sources worldwide. *Food Addit. Contam.* **1991**, *8*, 459–466.

60. Curtui, V.G.; Gareis, M.; Usleber, E.; Märtlbauer, E. Survey of Romanian slaughtered pigs for the occurrence of mycotoxins ochratoxins A and B, and zearalenone. *Food Addit. Contam.* **2001**, *18*, 730–738.

61. Resnik, S.; Neira, S.; Pacin, A.; Martinez, E.; Apro, N.; Latreite, S. A survey of the natural occurrence of aflatoxins and zearalenone in Argentine field maize: 1983–1994. *Food Addit. Contam.* **1996**, *13*, 115–120.

62. Schatzmayr, G.; Streit, E. Global occurrence of mycotoxins in the food and feed chain: Facts and figures. *World Mycotoxin J.* **2013**, *6*, 213–222.

63. Van Der Fels-Klerx, H.J.; Klemsdal, S.; Hietaniemi, V.; Lindblad, M.; Ioannou-Kakouri, E.; Van Asselt, E.D. Mycotoxin contamination of cereal grain commodities in relation to climate in North West Europe. *Food Addit. Contam. A* **2012**, *29*, 1581–1592.

64. Rodrigues, I.; Handl, J.; Binder, E.M. Mycotoxin occurrence in commodities, feeds and feed ingredients sourced in the Middle East and Africa. *Food Addit. Contam. B* **2011**, *4*, 168–179.

65. Vanheule, A.; Audenaert, K.; De Boevre, M.; Landschoot, S.; Bekaert, B.; Munaut, F.; Eeckhout, M.; Höfte, M.; De Saeger, S.; Haesaert, G. The compositional mosaic of Fusarium species and their mycotoxins in unprocessed cereals, food and feed products in Belgium. *Int. J. Food Microbiol.* **2014**, *181*, 28–36.

66. Han, R.W.; Zheng, N.; Wang, J.Q.; Zhen, Y.P.; Xu, X.M.; Li, S.L. Survey of aflatoxin in dairy cow feed and raw milk in China. *Food Control* **2013**, *34*, 35–39.

67. Van Asselt, E.D.; Azambuja, W.; Moretti, A.; Kastelein, P.; De Rijk, T.C.; Stratakou, I.; Van Der Fels-Klerx, H.J. A Dutch field survey on fungal infection and mycotoxin concentrations in maize. *Food Addit. Contam. A* **2012**, *29*, 1556–1565.

68. Anjum, M.A.; Khan, S.H.; Sahota, A.W.; Sardar, R. Assessment of aflatoxin B1 in commercial poultry feed and feed ingredients. *J. Anim. Plant Sci.* **2012**, *22*, 268–272.

69. Zhang, Y.; Caupert, J. Survey of mycotoxins in US distiller's dried grains with solubles from 2009 to 2011. *J. Agric. Food Chem.* **2012**, *60*, 539–543.

70. Martins, H.M.; Almeida, I.; Camacho, C.; Costa, J.M.; Bernardo, F. A survey on the occurrence of ochratoxin A in feeds for swine and laying hens. *Mycotoxin Res.* **2012**, *28*, 107–110.

71. Del Pilar Monge, M.; Magnoli, C.E.; Chiacchiera, S.M. Survey of Aspergillus and Fusarium species and their mycotoxins in raw materials and poultry feeds from Córdoba, Argentina. *Mycotoxin Res.* **2012**, *28*, 111–122.

72. Njobeh, P.B.; Dutton, M.F.; Åberg, A.T.; Haggblom, P. Estimation of multi-mycotoxin contamination in South African compound feeds. *Toxins* **2012**, *4*, 836–848.

73. Tabuc, C.; Taranu, I.; Calin, L. Survey of mould and mycotoxin contamination of cereals in South-Eastern Romania in 2008–2010. *Archiva Zootechnica* **2011**, *14*, 25.

74. Skrbić, B.; Malachova, A.; Zivancev, J.; Veprikova, Z.; Hajslová, J. Fusarium mycotoxins in wheat samples harvested in Serbia: A preliminary survey. *Food Control* **2011**, *22*, 1261–1267.

75. Guan, S.; Gong, M.; Yin, Y.; Huang, R.; Ruan, Z.; Zhou, T.; Xie, M. Occurrence of mycotoxins in feeds and feed ingredients in China. *J. Food Agric. Environ.* **2011**, *9*, 163–167.

76. Botana, L.M.; Sainz, M.J. *Climate Change and Mycotoxins*; Walter de Gruyter GmbH & Co KG: Berlin, Germany; Boston, MA, USA, 2015.

77. McGill, R.; Tukey, J.W.; Larsen, W.A. Variations of box plots. *Am. Stat.* **1978**, *32*, 12–16.

78. Tanaka, T.; Hasegawa, A.; Yamamoto, S.; Lee, U.S.; Sugiura, Y.; Ueno, Y. Worldwide contamination of cereals by the Fusarium mycotoxins nivalenol, deoxynivalenol, and zearalenone. 1. Survey of 19 countries. *J. Agric. Food Chem.* **1988**, *36*, 979–983.

79. Berthiller, F.; DallAsta, C.; Corradini, R.; Marchelli, R.; Sulyok, M.; Krska, R.; Adam, G.; Schuhmacher, R. Occurrence of deoxynivalenol and its 3-beta-D-glucoside in wheat and maize. *Food Addit. Contam.* **2009**, *26*, 507–511.

80. Pena, T.; Lara, P.; Castrodeza, C. Multiobjective stochastic programming for feed formulation. *J. Oper. Res. Soc.* **2009**, *60*, 1738–1748.

81. Van Egmond, H.P.; Schothorst, R.C.; Jonker, M.A. Regulations relating to mycotoxins in food. *Anal. Bioanal. Chem.* **2007**, *389*, 147–157.

82. Versilovskis, A.; Bartkevics, V.; Mickelsone, V. Sterigmatocystin presence in typical Latvian grains. *Food Chem.* **2008**, *109*, 243–248.

83. Kosiak, B.; Torp, M.; Skjerve, E.; Thrane, U. The prevalence and distribution of fusarium species in norwegian cereals: A Survey. *Acta Agric. Scand. Sect. B Soil Plant Sci.* **2003**, *53*, 168–176.

84. Sydenham, E.W.; Thiel, P.G.; Marasas, W.F.; Shephard, G.S.; Van Schalkwyk, D.J.; Koch, K.R. Natural occurrence of some Fusarium mycotoxins in corn from low and high esophageal cancer prevalence areas of the Transkei, Southern Africa. *J. Agric. Food Chem.* **1990**, *38*, 1900–1903.

85. Kabak, B.; Dobson, A.D.; Var, I. Strategies to prevent mycotoxin contamination of food and animal feed: A review. *Crit. Rev. Food Sci. Nutr.* **2006**, *46*, 593–619.

86. Naehrer, K.; Kovalsky, P. The Biomin Mycotoxin Survey, Identifying the threats in 2013. *Mycotoxins Science and Solutions*; April 2014; pp. 2–7. Available online: http://www.biomin.net/uploads/tx_news/MAG_SciSol_Special_MTX_EN_0414.pdf (accessed on 5 December 2016)

87. Malachová, A.; Sulyok, M.; Beltrán, E.; Berthiller, F.; Krska, R. Optimization and validation of a quantitative liquid chromatography–tandem mass spectrometric method covering 295 bacterial and fungal metabolites including all regulated mycotoxins in four model food matrices. *J. Chromatogr. A* **2014**, *1362*, 145–156.

88. Berendsen, B.; Meijer, T.; Wegh, R.; Mol, H.; Smyth, W.; Armstrong Hewitt, S.; van Ginkel, L.; Nielen, M. A critical assessment of the performance criteria in confirmatory analysis for veterinary drug residue analysis using mass spectrometric detection in selected reaction monitoring mode. *Drug Test. Anal.* **2016**, *8*, 477–490.

89. Pihlström, T. *Method Validation and Quality Control Procedures for Pesticide Residues Analysis in Food and Feed*; DG SANCO/12495/2011; European Commission Directorate-General for Health and Food Safety: Brussels, Belgium, 2011; pp. 1–41.

© 2016 by the authors. Licensee MDPI, Basel, Switzerland. This article is an open access article distributed under the terms and conditions of the Creative Commons Attribution (CC BY) license (http://creativecommons.org/licenses/by/4.0/).

toxins

MDPI

Article

Spotlight on the Underdogs—An Analysis of Underrepresented *Alternaria* Mycotoxins Formed Depending on Varying Substrate, Time and Temperature Conditions

Theresa Zwickel [1,3,*], Sandra M. Kahl [2,4], Horst Klaffke [1], Michael Rychlik [3] and Marina E. H. Müller [2,*]

[1]   Federal Institute for Risk Assessment (BfR), Max-Dohrn-Str 8-10, Berlin 10589, Germany; horst.klaffke@bfr.bund.de

[2]   Leibniz-Centre for Agricultural Landscape Research (ZALF), Institute of Landscape Biogeochemistry, Eberswalder Str. 84, Müncheberg 15374, Germany; sakahl@uni-potsdam.de

[3]   Technische Universität München, Chair of Analytical Food Chemistry, Alte Akademie 10, Freising 85354, Germany; michael.rychlik@tum.de

[4]   University of Potsdam, Maulbeerallee 1, Potsdam 14469, Germany

*   Correspondence: theresa.zwickel@bfr.bund.de (T.Z.); mmueller@zalf.de (M.E.H.M.); Tel.: +49-30-184120 (T.Z.); +49-33432-82420 (M.E.H.M.)

Academic Editor: Aldo Laganà

Received: 15 October 2016; Accepted: 13 November 2016; Published: 19 November 2016

Abstract: *Alternaria (A.)* is a genus of widespread fungi capable of producing numerous, possibly health-endangering *Alternaria* toxins (ATs), which are usually not the focus of attention. The formation of ATs depends on the species and complex interactions of various environmental factors and is not fully understood. In this study the influence of temperature (7 °C, 25 °C), substrate (rice, wheat kernels) and incubation time (4, 7, and 14 days) on the production of thirteen ATs and three sulfoconjugated ATs by three different *Alternaria* isolates from the species groups *A. tenuissima* and *A. infectoria* was determined. High-performance liquid chromatography coupled with tandem mass spectrometry was used for quantification. Under nearly all conditions, tenuazonic acid was the most extensively produced toxin. At 25 °C and with increasing incubation time all toxins were formed in high amounts by the two *A. tenuissima* strains on both substrates with comparable mycotoxin profiles. However, for some of the toxins, stagnation or a decrease in production was observed from day 7 to 14. As opposed to the *A. tenuissima* strains, the *A. infectoria* strain only produced low amounts of ATs, but high concentrations of stemphyltoxin III. The results provide an essential insight into the quantitative in vitro AT formation under different environmental conditions, potentially transferable to different field and storage conditions.

Keywords: *Alternaria infectoria; A. tenuissima*; mycotoxin profile; wheat; rice; *Alternaria* toxin sulfates; modified *Alternaria* toxins; altertoxins; altenuic acid; HPLC-MS/MS

1. Introduction

The genus *Alternaria* contains pathogenic and saprophytic fungi with a considerable number of host plants such as ornamental and crop plants, fruits and vegetables [1,2]. Saprophytic *Alternaria* strains can be found on grains in storage included in postharvest spoilage processes or in the soil as microbiome members in the rhizosphere [3,4]. Pathogenic *Alternaria* species affect healthy plants and seeds inducing necrosis on leaves, reducing grain germination and can also cause stem cancer, leaf blight or leaf spot diseases [3,5,6].

During its growth the fungus may produce a vast number of mycotoxins, host-specific as well as non-host specific, which are associated with the infection and colonization of its plant substrates [7,8]. So far it is known that *Alternaria* strains are capable of producing over seventy secondary metabolites. However, the chemical structures of only a few of them have been elucidated. *Alternaria* toxins (ATs) differ widely in their chemical structures and therefore can be divided into five groups [9]:

- Dibenzo-α-pyrone derivatives: e.g., alternariol (AOH), alternariol mono methylether (AME), altenuene (ALT), isoaltenuene (isoALT) and altenuisol (ATL) whose originally proposed structure [10] has been recently revised [11];
- Tetramic acid derivatives: e.g., tenuazonic acid (TeA);
- Perylene quinone derivatives: e.g., altertoxin I, II (ATX-I, -II) and stemphyltoxin III (STTX-III);
- Aminopentol esters: host-specific ATs that are produced by the fungus *Alternaria alternata* f. sp. *lycopersici* and collectively known as the AAL toxins, e.g., AAL TB1 and TB2;
- Miscellaneous structures: e.g., tentoxin (TEN) a cyclic tripeptide and altenuic acid III (AA-III) a resorcylic acid substituted with butenolide and a second carboxylic acid in the side-chain [12,13].

Recently, so-called modified mycotoxins [14] such as sulfates and glucosides of AOH and AME (e.g., AOH-3-sulfate, AME-3-β,D-glucoside) were identified and synthesized [15,16].

Due to the ubiquitous occurrence of *Alternaria* these mycotoxins can be present as natural contaminants throughout the entire food and feed chain. Among the various produced mycotoxins AOH, AME, ALT, TeA and TEN are the best examined ones and can be frequently found in a broad spectrum of foodstuff commodities such as cereal products, vegetables, fruits and oil seeds [3,4,9,17,18].

The occurrence as food and feed contaminants is a cause of concern as some of the ATs are suspected to pose a serious health risk to humans and animals. The European Food Safety Authority (EFSA) classified AOH and AME as genotoxic substances. TeA, however, was classified as non-genotoxic, but due to its acute oral toxicity in mice and rats it was classified as possibly harmful. TEN is considered as non-harmful [9]. According to Frizzell et al. [19] AOH has the potential to modulate the human endocrine system by altering the hormone production and interfering with gene and receptor expression. Furthermore, AOH and AME inhibit the progesterone secretion of porcine ovarian cells in vitro, which may affect reproductive performance in livestock [20]. Studies focusing on the toxicity of other *Alternaria* mycotoxins remain scarce: Alternariol 5-*O*-sulfate demonstrated lower bioactivity than free AOH, but higher bioactivity compared to free AME in in vitro screening for cytotoxic activity [15]. Mutagenic effects of ATX-I-III and STTX-III were determined by in vitro assays [21,22]. ATX-II also demonstrated the highest toxicity against HeLa cells followed by AOH and ATL [23]. Furthermore, ATX-II has a 50-fold higher mutagenic potency compared to AOH and has been reported to induce gene mutations and DNA strand breaks in V79 cells, however it does not interfere with the cell cycle [24]. The genotoxic effects of ATX-II in human cells also far exceeded the effects of the main toxins AOH and AME [25]. A carcinogenic potential has been proven as well for the occurrence of *A. alternata* and its mycotoxins, AOH and AME, causing esophageal cancer [8,26,27]. However, ATX-II and STTX-III have only recently been considered to be more likely responsible for esophageal cancer than AOH and AME. The genotoxic potential of the perylene quinones with an epoxide group is stated to be probably caused by the formation of DNA adducts. The DNA strand breaks induced by ATX-II and STTX-III were more persistent than the ones induced by AOH [28].

The few in vitro studies on the influence of environmental conditions on the fungus growth and the mycotoxin production capability have been reviewed [4,17,29]. Toxins most frequently analyzed were AOH, AME and TeA, followed by ALT and ATX-II all of which are produced by *A. alternata* or *A. tenuissima*. The temperature and water activity (a_w) level optima for these toxins were generally around 25 to 30 °C and 0.98 [4,17,29]. An Argentinean study showed that 75% of 87 *Alternaria* strains isolated from tomato, wheat, blueberry and walnut were able to produce mycotoxins. AOH and AME were the most common metabolites followed by TEN, ATX and TeA, and several qualitatively detected further metabolites [30]. The production of five different ATs by the same *Alternaria* isolates was compared in an in vitro assay, in cabbage, cultivated rocket *(Eruca sativa)* and cauliflower. AOH, AME, ATL and TEN showed a good correlation, while TeA was only produced in higher levels in liquid culture [31]. A difference in mycotoxin production between *Alternaria* species groups could be shown as well: strains characterized as *A. infectoria* isolated from wheat can be differentiated from other *Alternaria* species groups by low production of AOH, AME, ALT, ATX-I and TeA, by the color of cultivated colonies and by molecular classification [32].

The co-occurrence of several ATs is not only reported in in vitro studies but also in food and feed samples [9,17,18]. A few HPLC-MS/MS multi-methods have been published for the simultaneous quantification of the main ATs TeA, AOH, AME, ALT, TEN [33] and underrepresented ones such as ATX-I in tomato products [34], additionally ATX-II, STTX-III, ATL, isoALT, AA-III and AAL toxins TB1 and TB2 in fruit juices, vegetable juices and wine [13], in bakery products, sunflower seeds and vegetable oils [35], further including conjugated toxins of AOH and AME in cereal products [36] and also stable isotope dilution assay for several ATs in various commercial food samples [37].

To conclude, *Alternaria* species play not only an important role in plant pathology of agronomic crops with losses of harvest but also in food and feed quality and safety. For a reliable risk assessment of ATs in agricultural and processed food products it is also crucial to know more about the influencing factors during the mycotoxin producing processes.

The present study reports an overview of the quantitative production of AOH, AME, ALT, TeA, TEN, ATX-I and of the underrepresented toxins ATX-II, STTX-III, isoALT, ATL, AA-III and AAL TB1 and TB2 (Figure 1) under combinations of different growth conditions such as temperature, incubation time and nutrient media in in vitro cultures. Additionally, individual differences in mycotoxin production of three *Alternaria* strains from two species groups are examined as well. Furthermore, screening for modified ATs was carried out.

I. dibenzo-α-pyrone derivatives

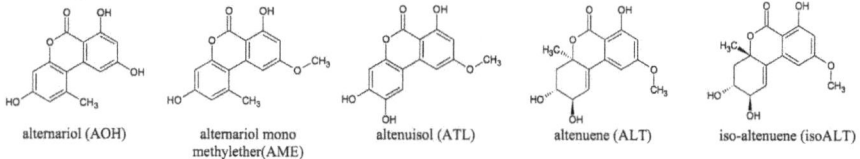

| alternariol (AOH) | alternariol mono methylether(AME) | altenuisol (ATL) | altenuene (ALT) | iso-altenuene (isoALT) |

II. tetramic acid derivatives

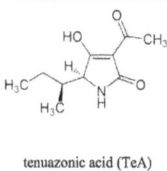

III. perylene quinone derivatives

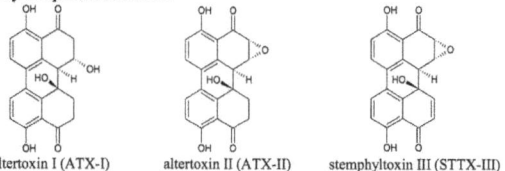

| tenuazonic acid (TeA) | altertoxin I (ATX-I) | altertoxin II (ATX-II) | stemphyltoxin III (STTX-III) |

IV. miscellaneous structures

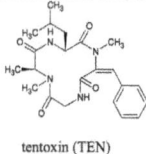

V. aminopentol esters

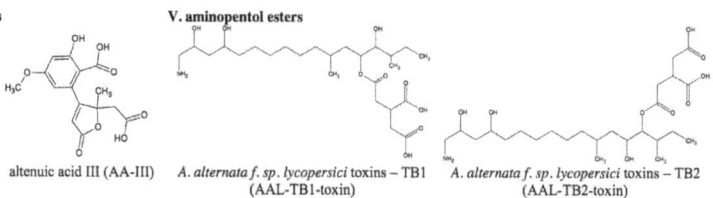

| tentoxin (TEN) | altenuic acid III (AA-III) | *A. alternata f. sp. lycopersici* toxins – TB1 (AAL-TB1-toxin) | *A. alternata f. sp. lycopersici* toxins – TB2 (AAL-TB2-toxin) |

Figure 1. Chemical structures of *Alternaria* toxins (ATs) determined in this study.

2. Results

The three *Alternaria* strains produced a highly variable mycotoxin cocktail on both substrates with up to eleven components in widely differing concentrations depending on the environmental conditions investigated in this study. Optimal conditions (rice as nutrient medium, 25 °C, 14 days of growth) trigger the production of a complex mixture of eleven or seven mycotoxins by both *A. tenuissima* strains or by the *A. infectoria* strain, respectively. Neither of the AAL-TB toxins were detected under either of the conditions. However, AOH-, AME- and ATL-sulfates could be tentatively identified at 25 °C in rice and in wheat. The composition of this cocktail and the amount of toxin produced was remarkably influenced by temperature, incubation time and the *Alternaria* species group and—to a lesser extent—by nutrient media used. In all cases, the underdogs within the ATs were analyzed as an essential part in these mixtures.

Results are presented as mean values of triplicates with logarithmic scaling because of high differences in concentration between the toxins produced by the strains in Figures 2–4 and are summarized as mean values in mg·kg^{-1} ± standard error of the mean in mg·kg^{-1} in Tables S1 and S2 available in the supplementary material. Percental distribution of the ATs depending on the strain are shown in Figure 5.

2.1. Mycotoxin Production Depending on the Incubation Time

To observe the effect of the incubation time on the toxin production, three different duration times (4, 7 and 14 days) were chosen to allow the fungus to grow and produce secondary metabolites (Figure 2a–f; Tables S1 and S2). In general, we observed a steady increase of AT concentrations from day 4 to day 14 with different rates depending on each secondary metabolite. Few exceptions could be observed with a decreasing phase between day 7 and day 14.

Most striking at 7 °C was the production of TeA by RN01Ct after 4 and 7 days in very low amounts in wheat and rice (0.046/1.8 and 1.6/0.41 mg·kg^{-1}) followed by a considerable increase up to 139 and 252 mg·kg^{-1} after 14 days. The same was observed for STTX-III, which firstly could not be detected after 4 days, and after 7 days only in small amounts (0.40 mg·kg^{-1} in rice by RN04Ci) but appeared after 14 days with 163 mg·kg^{-1} (Table S1).

Contrasting the results under low temperature conditions, an increasing amount of all examined secondary metabolites was found at 25 °C with advancing incubation time. TEN production increased constantly from day 4 to day 7 by a factor of 4.5 from 21 to 92 mg·kg^{-1} and finally doubled from day 7 to 14 up to 185 mg·kg^{-1} in wheat and similarly in rice (GH15t). For AA-III, ATX-I and the sum of ALT and isoALT (Σ(iso)ALT), a steady increase in production by the strains GH15t and RN01Ct in wheat and rice was observed over the whole period as well. The same was determined for ATX-II, ATL and AOH, except in rice (GH15t), the production of ATX-II and ATL stagnated and the AOH concentration decreased by 40% from day 7 to 14. Likewise, AME was formed in increasing amounts in wheat by the GH15t and RN01Ct strains (0.078/2.4/110 mg·kg^{-1} and 0.71/506/717 mg·kg^{-1}, respectively), however in rice a 60% (GH15t) or 70% (RN01Ct) decline in the total amount of AME was observed from day 7 to 14 (Figure 2a–d; Table S2).

In contrast to this, an increase of 100% and even 200% in TeA amount in wheat between day 4 and day 7 (2053 to 4309 mg·kg^{-1} GH15t; 1984 to 6098 mg·kg^{-1} RN01Ct) was detected while a 14% decline (3706 mg·kg^{-1}; GH15t) or a stagnation (5995 mg·kg^{-1} RN01Ct) in the total amount of TeA was observed from day 7 to day 14. In rice, TeA was formed in similar amounts over the whole incubation time period (3358, 2946 and 3327 mg·kg^{-1}; GH15t) or with a slight increase from day 7 to 14 (4637 to 5464 mg·kg^{-1}; RN01Ct) (Figure 2a–d; Table S2).

The STTX-III concentration increased from day 4 to 7 by a factor of 2 or 10 in rice or wheat (225 to 470 mg·kg^{-1}; 39 to 380 mg·kg^{-1}) and by another 30% from day 7 to 14 in rice, but stagnated in wheat (RN01Ct) (Figure 2c,d). Strain GH15t produced STTX-III in increasing amounts over the whole incubation time in wheat but showed a decrease of 18% from day 7 to 14 in rice (251 to 206 mg·kg^{-1}) (Figure 2a). The same can be stated for strain RN04Ci where a 50% decline in the total amount of STTX-III from day 7 to 14 in rice was determined (563 to 277 mg·kg^{-1}) (Figure 2f; Table S2).

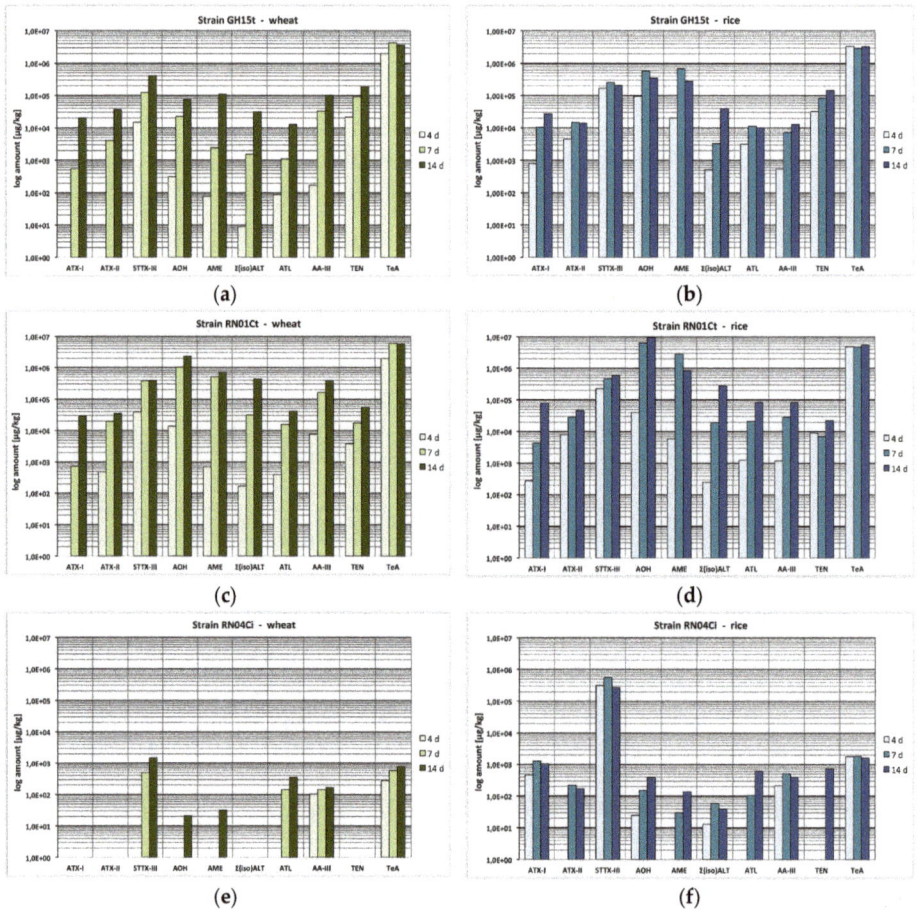

Figure 2. Production of altertoxin I (ATX-I), altertoxin II (ATX-II), stemphyltoxin III (STTX-III), alternariol (AOH), alternariol mono methylether (AME), sum of altenuene and isoaltenuene (Σ(iso)ALT), altenuisol (ATL), altenuic acid III (AA-III), tentoxin (TEN) and tenuazonic acid (TeA) after 4, 7 and 14 days at 25 °C; please note logarithmic scaling due to high difference in concentration between the toxins ($\mu g \cdot kg^{-1}$); (**a**) RN01Ct strain in wheat; (**b**) RN01Ct strain in rice; (**c**) GH15t strain in wheat; (**d**) GH15t strain in rice; (**e**) RN04Ci strain in wheat; (**f**) RN04Ci strain in rice.

2.2. Mycotoxin Production Depending on the Temperature

The samples were incubated at two different temperatures: at 7 °C and at 25 °C. Overall, the toxin production was heavily influenced by this factor resulting in either formation of all eleven ATs in huge amounts up to $g \cdot kg^{-1}$ at 25 °C or a strongly delayed formation of only a few ATs (TeA, AA-III, AOH, STTX-III) in small quantities ($\mu g \cdot kg^{-1}$) at 7 °C. None of the other investigated *Alternaria* toxins could be detected at 7 °C in rice or wheat. This was the case after 4, 7 and 14 days. Only in one case did the toxin concentration achieve a similar or even higher level at 7 °C compared to 25 °C after 14 days (STTX-III in rice by strain RN04Ci) (Figure 3c,d).

AA-III, TeA and AOH were produced at 7 °C in low amounts after 4 days (0.011 mg·kg^{-1}–1.58 mg·kg^{-1}). After 7 days, in addition to these three toxins, STTX-III was found (0.041 mg·kg^{-1}–0.24 mg·kg^{-1}). After 14 days, the RN01Ct strain had produced TeA in wheat and rice up to 139 and 252 mg·kg^{-1} (Figure 3a,b), respectively, whereas the RN04Ci strain had produced STTX-III up to 163 mg·kg^{-1} in rice (Figure 3c,d). However, AA-III and AOH could still only be determined in small amounts below 0.2 mg·kg^{-1} (Table S1).

In comparison to this, all examined toxins were detected at 25 °C, most of them in small amounts already after 4 days. Generally, it can be stated that ATX-I, ATX-II and ATL were produced in comparably small quantities over the whole incubation time followed by AA-III, TEN, Σ(iso)ALT, STTX-III and AME in significantly higher amounts, whereas TeA and AOH were produced in large quantities, with maximum amounts of 6098 mg TeA kg^{-1} after 7 days in wheat (RN01Ct) and 9693 mg AOH kg^{-1} after 14 days in rice (RN01Ct), respectively. Both *A. tenuissima* strains showed similar toxin formation capability, whereas the *A. infectoria* strain showed little toxin production at 25 °C for the whole incubation period, except for STTX-III, which could be found in amounts up to 563 mg·kg^{-1} after 7 days in rice. All other toxins were only produced between 0.01 and 1.9 mg·kg^{-1} (Figure 2e,f; Table S2).

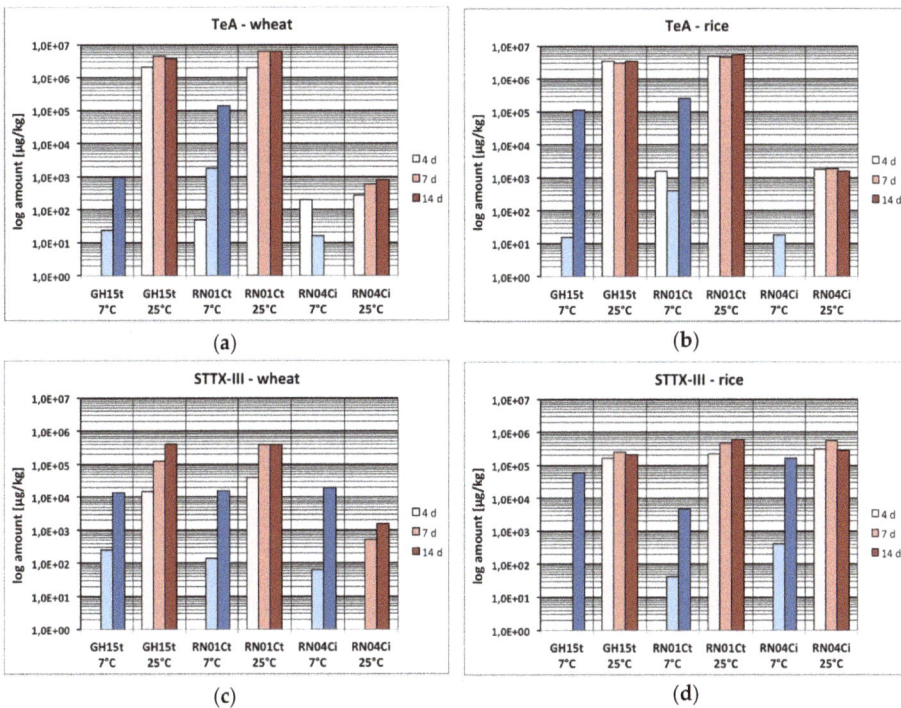

Figure 3. Production of tenuazonic acid (TeA) and stemphyltoxin III (STTX-III) at different temperatures after 4, 7 and 14 days by GH15t, RN01Ct and RN04Ci strains; please note logarithmic scaling due to high difference in concentration depending on the temperature (µg·kg^{-1}); (**a**) Comparison of TeA production at 7 °C and 25 °C in wheat; (**b**) Comparison of TeA production at 7 °C and 25 °C in rice; (**c**) Comparison of STTX-III production at 7 °C and 25 °C in wheat; (**d**) Comparison of STTX-III production at 7 °C and 25 °C in rice.

2.3. Mycotoxin Production Depending on the Substrate

To test the influence of different growth substrates, the three *Alternaria* strains were incubated in rice and wheat kernels. The substrate only marginally influenced the mycotoxin profile of the strains but the concentration of some toxins produced on the two different media varied up to ten-fold (Figure 4a,b).

For example, at 25 °C, AOH was produced by strain GH15t in higher concentrations in rice compared to in wheat after 4 (96 and 0.31 mg·kg^{-1}), 7 (570 and 22 mg·kg^{-1}) and 14 (346 and 74 mg·kg^{-1}) days (Figure 2a,b and Figure 4a; Table S2). The same was observed for strain RN01Ct, which produced AOH in significantly higher concentrations in rice than in wheat after 4 (40 and 14 mg·kg^{-1}), 7 (6452 and 1030 mg·kg^{-1}) and 14 (9693 and 2289 mg·kg^{-1}) days (Figure 2c,d and Figure 4b; Table S2). This trend was also determined for AME and ATX-I, but only with slightly higher amounts in rice for both *A. tenuissima* strains. In contrast to this, more AA-III could be detected in wheat than in rice after 4 (7.7 and 1.2 mg·kg^{-1}), 7 (163 and 29 mg·kg^{-1}) and 14 days (375 and 84 mg·kg^{-1}) for strain RN01Ct, which was similar for strain GH15t. The production of the other toxins (ATX-II, STTX-III Σ(iso)ALT, ATL, TEN, TeA) varied in both substrates, mostly starting on day 4 with higher amounts in rice and then switching to a comparable or slightly higher toxin production in wheat after 7 or 14 days. For example TEN was produced in rice and wheat after 4 (9.1 and 3.9 mg·kg^{-1}), 7 (7.1 and 18 mg·kg^{-1}) and 14 days (22 and 55 mg·kg^{-1}) by strain RN01Ct (Figure 2c,d and Figure 4b; Table S2). The toxin production capability of the *A. infectoria* strain RN04Ci was higher in rice than wheat both in concentrations and a number of ATs. In particular, STTX-III was produced in high concentrations in rice but not in wheat (Figure 2e,f; Table S2).

However, the mycotoxin profiles of the *A. tenuissima* strains GH15t and RN01Ct at 25 °C and after 14 days in rice reflect their profiles in wheat (Figure 4a,b).

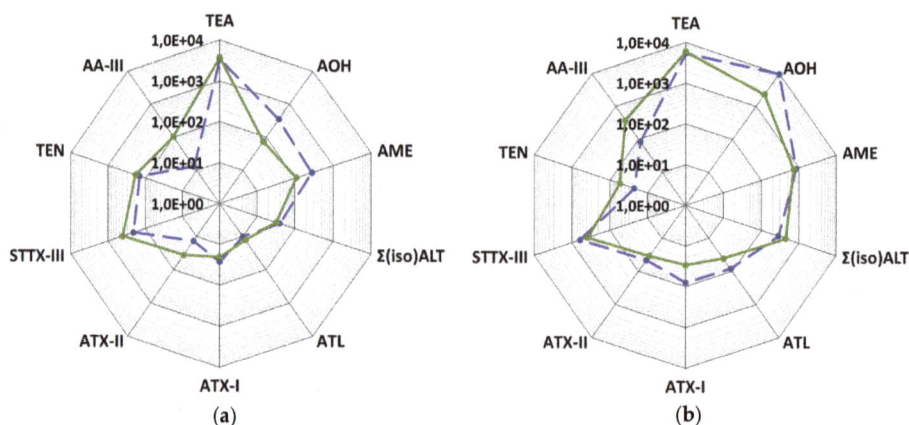

Figure 4. *Alternaria* toxin profiles of altertoxin I (ATX-I), altertoxin II (ATX-II), stemphyltoxin III (STTX-III), alternariol (AOH), alternariol mono methylether (AME), sum of altenuene and isoaltenuene (Σ(iso)ALT), altenuisol (ATL), altenuic acid III (AA-III), tentoxin (TEN) and tenuazonic acid (TeA) after 14 days at 25 °C in wheat and rice; please note logarithmic scaling of the contents (mg·kg^{-1}); solid green lines: rice kernels; dashed blue lines: wheat kernels; (**a**) *Alternaria* toxin profile of GH15t ; (**b**) *Alternaria* toxin profile of RN01Ct.

2.4. Mycotoxin Production Dependent on the Strain

Two strains from the *A. tenuissima* species group (GH15t and RN01Ct) and one strain from the *A. infectoria* species group (RN04Ci) were chosen for the study. The two *A. tenuissima* strains were able to produce all eleven ATs and some of them in high amounts at 25 °C: 9693 mg·kg^{-1} AOH in rice substrate after 14 days, 6098 mg·kg^{-1} TeA in wheat kernels after 7 days and 470 mg·kg^{-1} STTX-III in rice after 7 days (Figure 5a,b,d,e). In comparison, the *A. infectoria* strain could form all examined ATs, but only in low concentrations under the same conditions: 0.4 mg kg^{-1} AOH and 0.6 mg kg^{-1} TeA, but remarkably, 563 mg·kg^{-1} of STTX-III (Figure 5c,f).

The mycotoxin cocktails of the two *A.* species groups analyzed after 14 days at 25 °C showed remarkable differences: The main component of the cocktail produced by both *A. tenuissima* strains in wheat was TeA with 58% for strain RN01Ct followed by AOH (22%) and in descending order AME (7%), Σ(iso)ALT (4%), STTX-III (4%) and AA-III (4%) (Figure 5b). With regard to strain GH15t, the TeA fraction was even 79% followed by STTX-III (9%) and TEN (4%), AOH (2%), AME (2%) and AA-III (2%) (Figure 5a). Strain GH15t showed a similar toxin distribution in rice, whereas strain RN01Ct produced mostly AOH (56%) instead of TeA (32%) (Figure 5d,e).

In contrast, *A. infectoria* produced predominantly STTX-III (53%), followed by TeA (27%), ATL (12%), AA-III (6%), AME (1%) and AOH, all in low amounts up to 1.5 mg·kg^{-1} in wheat whereas in rice STTX-III was produced in higher amounts (277 mg·kg^{-1}) almost exclusively (Figure 5c,f).

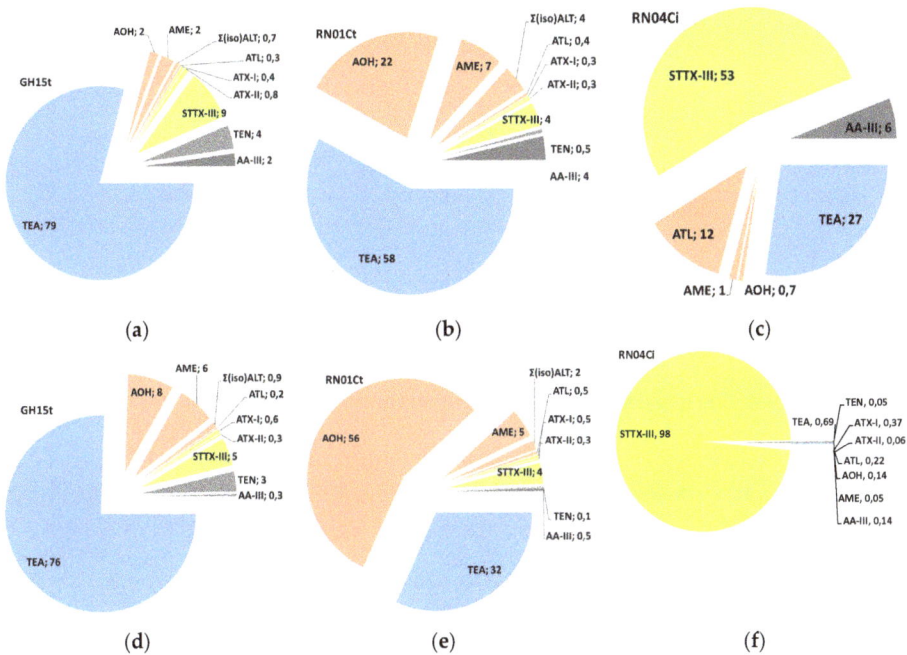

Figure 5. Distribution of *Alternaria* toxin production in percentage after 14 days at 25 °C in wheat kernels (**a–c**) and in rice kernels (**d–f**) by the *Alternaria tenuissima* strain GH15t (**a,d**), the *Alternaria tenuissima* strain RN01Ct (**b,e**) and the *Alternaria infectoria* strain RN04Ci (**c,f**).

2.5. Modified Toxins

Due to the lack of commercially available analytical calibration standards of modified *Alternaria* toxins several experiments were carried out for identification (see Section 5.7). These experiments included precursor ion scans, neutral loss scans and collision induced mass spectra (MS2) in high-resolution or tandem mass spectrometry. AOH-, AME- and ATL-sulfates could be tentatively detected in extracts from rice and wheat, which were inoculated with strain GH15t or RN01Ct at 25 °C. Sulfates of ALT, isoALT, ATX-I, ATX-II or STTX-III or conjugations with glucoside could not be detected with the applied tandem mass spectrometry or high-resolution MS methods. Multiple reaction monitoring (MRM) transitions were monitored and the ion ratios of the selected ion transitions displayed no significant variances (<5%) within all positive samples. The earlier retention times of the substances in comparison to the retention times of the respective ATs is a further indication for the more water-soluble modified ATs (Table S3). The retention times of all quantified ATs were within a permitted tolerance of ±0.2 min in solvent calibration samples compared to matrix samples [13] and retention times of the sulfates were within this permitted tolerance in all positive samples.

Neutral loss scanning on a triple quadrupole mass spectrometer for semi-targeted detection of sulfoconjugated metabolites has been described before as a powerful screening tool, because many sulfated compounds show a characteristic neutral loss of 80 Da (SO_3) [38].

The measured accurate masses of the sulfates were obtained with an accuracy of ±5 ppm compared to the calculated exact masses [16] (Table S4).

The appearance of AOH-sulfate was detected in rice after 7 days (strain GH15t). Strain RN01Ct appeared to produce AOH-sulfate over the whole incubation period in rice. The response of the signal increased by a factor of 6.5 from day 4 to day 7 and decreased by 65% from day 7 to day 14. On the contrary, the amount of AOH increased enormously (40/6452/9694 mg·kg^{-1}) over the whole period. Strain RN01Ct produced AOH-sulfate on day 7 and day 14 in wheat, too. The response of the signal decreased by 70%; conversely, the AOH concentration rose by a factor of 2 (Table S2).

AME-sulfate was formed by strain GH15t after 14 days in wheat and between 7 and 14 days in rice. The response of the AME-sulfate signal decreased by 80% from day 7 to 14 as well as the amount of AME (674 to 278 mg·kg^{-1}). Strain RN01Ct produced AME-sulfate after 4, 7 and 14 days in rice. At first the signal increased by a factor of 1.5, but showed a decline by 45% from days 7 to 14. Likewise, the AME concentration increased (6.0–2782 mg·kg^{-1}) and decreased by 70% after 14 days (831 mg·kg^{-1}). In wheat the AME-sulfate signal increased by a factor of 2.3 as the AME amount increased by a factor of 1.4 (506 to 717 mg·kg^{-1}) from day 7 to 14 (Table S2).

ATL-sulfate was found only after 14 days and only in rice (RN01Ct).

3. Discussion

Our results show that different *Alternaria* sp. isolates from wheat are able to produce up to eleven *Alternaria* toxins under varying in vitro conditions in wheat and rice substrate. TeA is the most extensively produced toxin under nearly all conditions. The host-specific AAL toxins TB1 and TB2 were not detected at all as assumed. The obtained results of the study are in line with results obtained in previous in vitro studies relating to the major examined ATs TeA, AOH, AME, ALT and TEN [17,30,31]. TeA has been indicated as a major mycotoxin in naturally infected tomatoes, wheat and cereal-based foodstuffs [9,18]. *A. alternata* and *A. tenuissima* are regarded as highly potential toxin producers: 100% of strains isolated in the Mediterranean regions produced TeA, 93% AOH and AME [39]; all strains isolated in China from weather-damaged wheat were able to accumulate AOH and AME [40] as were all isolates from Poland [41], Italy [42], Argentina [43] and Germany [32]. The concentrations of AOH, AME and TeA analyzed in different in vitro experiments are in line with those reported in our study.

The production of mycotoxins is dependent on the growth conditions of the fungus: humidity (water activity), temperature, composition of nutrient media and pH conditions. Lee et al. [17] revised the optimum conditions of some abiotic factors for the growth and mycotoxin production of several *Alternaria* strains isolated from different host plants. Fungal development and AOH, AME and TeA formation were found within a wide temperature range (5–35 °C), water activity range (0.88–0.999) and under different pH conditions [44–46] independently of the host plants. Our results show the production of not only TeA but also STTX-III, AOH and AA-III at 7 °C and indicate the high likelihood of accumulating these toxins at cooler temperatures during the vegetation period of different vegetables, fruits and crop plants and under storage conditions. This may impair the capability to store agricultural commodities such as wheat kernels and possibly leads to a misjudgement of mycotoxin contamination. The present study not only summarizes the significance of the major toxins AOH, AME and TeA, but also highlights the formation of a multicomponent mycotoxin mixture under various abiotic conditions. Shier et al. [47] and Vesonder et al. [48] already described the risk of additional or synergistic effects caused by the co-occurrence of different mycotoxins. The simultaneous exposure to diverse *Alternaria* toxins should be linked to a greater and stronger adverse impact on human and animal health than indicated by a single mycotoxin. Therefore, the knowledge of the mycotoxin cocktail production influenced by environmental conditions is important in developing forecasting risk models in natural habitats.

During the present study, temperature, nutrient media and incubation time have influenced the production of all mycotoxins evaluated depending on the strain and its belonging to a species group of *Alternaria*. The obtained data on mycotoxin production in rice and wheat substrate verify the observations made by Kahl et al. which revealed the segregation of twenty-nine *A. infectoria* strains from *A. alternata*, *A. tenuissima* and *A. arborescens* due to the non-production of some toxins or to a different toxigenic pattern [32]. The entirely different mycotoxin profile of *A. infectoria* detected in the present study facilitates the singular position of this species group within the genus *Alternaria*. Furthermore, stemphyltoxin III is the only mycotoxin whose concentration is comparable in the *A. infectoria* as well as in the two *A. tenuissima* cultures. But the large share of STTX-III in the mycotoxin mixture of *A. infectoria* could be an explicit attribute for this species-group and may differentiate between the species groups. Moreover, STTX-III is a widely neglected metabolite among the *Alternaria* toxins though it has mutagenic potential [21]. However, the instability of STTX-III in solvent solution has been shown in a previous stability study by Zwickel et al. [13]. Thus similar analyses dealing with the production of STTX-III in *Alternaria* species are not known.

The formation of the modified mycotoxin—AOH-sulfate—showed no unequivocal pattern with regard to the formation of AOH. Between day 4 and day 7, the more AOH that was produced the more sulfate was detected, which was demonstrated by a higher response of the signals. However, from day 7 to day 14 the AOH-sulfate signal diminished even though the AOH concentration rose further. By comparison, AME-sulfate formation showed an analogy compared to the AME formation. After an increase in AME production, an increase in the respective sulfate was also observed. Likewise, a decline in the AME amount correlated with a decrease in the response of the AME-sulfate signal. ATL-sulfate was only detected in one sample after 14 days at 25 °C. ATL itself was generally produced in low amounts but in this sample it was produced in higher amounts. Therefore, the conclusion can be drawn that the sulfoconjugation is triggered depending on the concentration of the respective toxin. AOH- and AME-sulfates were previously reported in *Alternaria* sp. isolates from *Polygonum senegalense* grown either in liquid culture or in solid rice [15]. Recently, the formation of these two and an additional six new sulfoconjugated metabolites in cultures of *Alternaria* sp. was described by Kelman et al. [49]. The *Alternaria* isolates hereby were inoculated in liquid culture as well as rice and Cheerios (American brand of breakfast cereals). For identification high-resolution neutral loss filtering was used as well in this study.

In our study the rice and wheat kernels were sterilized twice by autoclaving at 121 °C. This procedure destroyed all metabolic activity of the substrate. Therefore, the enzymatic activity for phase-I and phase-II reactions could only be derived by the fungus itself. Therefore, the conjugation of the toxins AOH, AME and ATL with sulfate to more water-soluble molecules is a fungal reaction during the later stages of growth. One reason for this might be that the self-detoxification of the fungus as AOH and AME may also affect the fungal development. One hypothesis regarding the question as to why secondary metabolites including mycotoxins are formed by fungi is that these substances have extrinsic functions and may give the producers a competitive advantage towards other microorganisms and during the infection processes of the host cells [44,50,51]. Therefore, we hypothesize that the formation of sulfate derivatives inactivates or decreases the toxicity of the basic substances and promotes the faster excretion of the more water-soluble conjugates as a self-detoxification process.

In vitro studies should provide an indication of the behavior of fungal strains in a natural complex environment. Abiotic driving factors recognized in laboratory experiments influence the fungal growth and mycotoxin production in natural habitats too but in a more complex and multifactorial situation. *Alternaria* strains are members of a microbial community and a biotic network in the phyllosphere (aerial plant parts) and rhizosphere of different crop plants. Interactions of fungi and bacteria with *Alternaria* could exist more often in competition and in antagonistic processes than in undisturbed coexistence. Few studies dealing with the co-occurrence of *Fusarium* and fluorescent pseudomonads with *Alternaria* suggest the implication of mycotoxins as antagonistic substances in these interactions [52,53] Simultaneously, competition processes between different fungal genera affect the production of mycotoxins or increase the metabolization of toxins [54,55]. That means that the production of mycotoxins in natural habitats is influenced by biotic and abiotic factors and could represent various toxin profiles and different concentrations of mycotoxins [56,57].

In this cooperation work we conducted a further experiment with strains isolated from different habitats [32]. Also, the production of STTX-III in remarkably high amounts by *A. infectoria* strain will be monitored by investigation of several other isolated *A. infectoria* strains. The results of this next experiment will soon be published.

4. Conclusions

In general, fungi of the *Alternaria* species-group *tenuissima* and *infectoria* are able to produce a complex multicomponent mycotoxin cocktail. Temperature, nutrient media, incubation time and the fungal strain had a strong influence on what toxins were formed, the percentage they made up of the total mixture and the concentrations they were produced in. Here, we present the first in vitro study on the formation of underrepresented *Alternaria* toxins such as STTX-III, ATX-II, ATL, AA-III, AOH-sulfate, AME-sulfate and ATL-sulfate along with the better investigated ATs AOH, AME, ALT, TeA and TEN, depending on varying substrate, time and temperature conditions. It can thus be concluded that these underrepresented ATs could also be harmful substances in naturally contaminated crops, fruits and vegetables. A monitoring assessment of food and feeds should include the analyses of these underdogs. To our best knowledge this is the first time that the occurrence of sulfoconjugated altenuisol has been described.

5. Materials and Methods

5.1. Reagents, Solvents and Equipment

Analytical solid calibrants of AOH, AME, TEN, TeA and a mixture of AAL-TB1 and TB2 toxin, all >99% purity were purchased from Sigma-Aldrich, St. Louis, MO, USA. ATX-I, ATX-II and STTX-III were provided by the group of Professor Metzler at the Institute of Applied Bioscience (Karlsruher Institut für Technologie (KIT), Karlsruhe, Germany) via preparative HPLC of *Alternaria* isolates [58]. Crystalline ALT, iso(ALT), AA-III, and ATL were synthesized and provided by the group of Professor Podlech at the Institute of Organic Chemistry (KIT) [11,12]. Analytical reagent quality ammonium acetate, ammonium hydroxide solution (25%), LiChrosolv LC-MS hypergrade quality methanol and acetonitrile were purchased from Merck, Darmstadt, Germany. Analytical grade water (0.055 $\mu S \cdot cm^{-1}$) was generated from a Milli-Q system (Merck, Darmstadt, Germany).

5.2. Preparation of Standard Solutions

Stock solutions were prepared by dissolving 1 mg of the solid material in 1 mL methanol to obtain 1 $mg \cdot mL^{-1}$ stock solutions. Dried down standards (ATX-I, ATX-II and STTX-III) with a proposed absolute amount of 100 µg of the toxin were redissolved in 1 mL methanol. All stock solutions were kept at -30 °C. The actual concentrations were determined by ultraviolet–visible spectroscopy (UV/Vis) using a UV-1700 Pharma Spec (Shimadzu, Kyoto, Japan). Used wavelength and extinction coefficients for each toxin are described in a published work [13]. Mixed stock solutions with a concentration of 1 $\mu g \cdot mL^{-1}$ of each toxin were prepared beforehand in a mixture of 75% 1 mM ammonium acetate (pH 9) and 25% methanol. Calibration mix solutions were always prepared freshly in a range from 1 $ng \cdot mL^{-1}$ to 1000 $ng \cdot mL^{-1}$.

5.3. Instrumentation and Equipment

Analyses were performed using a 1100 HPLC system from Agilent Technologies (Santa Clara, CA, USA) coupled to an API 4000 (SCIEX, Foster City, CA, USA) triple quadrupole mass spectrometer. The system was equipped with an electrospray (ESI) interface (Turbo V^{TM}, SCIEX, Foster City, CA, USA) and negative and positive ionization was used during acquisition. Chromatographic separation was performed on a 5 μm particle size, 100 mm × 2 mm i.d Gemini NX-C18 HPLC column from Phenomenex (Aschaffenburg, Germany). Data acquisition and evaluation were performed with Analyst version 1.6.2 (SCIEX, Foster City, CA, USA, 2013).

High-resolution mass spectrometry (HRMS) analyses to confirm the modified ATs (AOH-, AME- and ATL-sulfate) were performed on an Accela HPLC system coupled to an Exactive^{TM} (orbitrap) HCD (higher energy collisional dissociation) system fitted with a HESI II (heated-electrospray ionization) source (Thermo Fisher Scientific Inc., Waltham, MA, USA).

5.4. Fungal Isolates Used in this Study

Alternaria-infected grain samples were collected from winter wheat fields in the northeast Germany (Uckermark region) and central Russia (Novosibirsk region) in 2012. The fields underlie common agricultural practices and German sites take part in long-term examinations by the Leibniz-Centre for Agricultural Landscape Research Müncheberg, Germany (ZALF). *Alternaria* was isolated from grains after incubation on nutrient agar (potato dextrose agar, PDA, Carl Roth, Karlsruhe, Germany), morphologically identified and classified as belonging to the *A. infectoria* or *A. tenuissima* species group [32]. Single spore stock cultures of the isolates are maintained in sterile wheat mixtures at −20 °C. Three *Alternaria* isolates were chosen for the present mycotoxin analysis on the basis of previous mycotoxin measurements of ALT, AOH, AME, ATX-I and TeA by means of HPLC with diode-array detection (HPLC-DAD) analysis. A high (GH15t) and a medium *Alternaria* mycotoxin producer (RN01t) of the *A. tenuissima* species group were chosen additionally to one *A. infectoria* strain (RN04Ci), which could only produce low amounts of ATs [32].

5.5. Growth Conditions for Fungal Isolates

A spore suspension with 1×10^5 spores mL^{-1} was produced for each isolate after incubation on potato-carrot-agar described by Kahl et al. [32]. A total of 114 centrifuge tubes (15 mL) were filled with 0.2 g of either rice (*n* = 57) or wheat kernels (*n* = 57) and 250 μL of demineralized water. Tubes were sterilized twice before inoculation. Three wheat and three rice samples as controls were not inoculated but incubated at 25 °C for 14 days to prove that there was no contamination. For each of the three *Alternaria* isolates 36 tubes—half of them containing wheat kernels the other half rice kernels—were inoculated with 50 μL of the respective spore suspension. Nine samples for each substrate group were incubated at 7 °C to simulate standard storage conditions and the other nine samples at 25 °C to simulate the conditions during the ripening process of cereals. Three samples of each temperature were incubated in darkness for 4, 7 and 14 days, respectively. After the respective incubation time samples were frozen at −20 °C to stop the growth of the fungus and any further mycotoxin production.

5.6. Extraction of Alternaria Toxins from Fungal Cultures

After defrosting, 5 mL of an acetonitrile/water/acetic acid mixture (79:20:1; *v/v/v*) were added to each sample tube. The fungus-kernel complex was mechanically pounded with a spatula. Ultrasonic extraction was carried out for 15 min. Subsequent tubes were roughly shaken at 25 °C for 60 min. The two steps were repeated one more time and after that the sample tubes were centrifuged at 4000 rpm for 6 min. Then, 2 mL of the supernatant were filled in 2 mL glass vials and stored at −30 °C until further processing.

5.7. HPLC-MS/MS Analysis of Alternaria Toxins

The ATs were simultaneously separated on a Gemini NX-C18 (2.1 × 100 mm, 5 μm) HPLC column equipped with a C18 SecurityGuard™ cartridge system (0.3 mm) using binary linear gradient elution. Column oven temperature was set to 40 °C. Solvent A contained 1 mM of ammonium acetate and a pH adjusted to 9 with ammonium hydroxide solution (25%), while solvent B was methanol. The flow rate was set to 0.3 mL·min⁻¹. Solvent B was 0% at 0 min, 0% at 1.0 min, 95% at 1.2 min, 95% at 6.0 min and 0% at 7.0 min. Equilibration time was set to 5 min with 100% of solvent A. Total run time was 12 min. The autosampler was operated at 10 °C and the injection volume was set to 5 μL. 2 mM ammonium acetate in methanol was used as post column solvent with a constant flow rate of 0.2 mL·min⁻¹. The HPLC method has been precisely published by Zwickel et al. [13].

The mass spectrometer was operated in the negative electrospray ionization mode using multiple reaction monitoring (MRM) mode. Two ion transitions were scanned for each target compound. After the last analyte was detected, the polarity was switched to positive mode for 3 min to prevent the MS from possible contamination with negatively loaded matrix compounds [13]. The selected ion transitions with the optimized collision energies (CE), collision cell exit potential (CXP) and declustering potential (DP) for each analyte [13] are summarized in Table S3, entrance potential (EP) was set to −10 V and the gas for the collisionally activated dissociation (CAD) to 6 (arbitrary unit). Both quadrupoles (Q1 and Q3) were set to unit resolution. The source parameters were set as follows: curtain gas, 40 psi, ion spray voltage, −3500 V (0–7 min); +3500 (7.01–10 min); temperature 550 °C; spray gas (GS1), 35 psi; dry gas (GS2) 50 psi. Ultra high purity nitrogen (99.999%) was used as gas.

5.8. Neutral Loss Scan and HRMS Analysis of Modified Alternaria Toxins

Sulfoconjugates were detected in negative mode by monitoring constant losses of 80 Da (SO₃) and precursors of m/z 80. Subsequently product ion scans of the detected precursors were obtained by collision-induced dissociation (CID) experiments. For each detected AT-sulfate two ion transition were selected—AOH-sulfate (m/z 337.1–m/z 257; m/z 337.1–m/z 213); AME-sulfate (m/z 351.1–m/z 271; m/z 351.1–m/z 256) and ATL-sulfate (m/z 353.1–m/z 273; m/z 353.1–m/z 230) (Table S3)—and integrated into the original method [13]. AOH- and AME-sulfate ion transitions correspond with the obtained transitions from analytical standard substances [36]. Ion ratios and retention times of detected sulfates in samples were monitored.

The detected precursor ions of AOH-, AME-sulfate were confirmed via high-resolution MS by comparing their calculated exact masses with the actual detected masses and the presented masses from the synthesized molecules [16] (Table S4). HESI parameters were set as follows: sheath gas flow 20 psi; spray voltage 4 kV, capillary temperature 350 °C; capillary voltage −60 V, tube lens voltage −120 V; skimmer voltage −25 V; heater temperature 350 °C. Ultra high purity nitrogen (99.999%) was used as gas. HRMS scan parameters were set as follows: scan range 150.00–1000.00 m/z; resolution was set to ultra-high; polarity was negative; AGC target was balanced.

5.9. Validation Parameters and Sample Preparation for HPLC-MS/MS Measurement

Limits of detection (LODs) and limits of quantification (LOQs) were determined according to DIN EN standard 32645 in extracts from not inoculated but incubated samples for all *Alternaria* toxins [13]. LODs ranged from 0.5 to 3.0 μg·kg⁻¹ and LOQ ranged from 1.8–9.5 μg·kg⁻¹ (see table S5).

External calibration curves were prepared in injection solution (1 mM ammonium acetate (pH 9): methanol; 75:25; v/v). Linear calibration curves ranged between 1.0–1000 ng·mL⁻¹ for all analytes because of high differences in concentration between the toxins in the samples. Due to inhomogeneity of variance, what is almost inevitable for LC-ESI-MS/MS measurement when using a large calibration range is that the accuracy in the lower end of the range is not confidently given. Next to using weighted least squares linear regression (1/x) another alternative to counteract inhomogeneity of variance is to stagger calibration solutions and up to 20 samples and calculate the concentrations of the toxins with the preceding calculation curve. Both methods were compared and led to similar results.

Samples had to be diluted 1:2000 (14 day samples at 25 °C), 1:1000 (7 and 14 day samples at 25 °C), 1:500 and at least 1:100 (all other samples) with injection solution (1 mM ammonium acetate (pH 9): methanol; 75:25; v/v) due to high differences in concentration between the ATs. Every dilution approach was tested for matrix effects (ME), which may lead to ion suppression or ion enhancement. Therefore, the blank samples (not inoculated) were diluted equal to the inoculated samples and spiked with AT mix solution in three concentrations (1, 200 and 1000 ng·mL^{-1}). The effect of the presence of matrix on each analyte was calculated by dividing the analyte peak areas of matrix-matched standard samples through the peak areas of standard samples in pure injection solution: ME (%) = peak area of analyte in matrix solution/peak area of analyte in neat solution – 1) × 100 [13]. A negative ME indicates ion suppression and a positive one ion enhancement. For each analyte in each dilution approach a ME < ±10% was calculated. Additionally, samples with high toxin contents were diluted 2000, 1000, 500 and 100 times. Calculation via external calibration led to comparable results. Hence, the dilution of the samples resulted in a sufficient dilution of the matrix and no further compensation measures were needed.

Supplementary Materials: The following are available online at www.mdpi.com/2072-6651/8/11/344/s1, Table S1: Mean content ($n = 3$) of produced *Alternaria* toxins at 7 °C after 4, 7 and 14 days in wheat and in rice in mg kg^{-1} (± standard error of the mean; n.d. not detected), Table S2: Mean content ($n = 3$) of produced *Alternaria* toxins at 25 °C after 4, 7 and 14 days in wheat and in rice in mg kg^{-1} (± standard error of the mean; n.d. not detected), Table S3: Selected ion transitions with optimized collision energies (CE), collision cell exit potential (CXP), declustering potential (DP), quantifier A(1) to qualifier A(2) ratio and retention time (Rt) for each analyte, Table S4: Monoisotopic calculated exact masses (EM) and measured accurate masses (AM) of negatively loaded alternariol sulfate ion, alternariol mono methylether sulfate ion and altenuisol sulfate ion, Table S5: Limits of detection (LODs) and limits of quantification (LOQs) determined according to DIN EN ISO 32645.

Acknowledgments: The authors gratefully acknowledge the technical assistance by Grit van der Waydbrink and Martina Peters.

Author Contributions: M.E.H.M., T.Z. and S.M.K. conceived and designed the experiments. T.Z. prepared the samples for measurement, performed the chemical analysis and analysed the data; S.M.K. provided the in vitro assay and the extraction of mycotoxins. T.Z., S.M.K. and M.E.H.M. wrote the paper. M.E.H.M., H.K. and M.R. performed a scientific supervision and manuscript revising.

Conflicts of Interest: The authors declare no conflict of interest.

References

1. Thomma, B.P. *Alternaria* spp.: From general saprophyte to specific parasite. *Mol. Plant Pathol.* **2003**, *4*, 225–236. [CrossRef] [PubMed]

2. Vučković, J.N.; Brkljača, J.S.; Bodroža-Solarov, M.I.; Bagi, F.F.; Stojšin, V.B.; Ćulafić, J.N.; Aćimović, M.G. *Alternaria* spp. on small grains. *Food Feed Res.* **2012**, *39*, 79–88.

3. Logrieco, A.; Moretti, A.; Solfrizzo, M. *Alternaria* toxins and plant diseases: An overview of origin, occurrence and risks. *World Mycotoxin J.* **2009**, *2*, 129–140. [CrossRef]

4. Ostry, V. *Alternaria* mycotoxins: An overview of chemical characterization, producers, toxicity, analysis and occurrence in foodstuffs. *World Mycotoxin J.* **2008**, *1*, 175–188. [CrossRef]

5. Hasan, H.A. Phytotoxicity of pathogenic fungi and their mycotoxins to cereal seedling viability. *Acta Microbiol. Immunol. Hung.* **2001**, *48*, 27–37. [CrossRef] [PubMed]

6. Mercado Vergnes, D.; Renard, M.E.; Duveiller, E.; Maraite, H. Identification of *Alternaria* spp. on wheat by pathogenicity assays and sequencing. *Plant Pathol.* **2006**, *55*, 485–493. [CrossRef]

7. Rotem, J. *The genus Alternaria: Biology, Epidemiology, and Pathogenicity*; American Phytopathological Society: St. Paul, MN, USA, 1994.

8. Graf, E.; Schmidt-Heydt, M.; Geisen, R. Hog map kinase regulation of alternariol biosynthesis in *Alternaria alternata* is important for substrate colonization. *Int. J. Food Microbiol.* **2012**, *157*, 353–359. [CrossRef] [PubMed]

9. EFSA. Scientific opinion on the risks for animal and public health related to the presence of *Alternaria* toxins in feed and food. *EFSA J.* **2011**, *9*, 2407–2504.

10. Pero, R.W.; Harvan, D.; Blois, M.C. Isolation of the toxin, altenuisol, from the fungus, *Alternaria tenuis* Auct. *Tetrahedron Lett.* **1973**, *14*, 948–1973.

11. Nemecek, G.; Cudaj, J.; Podlech, J. Revision of the structure and total synthesis of altenuisol. *Eur. J. Org. Chem.* **2012**, *2012*, 3863–3870. [CrossRef]

12. Nemecek, G.; Thomas, R.; Goesmann, H.; Feldmann, C.; Podlech, J. Structure elucidation and total synthesis of altenuic acid III and studies towards the total synthesis of altenuic acid II. *Eur. J. Org. Chem.* **2013**, *2013*, 6420–6432. [CrossRef]

13. Zwickel, T.; Klaffke, H.; Richards, K.; Rychlik, M. Development of a high performance liquid chromatography tandem mass spectrometry based analysis for the simultaneous quantification of various *Alternaria* toxins in wine, vegetable juices and fruit juices. *J. Chromatogr. A* **2016**, *1455*, 74–85. [CrossRef] [PubMed]

14. Rychlik, M.; Humpf, H.-U.; Marko, D.; Danicke, S.; Mally, A.; Berthiller, F.; Klaffke, H.; Lorenz, N. Proposal of a comprehensive definition of modified and other forms of mycotoxins including "masked" mycotoxins. *Mycotoxin Res.* **2014**, *30*, 197–205. [CrossRef] [PubMed]

15. Aly, A.H.; Edrada-Ebel, R.; Indriani, I.D.; Wray, V.; Müller, W.E.; Totzke, F.; Zirrgiebel, U.; Schachtele, C.; Kubbutat, M.H.; Lin, W.H.; et al. Cytotoxic metabolites from the fungal endophyte *Alternaria* sp. and their subsequent detection in its host plant *Polygonum senegalense*. *J. Nat. Prod.* **2008**, *71*, 972–980. [CrossRef] [PubMed]

16. Mikula, H.; Skrinjar, P.; Sohr, B.; Ellmer, D.; Hametner, C.; Frohlich, J. Total synthesis of masked *Alternaria* mycotoxins-sulfates and glucosides of alternariol (AOH) and alternariol-9-methyl ether (AME). *Tetrahedron* **2013**, *69*, 10322–10330. [CrossRef]

17. Lee, H.B.; Patriarca, A.; Magan, N. *Alternaria* in food: Ecophysiology, mycotoxin production and toxicology. *Mycobiology* **2015**, *43*, 93–106. [CrossRef] [PubMed]

18. Müller, M.E.H.; Korn, U. *Alternaria* mycotoxins in wheat—A 10 years survey in the northeast of germany. *Food Control* **2013**, *34*, 191–197. [CrossRef]

19. Frizzell, C.; Ndossi, D.; Kalayou, S.; Eriksen, G.S.; Verhaegen, S.; Sorlie, M.; Elliott, C.T.; Ropstad, E.; Connolly, L. An in vitro investigation of endocrine disrupting effects of the mycotoxin alternariol. *Toxicol. Appl. Pharmacol.* **2013**, *271*, 64–71. [CrossRef] [PubMed]

20. Tiemann, U.; Tomek, W.; Schneider, F.; Müller, M.; Pohland, R.; Vanselow, J. The mycotoxins alternariol and alternariol methyl ether negatively affect progesterone synthesis in porcine granulosa cells in vitro. *Toxicol. Lett.* **2009**, *186*, 139–145. [CrossRef] [PubMed]

21. Davis, V.M.; Stack, M.E. Mutagenicity of stemphyltoxin III, a metabolite of *Alternaria alternata*. *Appl. Eviron. Microbiol.* **1991**, *57*, 180–182.

22. Stack, M.E.; Prival, M.J. Mutagenicity of the *Alternaria* metabolites altertoxins I, II, and III. *Appl. Environ. Microbiol.* **1986**, *52*, 718–722. [PubMed]

23. Pero, R.W.; Posner, H.; Blois, M.; Harvan, D.; Spalding, J.W. Toxicity of metabolites produced by the "*Alternaria*". *Environ. Health Perspect.* **1973**, *4*, 87–94. [CrossRef] [PubMed]

24. Fleck, S.C.; Burkhardt, B.; Pfeiffer, E.; Metzler, M. *Alternaria* toxins: Altertoxin II is a much stronger mutagen and DNA strand breaking mycotoxin than alternariol and its methyl ether in cultured mammalian cells. *Toxicol. Lett.* **2012**, *214*, 27–32. [CrossRef] [PubMed]

25. Schwarz, C.; Tiessen, C.; Kreutzer, M.; Stark, T.; Hofmann, T.; Marko, D. Characterization of a genotoxic impact compound in *Alternaria alternata* infested rice as altertoxin II. *Arch. Toxicol.* **2012**, *86*, 1911–1925. [CrossRef] [PubMed]

26. Brugger, E.-M.; Wagner, J.; Schumacher, D.M.; Koch, K.; Podlech, J.; Metzler, M.; Lehmann, L. Mutagenicity of the mycotoxin alternariol in cultured mammalian cells. *Toxicol. Lett.* **2006**, *164*, 221–230. [CrossRef] [PubMed]

27. Liu, G.T.; Qian, Y.Z.; Zhang, P.; Dong, Z.M.; Shi, Z.Y.; Zhen, Y.Z.; Miao, J.; Xu, Y.M. Relationships between *Alternaria alternata* and oesophageal cancer. *IARC Sci. Publ.* **1991**, *105*, 258–262.

28. Fleck, S.C.; Sauter, F.; Pfeiffer, E.; Metzler, M.; Hartwig, A.; Koberle, B. DNA damage and repair kinetics of the *Alternaria* mycotoxins alternariol, altertoxin II and stemphyltoxin III in cultured cells. *Mutat. Res. Genet. Toxicol. Environ. Mutagen* **2016**, *798–799*, 27–34. [CrossRef] [PubMed]

29. Vaquera, S.; Patriarca, A.; Fernandez Pinto, V. Influence of environmental parameters on mycotoxin production by *Alternaria arborescens*. *Int. J. Food Microbiol.* **2016**, *219*, 44–49. [CrossRef] [PubMed]

30. Andersen, B.; Nielsen, K.F.; Fernandez Pinto, V.; Patriarca, A. Characterization of *Alternaria* strains from Argentinean blueberry, tomato, walnut and wheat. *Int. J. Food Microbiol.* **2015**, *196*, 1–10. [CrossRef] [PubMed]

31. Siciliano, I.; Ortu, G.; Gilardi, G.; Gullino, M.L.; Garibaldi, A. Mycotoxin production in liquid culture and on plants infected with *Alternaria spp.* isolated from rocket and cabbage. *Toxins* **2015**, *7*, 743–754. [CrossRef] [PubMed]

32. Kahl, S.M.; Ulrich, A.; Kirichenko, A.A.; Müller, M.E. Phenotypic and phylogenetic segregation of *Alternaria infectoria* from small-spored *Alternaria* species isolated from wheat in Germany and Russia. *J. Appl. Microbiol.* **2015**, *119*, 1637–1650. [CrossRef] [PubMed]

33. Tolgyesi, A.; Stroka, J.; Tamosiunas, V.; Zwickel, T. Simultaneous analysis of *Alternaria* toxins and citrinin in tomato: An optimised method using liquid chromatography-tandem mass spectrometry. *Food Addit. Contam. A* **2015**, *32*, 1512–1522. [CrossRef] [PubMed]

34. Noser, J.; Schneider, P.; Rother, M.; Schmutz, H. Determination of six *Alternaria* toxins with UPLC-MS/MS and their occurrence in tomatoes and tomato products from the Swiss market. *Mycotoxin Res.* **2011**, *27*, 265–271. [CrossRef] [PubMed]

35. Hickert, S.; Bergmann, M.; Ersen, S.; Cramer, B.; Humpf, H.-U. Survey of *Alternaria* toxin contamination in food from the German market, using a rapid HPLC-MS/MS approach. *Mycotoxin Res.* **2016**, *32*, 7–18. [CrossRef] [PubMed]

36. Walravens, J.; Mikula, H.; Rychlik, M.; Asam, S.; Ediage, E.N.; Di Mavungu, J.D.; Van Landschoot, A.; Vanhaecke, L.; De Saeger, S. Development and validation of an ultra-high-performance liquid chromatography tandem mass spectrometric method for the simultaneous determination of free and conjugated *Alternaria* toxins in cereal-based foodstuffs. *J. Chromatogr. A* **2014**, *1372C*, 91–101. [CrossRef] [PubMed]

37. Asam, S.; Rychlik, M. Recent developments in stable isotope dilution assays in mycotoxin analysis with special regard to *Alternaria* toxins. *Anal. Bioanal. Chem.* **2015**, *407*, 7563–7577. [CrossRef] [PubMed]

38. Lafaye, A.; Junot, C.; Ramounet-Le Gall, B.; Fritsch, P.; Ezan, E.; Tabet, J.C. Profiling of sulfoconjugates in urine by using precursor ion and neutral loss scans in tandem mass spectrometry. Application to the investigation of heavy metal toxicity in rats. *J. Mass Spectrom.* **2004**, *39*, 655–664. [CrossRef] [PubMed]

39. Logrieco, A.; Bottalico, A.; Mule, G.; Moretti, A.; Perrone, G. Epidemiology of toxigenic fungi and their associated mycotoxins for some mediterranean crops. *Eur. J. Plant Pathol.* **2003**, *109*, 645–667. [CrossRef]

40. Li, F.Q.; Toyazaki, N.; Yoshizawa, T. Production of *Alternaria* mycotoxins by *Alternaria alternata* isolated from weather-damaged wheat. *J. Food Prot.* **2001**, *64*, 567–571. [PubMed]

41. Grabarkiewicz-Szczesna, J.; Chelkowski, J. Metabolites produced by *Alternaria* species and their natural occurrence in Poland. In *Alternaria Biology, Plant Diseases and Metabolites*; Chelkowski, J., Visconti, A., Eds.; Elsevier: Amsterdam, The Netherlands, 1992; pp. 363–380.

42. Visconti, A.; Sibilia, A.; Sabia, C. *Alternaria alternata* from oilseed rape: Mycotoxin production, and toxicity to *Artemia salina* larvae and rape seedlings. *Mycotoxin Res.* **1992**, *8*, 9–16. [CrossRef] [PubMed]

43. Patriarca, A.; Azcarate, M.P.; Terminiello, L.; Fernandez Pinto, V. Mycotoxin production by *Alternaria* strains isolated from Argentinean wheat. *Int. J. Food Microbiol.* **2007**, *119*, 219–222. [CrossRef] [PubMed]

44. Magan, N.; Aldred, D. Why do fungi produce mycotoxins? In *Food Mycology: A Multifaceted Approach to Fungi and Food*; Dijksterhuis, J., Samson, R.A., Eds.; CRC Press: Boca Raton, FL, USA, 2007; Volume 25, pp. 121–133.

45. Oviedo, M.S.; Ramirez, M.L.; Barros, G.G.; Chulze, S.N. Effect of environmental factors on tenuazonic acid production by *Alternaria alternata* on soybean-based media. *J. Appl. Microbiol.* **2009**, *107*, 1186–1192. [CrossRef] [PubMed]

46. Pose, G.; Patriarca, A.; Kyanko, V.; Pardo, A.; Fernández Pinto, V. Effect of water activity and temperature on growth of *Alternaria alternata* on a synthetic tomato medium. *Int. J. Food Microbiol.* **2009**, *135*, 60–63. [CrossRef] [PubMed]

47. Shier, W.T.; Abbas, H.K.; Mirocha, C.J. Toxicity of the mycotoxins fumonisins b1 and b2 and *Alternaria alternata* f. sp. *lycopersici* toxin (AAL) in cultured mammalian cells. *Mycopathologia* **1991**, *116*, 97–104. [PubMed]

48. Vesonder, R.F.; Gasdorf, H.; Peterson, R.E. Comparison of the cytotoxicities of *Fusarium* metabolites and *Alternaria* metabolite AAL-toxin to cultured mammalian cell lines. *Arch. Environ. Contam. Toxicol.* **1993**, *24*, 473–477. [CrossRef] [PubMed]

49. Kelman, M.J.; Renaud, J.B.; Seifert, K.A.; Mack, J.; Sivagnanam, K.; Yeung, K.K.; Sumarah, M.W. Identification of six new *Alternaria* sulfoconjugated metabolites by high-resolution neutral loss filtering. *Rapid Commun. Mass Spectrom.* **2015**, *29*, 1805–1810. [CrossRef] [PubMed]

50. Proctor, R.H.; Desjardins, A.E.; McCormick, S.P.; Plattner, R.D.; Alexander, N.J.; Brown, D.W. Genetic analysis of the role of trichothecene and fumonisin mycotoxins in the virulence of *Fusarium*. *Eur. J. Plant Pathol.* **2002**, *108*, 691–698. [CrossRef]

51. Kang, Z.; Buchenauer, H. Studies on the infection process of *Fusarium culmorum* in wheat spikes: Degradation of host cell wall components and localization of trichothecene toxins in infected tissue. In *Mycotoxins in Plant Disease: Under the Aegis of Cost Action 835 'Agriculturally Important Toxigenic Fungi 1998–2003', eu Project (qlk 1-ct-1998-01380), and Ispp 'Fusarium Committee'*; Logrieco, A., Bailey, J.A., Corazza, L., Cooke, B.M., Eds.; Springer: Dordrecht, The Netherlands, 2002; pp. 653–660.

52. Müller, T.; Behrendt, U.; Ruppel, S.; von der Waydbrink, G.; Müller, M.E. Fluorescent pseudomonads in the phyllosphere of wheat: Potential antagonists against fungal phytopathogens. *Curr. Microbiol.* **2016**, *72*, 383–389. [CrossRef] [PubMed]

53. Saß, V.; Milles, J.; Krämer, J.; Prange, A. Competitive interactions of *Fusarium graminearum* and *Alternaria alternata* in vitro in relation to deoxynivalenol and zearalenone production. *J. Food Agric. Environ.* **2007**, *5*, 257–261.

54. Müller, M.E.; Steier, I.; Koppen, R.; Siegel, D.; Proske, M.; Korn, U.; Koch, M. Cocultivation of phytopathogenic *Fusarium* and *Alternaria* strains affects fungal growth and mycotoxin production. *J. Appl. Microbiol.* **2012**, *113*, 874–887. [CrossRef] [PubMed]

55. Müller, M.E.H.; Urban, K.; Köppen, R.; Siegel, D.; Korn, U.; Koch, M. Mycotoxins as antagonistic or supporting agents in the interaction between phytopathogenic *Fusarium* and *Alternaria* fungi. *World Mycotoxin J.* **2015**, *8*, 311–321. [CrossRef]

56. Azcarate, M.P.; Patriarca, A.; Terminiello, L.; Pinto, V.F. *Alternaria* toxins in wheat during the 2004 to 2005 Argentinean harvest. *J. Food Prot.* **2008**, *71*, 1262–1265. [PubMed]

57. Webley, D.J.; Jackson, K.L.; Mullins, J.D.; Hocking, A.D.; Pitt, J.I. *Alternaria* toxins in weather-damaged wheat and sorghum in the 1995–1996 australian harvest. *Aust. J. Agric. Res.* **1997**, *48*, 1249–1255. [CrossRef]

58. Fleck, S.C.; Pfeiffer, E.; Metzler, M. Permeation and metabolism of *Alternaria* mycotoxins with perylene quinone structure in cultured Caco-2 cells. *Mycotoxin Res.* **2014**, *30*, 17–23. [CrossRef] [PubMed]

© 2016 by the authors. Licensee MDPI, Basel, Switzerland. This article is an open access article distributed under the terms and conditions of the Creative Commons Attribution (CC BY) license (http://creativecommons.org/licenses/by/4.0/).

toxins

Article

Development and Validation of a LC-ESI-MS/MS Method for the Determination of *Alternaria* Toxins Alternariol, Alternariol Methyl-Ether and Tentoxin in Tomato and Tomato-Based Products

Yelko Rodríguez-Carrasco *, Jordi Mañes, Houda Berrada and Cristina Juan

Department of Food Chemistry and Toxicology, Faculty of Pharmacy, University of Valencia,
Av. Vicent Andrés Estellés s/n, 46100 Burjassot, Spain; jordi.manes@uv.es (J.M.);
houda.berrada@uv.es (H.B.); cristina.juan@uv.es (C.J.)
* Correspondence: yelko.rodriguez@uv.es; Tel.: +34-96-354-4117; Fax: +34-96-354-4954

Academic Editor: Aldo Laganà
Received: 29 September 2016; Accepted: 9 November 2016; Published: 11 November 2016

Abstract: *Alternaria* species are capable of producing several secondary toxic metabolites in infected plants and in agricultural commodities, which play important roles in food safety. *Alternaria alternata* turn out to be the most frequent fungal species invading tomatoes. Alternariol (AOH), alternariol monomethyl ether (AME), and tentoxin (TEN) are some of the main *Alternaria* mycotoxins that can be found as contaminants in food. In this work, an analytical method based on liquid chromatography (LC) tandem mass spectrometry (MS/MS) detection for the simultaneous quantification of AOH, AME, and TEN in tomato and tomato-based products was developed. Mycotoxin analysis was performed by dispersive liquid-liquid microextraction (DLLME) combined with LC-ESI-MS/MS. Careful optimization of the MS/MS parameters was performed with an LC/MS system with the ESI interface in the positive ion mode. Mycotoxins were efficiently extracted from sample extract into a droplet of chloroform (100 µL) by DLLME technique using acetonitrile as a disperser solvent. Method validation following the Commission Decision No. 2002/657/EC was carried out by using tomato juice as a blank matrix. Limits of detection and quantitation were, respectively, in the range 0.7 and 3.5 ng/g. Recovery rates were above 80%. Relative standard deviations of repeatability (RSDr) and intermediate reproducibility (RSD$_R$) were $\leq$ 9% and $\leq$ 15%, respectively, at levels of 25 and 50 ng/g. Five out of 30 analyzed samples resulted positive to at least one *Alternaria* toxin investigated. AOH was the most common *Alternaria* toxin found, but at levels close to LOQ (average content: 3.75 ng/g).

Keywords: *Alternaria*; LC-MS/MS; dispersive liquid-liquid microextraction; tomato

---

1. Introduction

Alternaria is a widely distributed fungal genus frequently isolated from different plant crops, and has been documented as a pre- and post-harvest pathogen causing decay [1]. The fungal species of *Alternaria* are considered relevant contaminants of refrigerated fruits, vegetables, and stored foodstuffs, mainly as a consequence of their occurrence and the ability to grow and produce toxins even at low temperatures and low water activity [2]. *Alternaria* species produce a liarge variety of secondary metabolites capable of causing several health problems in humans and animals. The most relevant mycotoxins produced by *Alternaria* spp. are alternariol (AOH), alternariol monomethyl ether (AME), tentoxin (TEN), tenuazonic acid (TeA), altenuene (ALT), and altertoxins (ATXs).

The toxic effects of *Alternaria* toxins are wide-ranging. To date some of these mycotoxins have shown to be teratogenic in vivo. Genotoxic effects of AOH and AME in vitro have also been described [3]. Recently, some authors reported that AOH and AME are able to induce cell cycle arrest, apoptosis of cells, and DNA damaging effects [4–7]. In spite of the before-mentioned, there are currently no guideline limits set for *Alternaria* mycotoxins by regulatory authorities yet. The European Food Safety Agency (EFSA) provided a scientific opinion on the risks for animal and public health related to the presence of *Alternaria* toxins in feed and food [3]. EFSA evidenced a lack of robust occurrence data of *Alternaria* toxins in food and processed products, and recommended the collection of representative data across Europe to enable a proper risk assessment.

Tomatoes and many other soft-skinned vegetables and fruits can be easily infected by fungi and *Alternaria* is the main fungus responsible for spoilage. Tomato (*Solanum lycopersicum* L., syn. *Lycopersicon esculentum* Mill.) is considered to be one of the main important vegetable crops worldwide. Although tomatoes are commonly consumed fresh, over 80% of the tomato consumption comes from processed products, such as tomato juice, paste, puree, ketchup, and soup, such as gazpacho, a traditional Spanish, ready-to-serve cold vegetable soup, which contain fresh tomato (> 50%) and other ingredients, such as cucumber, pepper, olive oil, and other minor constituents, such as onion, garlic, wine vinegar, salt, and water.

Based on the increasing need for incidence data, a bunch of new analytical methods are demanded for detection and quantification of *Alternaria* toxins in foods. The coupling of both liquid (LC) and gas chromatography (GC) to tandem mass spectrometry (MS/MS) has enabled the development of highly selective, sensitive, and accurate methods for mycotoxin determination in both biological [8,9] and food samples [10,11]. For the analysis of mycotoxins in various food matrices, the traditional liquid-liquid extraction (LLE), solid phase extraction (SPE), and combinations of LLE and SPE have been commonly used as sample preparation procedures as recently reviewed by Turner et al. [12]. An ideal sample preparation procedure should be straightforward and rapid with low operational cost, as well as efficient in sample clean-up. Furthermore, to allow the trace-analysis of various compounds, a high enrichment factor could be of interest. To fulfill these ideal requirements, in 2006 it was proposed a novel very attractive dispersive liquid-liquid microextraction technique (DLLME) for the treatment of liquid samples, and has been recently used in the determination of some mycotoxins in several food samples [13–17]. Basically, DLLME consists of the formation of a cloudy solution promoted by the fast addition of a mixture of extraction and disperser solvents to an aqueous sample. The tiny droplets formed and dispersed among the aqueous sample solution are further joined by centrifugation. DLLME has been proved to be a powerful cleaning and preconcentration technique. Other benefits are its high speeds, the low solvent use, and final disposal.

The objective of the present work was to develop and validate a reliable DLLME-LC-ESI-MS/MS method for simultaneous determination of some *Alternaria* toxins in tomato and tomato-based products. Special attention was given on the optimization of the MS/MS parameters to attain the best response. Additionally, optimization of the DLLME procedure was also assessed, by careful evaluation of the nature and amount of extraction and disperser solvents as well as the amount of sample. The validated method was used to assess the occurrence of AOH, AME, and TEN in 30 tomato and tomato-based samples commercialized in Valencia, Spain.

2. Results and Discussion

2.1. MS/MS Optimization

A preliminary study was conducted in order to obtain the best instrumental conditions affording high resolution and short analysis time with a suitable analyte separation. The optimization of the analyte-dependent MS/MS parameters was performed via direct infusion of standards (diluted in a 1:1 mixture of eluent A and B) into the MS source using a syringe injection at a flow rate of 10 μL/min. Positive and negative ionization modes were tested, obtaining a better response in positive ionization mode for the studied *Alternaria* toxins. Compound-dependent parameters of quadrupole mode scans including declustering potential (DP), entrance potential (EP), collision cell entrance potential (CEP), collision cell exit potential (CXP), and collision energies (CE) were also evaluated and optimized to provide the best combination of efficiency and finding the optimal response value for each analyte. The acquisition of two single reaction monitoring transitions per analyte allowed the confirmation of the identity of the positive results according to the criteria established in Commission Decision No. 2002/657/EC [18]. The product ion with the highest intensity was selected as a quantifier, whereas the other was used as a qualifier. Table 1 lists the characteristic ions and the optimized mass spectrometry parameters for each compound during multiple reaction monitoring (MRM) acquisitions.

Table 1. Retention times, main transitions, collision energies (CE), declustering potential (DP), entrance potential (EP), collision cell entrance potential (CEP), and collision cell exit potential (CXP) for the *Alternaria* toxins analyzed in this study.

Analyte	Rt (min)	Parent Ion Q1 (m/z)	Product Ion Q3 (m/z)	CE (V)	DP (V)	EP (V)	CEP (V)	CXP (V)
TEN	8.1	415 [M + H]$^+$	312 ^{Q} 256 ^{q}	29 39	55	8	21	2
AOH	8.3	259 [M + H]$^+$	128 ^{Q} 184 ^{q}	65 42	39	10	16	3
AME	9.1	273 [M + H]$^+$	128 ^{Q} 228 ^{q}	60 40	32	10	16	13

^{Q} Quantification transition; ^{q} Qualification transition.

2.2. DLLME Optimization

DLLME has been proved to be a powerful cleaning and preconcentration technique. Other benefits, such as its high speeds and the low solvent use (if compared with traditional sample preparation procedures) and final disposal, have caused it to fast become one of the most popular analytical sample preparations in developing reliable quantitative multi-analyte methods. Other promising techniques, such as dilute-and-shoot approaches, only requires the dilution of the sample, however, it does not remove any interference which can cause chromatographic troubles, such as carryover, loss of sensitivity, increase of background interferences, or co-eluting peaks. Therefore, its use as a routine procedure may cause a time consuming drawback coming from the troubleshooting and cleaning of the system.

Hence, the DLLME sample preparation procedure was selected here based on its benefit in routine on a daily basis. The effect of the following parameters affecting the extraction efficiency was evaluated: (i) the type of extraction and disperser solvents; (ii) the extraction-disperser solvent ratios; and (iii) the amount of sample. The method of optimization was performed by recovery experiments in three replicates using tomato extract blank samples (5 mL) spiked with 25 ng/g of each targeted mycotoxin.

2.2.1. Influence of the Type of Extraction and Disperser Solvents

Selection of both an appropriate extraction and disperser solvent is very crucial to achieve good performance. The extraction solvent must have properties, such as a greater density than water, high extraction capability of the analytes, as well as low solubility in water. The role of the disperser solvent is to increase the dispersion of the extraction solvent as tiny droplets in an aqueous medium solution resulting in a large contact area between the extraction solvent and aqueous solution, thus improving the extraction efficiency. In this study, three common halogenated solvents, including CCl_4, CH_2Cl_2, and $CHCl_3$ were tested for extraction, whereas acetonitrile, acetone, and methanol were selected as disperser solvents.

Mixtures of 1.0 mL of different disperser solvents and 100 μL of extraction solvent were injected to 5 mL of tomato sample extracts spiked with the standard solution at 50 ng/g. Furthermore, 1 g of NaCl was added to the sample extract to improve the extraction efficiency, as well as to facilitate the phase's separation [19]. Extraction efficiency was evaluated by comparing the recoveries of the analytes. Results indicated that the best conditions were accomplished with the AcN-CHCl$_3$ pair, with satisfactory recoveries between 81%–94% (Table 2). There were no differences in extraction for the studied analytes.

Table 2. Recovery range of *Alternaria* toxins obtained by using different combinations of extraction (Ac, AcN and MeOH) and disperser solvents (CCl_4, CH_2Cl_2, and $CHCl_3$).

	Recovery Range (%) [a]		
Disperser Solvent	**Extraction Solvent**		
	CCl_4	CH_2Cl_2	$CHCl_3$
Ac	45–67	35–56	69–78
AcN	71–86	47–66	81–94
MeOH	58–81	53–78	65–83

[a] spiked level: 25 ng/g of each target mycotoxin.

2.2.2. Influence of the Extraction-Disperser Solvent Ratio

To evaluate the influence of the extraction-disperser solvent ratios on the extraction efficiency different volumes of chloroform (60, 80, 100, and 120 μL) and acetonitrile (0.5, 1.0, and 1.5 mL) were used. The optimal volumes of AcN and $CHCl_3$ were evaluated with the MATrix LABoratory (MATLAB)-based surface response design (Figure 1).

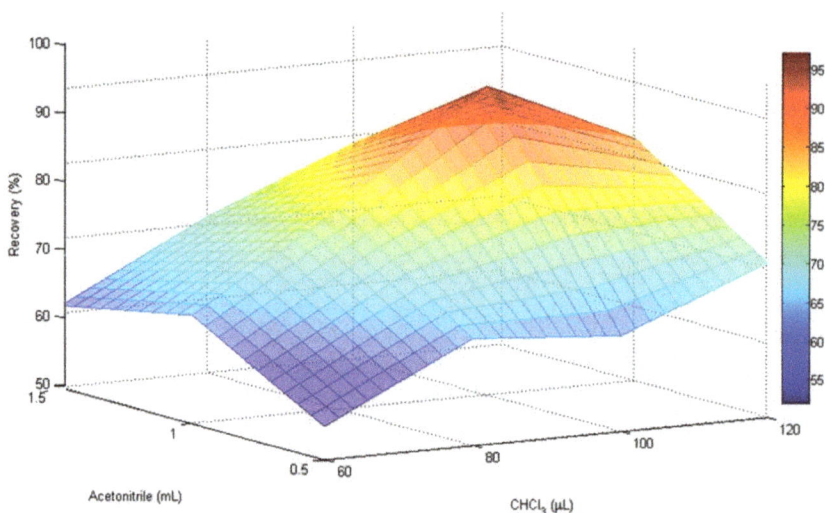

Figure 1. MATLAB-based surface response design showing the influence of AcN and $CHCl_3$ ratio in the extraction efficiency of AOH. Recovery experiments were conducted by spiking tomato extract blank samples with 25 ng/g of each targeted mycotoxin.

The enrichment factor improved with the lower volumes, but the lower the volumes the lesser the volume of the sedimented phase. Despite this, when the volume of extraction solvent was increased from 60 to 100 μL, the recoveries of the mycotoxins rose significantly. However, it should be noted that with the following combinations of AcN-$CHCl_3$: 1.50 mL–120 μL, 1.50 mL–100 μL, and 1 mL–120 μL, the matrix effect increased with respect to the other tested ratios. Hence, the combination of 1 mL of acetonitrile containing 100 μL of $CHCl_3$ was selected as a good compromise to reach the best DLLME conditions.

2.2.3. Influence of the Amount of Sample

The influence of the amount of sample was evaluated by testing different amounts of sample extract (5, 7.5 and 10 mL). Results showed that recoveries below 80% were obtained with 7.5 and 10 mL of sample, whereas recoveries greater than 80% were achieved with 5 mL. Thus, 5 mL of tomato extract was selected as the optimum amount of sample for a reliable and efficient extraction (Figure 2).

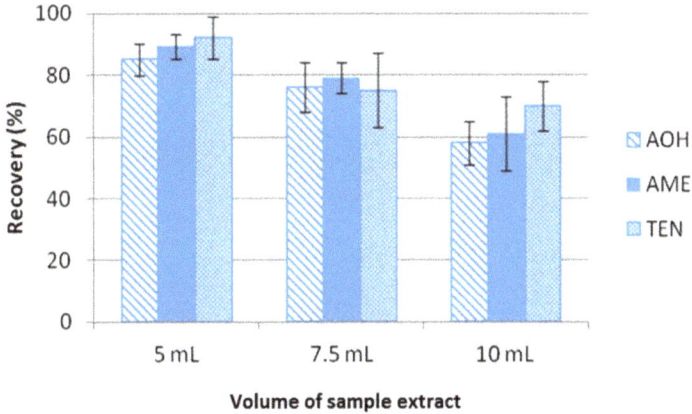

Figure 2. Influence of the volume of sample extract in the extraction efficiency of the *Alternaria* toxins studied. Recovery experiments were conducted by spiking tomato extract blank samples with 25 ng/g of each target mycotoxin.

2.3. Analytical Method Validation

Good linearity was achieved in all cases with regression coefficients higher than 0.990. Significant signal suppression was observed (from 65%–80%) between the slopes of the calibration lines meaning that the matrix effect is present (Table 3). Therefore, matrix-matched calibration curves were used for effective quantification in tomato samples.

Table 3. Overview of the correlation coefficient, extraction recovery, repeatability, and reproducibility (Rec (RSD), %), limits of detection (LODs) and quantitation (LOQ), and signal suppression/enhancement (SSE) for the studied analytes.

Mycotoxin	Correlation Coefficient (r)	Repeatability (RSD$_r$, %) [a]		Reproducibility (RSD$_R$, %) [a]		LOD (ng/g)	LOQ (ng/g)	SSE (%)
		25 ng/g [b]	50 ng/g [b]	25 ng/g [b]	50 ng/g [b]			
AOH	0.998	81 (6)	82 (4)	84 (8)	89 (6)	1.40	3.50	65
AME	0.996	86 (4)	89 (7)	90 (7)	93 (10)	1.40	3.50	80
TEN	0.995	91 (9)	94 (6)	94 (15)	90 (12)	0.70	1.75	78

[a] $n = 3$; [b] spiked level.

Satisfactory results in terms of recoveries were found (recovery range from 81%–94% for both spiking levels). Precision studies showed that the method was repeatable (RSD$_r$ < 9%) and reproducible (RSD$_R$ < 15%) (Table 3).

LOQs were 3.5 ng/g for AOH and AME, while the LOQ for TEN was 1.75 ng/g. LODs were 1.40 ng/g for AOH and AME, while the LOD for TEN was 0.70 ng/g (Table 3). These results showed the suitability of the developed method for the determination of trace amounts of the selected mycotoxins in tomato samples. No obvious interfering peak from blank samples was detected. MRM chromatograms of tomato juice spiked at 10 µg/L of AOH, AME, and TEN are shown in Figure 3.

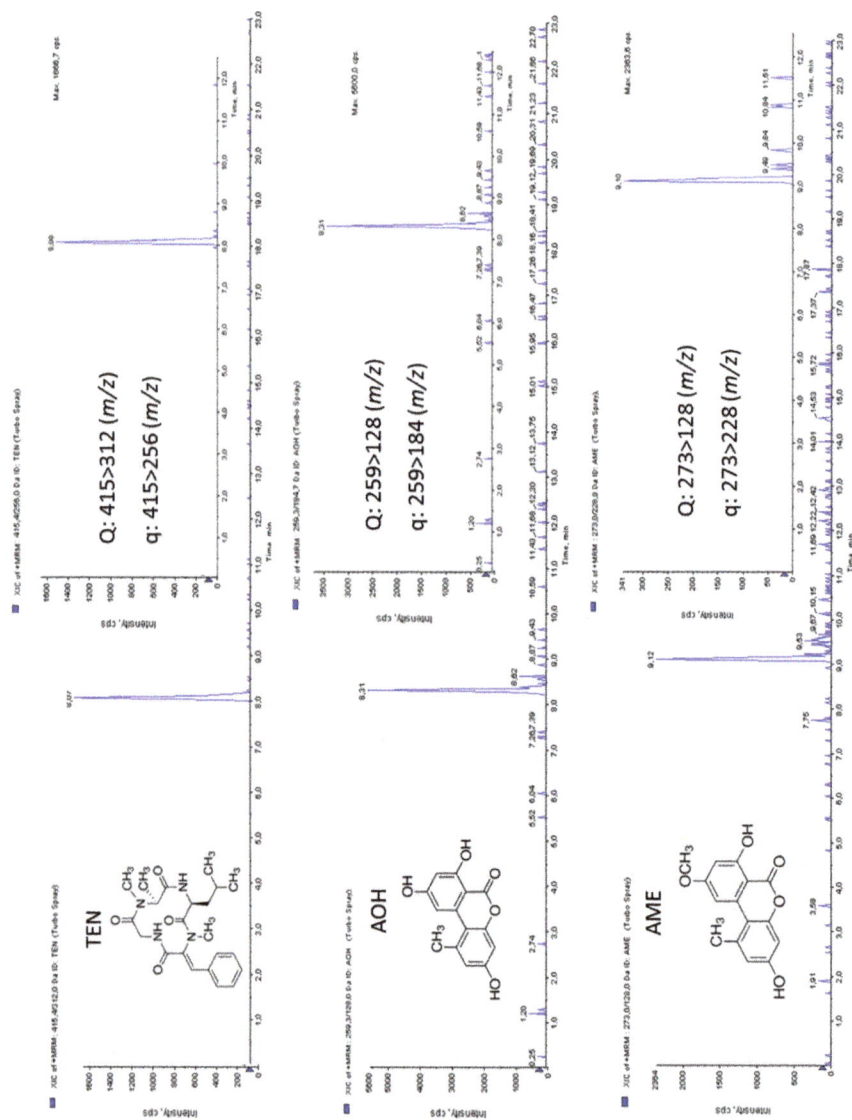

Figure 3. MRM chromatograms of tomato juice spiked with 3.5 µg/L of AOH, AME, and TEN (corresponding to 1 ng of each injected toxin).

The results obtained in the present study were within the limits set by Commission Decision, No. 2002/657/EC. According to Commission Decision, No. 2002/657/EC, it is acceptable that trueness of measurements is assessed through recovery of additions of known amounts of the analytes to a blank matrix and the guideline ranges for the deviation of the experimentally-determined recovery should be between 80% and 110% for a mass fraction ≥ 10 µg/kg. In the case of repeated analysis of a sample carried out under within-laboratory reproducibility conditions, the intra-laboratory relative standard deviation should not exceed 20% for a mass fraction of ≥ 10 µg/kg to 100 µg/kg. For analyses carried out under within-laboratory reproducibility conditions, the within-laboratory RSD shall not be greater than the reproducibility RSD.

2.4. Application to Samples

The developed method was evaluated carrying out a survey of AOH, AME, and TEN in 30 tomato and tomato-based products purchased in several Valencian supermarkets (Spain). Neither in tomato juice ($n = 5$) nor in gazpacho ($n = 5$) samples was the occurrence of target mycotoxins detected. However, five out of 20 fresh tomato samples (25%) resulted positive to at least one *Alternaria* toxin. AOH was detected in four out of the five contaminated samples but at levels close to LOQ (mean: 3.75 ng/g), whereas AME was identified in two fresh tomato samples but at levels between LOD and LOQ. TEN was not found in any analyzed sample. A MRM chromatogram of a naturally-contaminated tomato sample with AOH at 5.8 ng/g is shown in Figure 4.

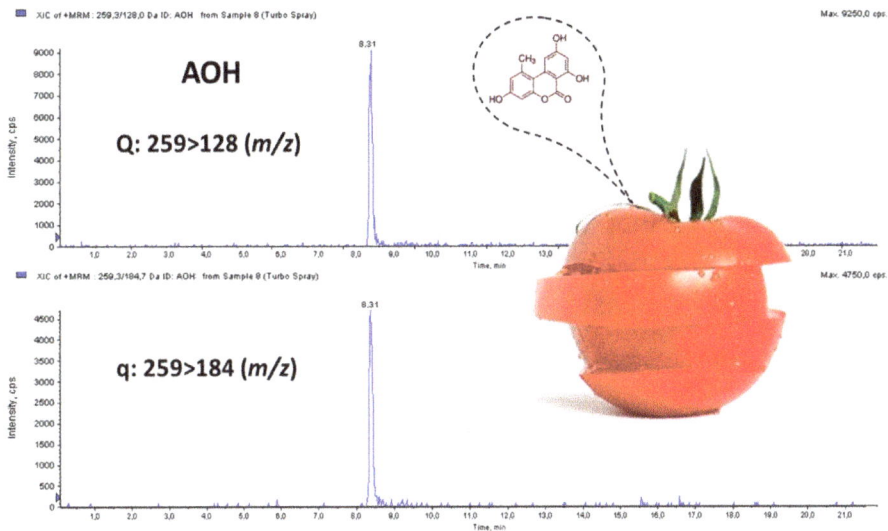

Figure 4. MRM chromatogram of a naturally contaminated tomato sample with AOH at 5.8 ng/g.

These findings are in agreement with the results reported in some European studies. In a study conducted in The Netherlands, AOH was detected in three out of 10 fresh tomato samples (levels ranging between 2–15 ng/g) but AME and TEN were not detected [20]. However, in the same study, AOH was quantified (at levels from 2–11 ng/g) in four out of 14 tomato juice samples. Similarly, in Switzerland, no *Alternaria* mycotoxins were found in fresh and whole tomato samples ($n = 4$) but AOH was detected in eight out of 24 tomato soup samples (levels from 4–10 ng/g). In the same products AME was detected in seven out of 24 samples but at lower concentration (from 1–4 ng/g). Only a few samples were positive for TEN (tomato puree, concentrated, and dried tomatoes) [21]. Those findings are also in line with the results reported in Italian tomato-based products ($n = 10$).

Occurrence of AOH was detected in five out of 10 samples (levels from 4–6.8 ng/g) and TEN in one sample (4.4 ng/g) whereas AME was not found [22]. In Germany, a higher occurrence of *Alternaria* mycotoxins in tomato products (*n* = 34) was recently reported [23]. AME, AOH, and TEN were detected in 100, 70, and 26% of samples, respectively. AOH was found at levels from 6.1–25 ng/g and AME from 1.2–7.4 ng/g, whereas TEN contamination was set at levels from LOD and LOQ (<6.6 ng/g).

In China, no *Alternaria* mycotoxins were found in a study conducted in 70 fresh tomato samples [24]. In contrast, contamination of 26.2% of AME (up to 1734 µg/kg) and 6.2% of AOH (up to 8756 µg/kg) was reported in 80 tomato purees processed and sold in Argentina [25]. Similar results were also reported by Van de Perre et al. [26] who analyzed a total of 144 samples of derived tomato products such as ketchups, concentrates, pulp, dried tomatoes, and juices, which were collected from local markets in different countries (i.e., Belgium, Spain, Egypt, Brazil, and South Africa). Puree and concentrate tomato samples showed the highest occurrence of AOH and AME whereas, in tomato juice samples, none of the studied toxins were detected.

3. Conclusions

A rapid, straight-forward, robust, sensitive and accurate analytical method based on DLLME-LC-ESI-MS/MS for determining various *Alternaria* toxins in tomato and tomato-based products was developed. Careful optimization of the MS/MS parameters was performed to reach the best analytical conditions. Additionally, parameters affecting the extraction efficiency of the DLLME were also evaluated and optimized. The method performance fulfilled the EU guideline standardized in the Commission Decision, NO. 2002/657/EC. The recoveries were greater than 80% and relative standard deviations of repeatability and intermediate reproducibility were ≤9% and ≤15%, respectively, at levels of 25 and 50 ng/g. Under the optimized conditions LODs and LOQs were in the range 0.7–3.5 ng/g, respectively. Significant signal suppression was observed and matrix-matched calibrations were used for quantitation purpose. The developed method was successfully applied to 30 commercially available tomato and tomato-based products acquired in Valencia, showing the occurrence of various *Alternaria* toxins, at levels of few nanograms per gram in 20% of samples, being AOH the most commonly mycotoxin found. Due to its simplicity, and by allowing a faster extraction, the proposed methodology is proposed as a reliable analytical tool. Furthermore, this method could be applied to gather data on the presence of *Alternaria* toxins in foodstuffs, which are highly recommended by EFSA to enable a proper risk assessment of these toxins.

4. Materials and Methods

4.1. Chemicals and Reagents

Acetonitrile (AcN), methanol (MeOH), acetone (Ac), chloroform (CHCl$_3$), dichloromethane (CH$_2$Cl$_2$), and carbone tetrachloride (CCl$_4$) were supplied by Merck (Darmstadt, Germany). Ammonium formate (99%) was supplied by Panreac Quimica S.A.U. (Barcelona, Spain). Deionized water was obtained in the laboratory using a Milli-Q SP® Reagent Water System (Millipore, Bedford, MA, USA).

Certified standards of AOH, AME and TEN were purchased from Sigma-Aldrich (Madrid, Spain). Standard solutions of AOH, AME, and TEN were prepared by dissolving 10 mg of each compound in 10 mL of MeOH. Stock solutions were diluted with pure MeOH afterwards order to get the appropriate working solutions. A multi-mycotoxin working standard solution was prepared by combining aliquots of each individual working solution and diluting with MeOH to obtain the final concentration of 0.02 mg/L for AOH, AME, and TEN. All solutions were stored at −20 °C in amber glass vials and darkness before use.

4.2. LC-MS/MS Analysis

The determination was performed using a system LC-MS/MS triple quadrupole, consisted of a LC Agilent 1200 (Agilent Technologies, Santa Clara, CA, USA)using a binary pump and automatic injector and coupled to a 3200 QTRAP® AB SCIEX (Applied Biosystems, Foster City, CA, USA). The chromatographic separation of the analyte was conducted at 25 °C with a reverse phase analytical column Gemini® C18 (3 µM, 150 × 2 mm ID) and a guard-column C18 (4 × 2 mm, ID; 3 µM) from Phenomenex (Madrid, Spain).

Mobile phase was a time programmed gradient using as phase A methanol (1% formic acid and 5 mM ammonium formate), and as phase B water (1% formic acid and 5 mM ammonium formate). The following gradient was employed: equilibration during 2 min at 10% A at 0.25 mL/min, 10%–80% A in 3 min at 0.25 mL/min, 80% A for 1 min at 0.25 mL/min, 80%–90% A in 2 min, 90% A for 6 min at 0.25 mL/min, 90%–100% A in 3 min at 0.25 mL/min, 100% for 1 min at 0.35 mL/min, 100%–50% in 3 min at 0.4 mL/min at, return to initial conditions in 2 min and maintain during 2 min. Total run time was 21 min. The injection volume was 20 µL.

To analyze the mycotoxins, a triple quadrupole mass spectrometry detector (MS/MS) 3200 QTRAP® System AB SCIEX (Applied Biosystems, Concord, ON, Canada) was used. Electrospray ionization (ESI) interfaces were used to analyze these mycotoxins with the following settings for source/gas parameters: curtain gas (CUR) 20, ionspray voltage (IS) 5500 V, source temperature (TEM) 450 °C, ion source gas 1 (GS1), and ion source gas 2 (GS2) 50. Therefore, in this study, the optimization of the MS/MS parameters was performed with an LC/MS system with the ESI interface in the positive ion mode using a mycotoxin standard mixture. The precursor ions (Q1) of each mycotoxin were confirmed in product ion (Q3) scan mode. As shown in Table 1, a protonated molecule was observed as the base peak ion in the mass spectra of AOH, AME, and TEN. Hence, these ions were selected as precursor ions (Q1) for each mycotoxin. The optimization of product ions (Q3) and their collision energy were performed in the product ion scan mode. The final selection of multiple reaction monitoring (MRM) transitions in positive ion mode for each compound, the optimal declustering potential (DP), entrance potential (EP), collision cell entrance potential (CEP), collision cell exit potential (CXP), and collision energies (CE) are shown in Table 1. Data acquisition and processing were performed using Analyst® software version 1.5.2. (MDS Analytical Technologies, 2008 MDS Inc, ON, Canada,).

4.3. Sampling and Sample Preparation

Thirty tomato and tomato-based samples were purchased from local supermarkets located in Valencia (Spain). Unwashed fresh tomato samples (n = 20) were chopped using a blender immediately after reception of samples in the laboratory. After blending, homogenization, and centrifugation (3000 rpm, 4° C for 5 min), the tomato samples were placed in 100 mL closed polyethylene flasks before storing at 4 °C until analysis. Analyses were carried out within the following two days after reception. Tomato-based samples consisted of gazpacho samples (n = 5) and tomato juice (n = 5).

A modified version of the DLLME method for sample preparation of fruit juices was used [27]. A mixture of 1 mL of AcN (as disperser solvent) and 100 µL of CHCl$_3$ (as extraction solvent), was rapidly injected into 5 mL of centrifuged tomato extract containing 1 g of NaCl. The mixture was vortexed for 1 min and centrifuged at 4000 rpm for 5 min, and the droplet formed was collected by a 100 µL syringe and transferred to a chromatography vial. Then, the droplet was evaporated to dryness under a gentle stream of N$_2$ and reconstituted with 1 mL of MeOH:H$_2$O (50:50, *v/v*). The solution was filtered through 13 mm/0.20 µm nylon filter and injected into the LC–MS/MS system for mycotoxin analysis.

4.4. Method Performance

The method performance was performed under optimized conditions following the Commission Decision No. 2002/657/EC. The method validation included the evaluation of linearity, limits of detection (LODs), limits of quantification (LOQs), recoveries, repeatability (intra-day precision), and intermediate reproducibility (inter-day precision). All of the parameters were evaluated by spiking blank tomato juice samples at 25 and 50 ng/g. Samples were spiked and left to equilibrate over night before the analysis.

Linearity was assessed through six concentration levels in a linear range between LOQ and 100 × LOQ in triplicate. The correlation coefficient was obtained by plotting the signal intensity against analyte concentrations. A calibration curve was injected at the end of each batch to assess the response drift of the method. Components from matrix can negatively influence the method performance if they co-elute with the analyte of interest and can cause ion suppression or enhancement in the ion source. Therefore, matrix effect was also evaluated. The matrix effect (ME), is defined as the ratio between the slopes of the matrix-matched calibration and the solvent calibration one, and it was calculated as follows:

$$ME(\%) = \frac{Slope_{matrix-matched}}{Slope_{solvent}} \times 100$$

Matrix-matched calibration curves were built by spiking blank sample extracts with the studied analytes at the same concentration levels than those used in solvent standard calibration curves.

The accuracy was evaluated through recovery studies using spiked blank samples at 25 ng/g and 50 ng/g concentration levels. Recovery studies were performed in triplicate in the same day, as well as in three different days. Precision (expressed as %RSD) of the method was determined by repeatability (intraday precision, RSDr) and intermediate reproducibility (interday precision, RSDR). Intraday variation was evaluated in three determinations per concentration in a single day, whereas interday variation was tested on three different working days within 20 days. RSDr and RSDR were determined by spiking blank samples at the 25 ng/g and 50 ng/g concentration levels.

Limits of detection (LODs) and limits of quantitation (LOQs) were estimated from a blank juice tomato sample fortified with decreasing concentrations of the analytes. LODs were calculated using a signal-to-noise ratio of 3:1. LOQs Results were calculated using a signal-to-noise ratio of 10:1. The specificity of the method was evaluated with respect to interferences from endogenous compounds. Five samples of blank tomato juice samples were analyzed and compared with the corresponding spiked samples at the LOQ level to check for possible interference with the detection of the analytes.

4.5. Confirmation Criteria

Confirmation criteria were based on the following items; (i) chromatographic separation: the retention time of the analyte in the extract should correspond to that of the matrix-matched calibration within a ± 2.5% interval of the retention time; (ii) mass spectrometric detection: extracted ion chromatograms of sample extracts should have peak shapes and response ratios to those obtained from calibration standards analyzed at comparable concentrations in the same batch. The relative intensities or ratios of selective ions, expressed as a ratio relative to the most intense ion used for identification, should correspond to those of the calibration standard solutions. The ion ratio should not deviate more than 30% (relative).

Acknowledgments: This work was supported by the Spanish Ministry of Economy and Competitiveness (AGL2013-43194-P) and Ministry of Education (No. AP2010-2940) F.P.U. Grant. In addition, we would like to thank to J.J. Sánchez for the processing of data by MATrix LABoratory (MATLAB).

Author Contributions: Houda Berrada and Jordi Mañes conceived and designed the experiments; Yelko Rodríguez-Carrasco, Cristina Juan performed the experiments, analyzed the data and wrote the paper.

Conflicts of Interest: The authors declare no conflict of interest.

References

1. Logrieco, A.; Moretti, A.; Solfrizzo, M. *Alternaria* Toxins and Plant Diseases: An Overview of Origin, Occurrence and Risks. *World Mycotoxin J.* **2009**, *2*, 129–140. [CrossRef]
2. Siciliano, C.I.; Ortu, G.; Gilardi, G.; Gullino, M.L.; Garibaldi, A. Mycotoxin Production in Liquid Culture and on Plants Infected with *Alternaria* spp. Isolated from Rocket and Cabbage. *Toxins* **2015**, *7*, 743–754. [CrossRef] [PubMed]
3. EFSA Panel on Contaminants in the Food Chain (CONTAM). Scientific Opinion on the Risks for Animal and Public Health Related to the Presence of *Alternaria* toxins in Feed and Food. *EFSA J.* **2011**. [CrossRef]
4. Bensassi, F.; Gallerne, C.; Dein, O.S.E.; Hajlaoui, M.R.; Bacha, H.; Lemaire, C. Mechanism of Alternariol Monomethyl Ether-Induced Mitochondrial Apoptosis in Human Colon Carcinoma Cells. *Toxicology* **2011**, *290*, 230–240. [CrossRef] [PubMed]
5. Bensassi, F.; Gallerne, C.; Sharaf El Dein, O.; Hajlaoui, M.R.; Bacha, H.; Lemaire, C. Cell Death Induced by the *Alternaria* Mycotoxin Alternariol. *Toxicol. In Vitro* **2012**, *26*, 915–923. [CrossRef] [PubMed]
6. Fernández-Blanco, C.; Font, G.; Ruiz, M. Oxidative Stress of Alternariol in Caco-2 Cells. *Toxicol. Lett.* **2014**, *229*, 458–464. [CrossRef] [PubMed]
7. Fernández-Blanco, C.; Juan-García, A.; Juan, C.; Font, G.; Ruiz, M. Alternariol Induce Toxicity via Cell Death and Mitochondrial Damage on Caco-2 Cells. *Food Chem. Toxicol.* **2016**, *88*, 32–39. [CrossRef] [PubMed]
8. Rodríguez-Carrasco, Y.; Moltó, J.C.; Mañes, J.; Berrada, H. Development of a GC-MS/MS Strategy to Determine 15 Mycotoxins and Metabolites in Human Urine. *Talanta* **2014**, *128*, 125–131. [CrossRef] [PubMed]
9. Rodríguez-Carrasco, Y.; Heilos, D.; Richter, L.; Süssmuth, R.D.; Heffeter, P.; Sulyok, M.; Kenner, L.; Berger, W.; Dornetshuber-Fleiss, R. Mouse Tissue Distribution and Persistence of the Food-Born Fusariotoxins Enniatin B and Beauvericin. *Toxicol. Lett.* **2016**, *247*, 35–44. [CrossRef] [PubMed]
10. Rodríguez-Carrasco, Y.; Berrada, H.; Font, G.; Mañes, J. Multi-Mycotoxin Analysis in Wheat Semolina using an Acetonitrile-Based Extraction Procedure and Gas Chromatography–Tandem Mass Spectrometry. *J. Chromatogr. A* **2012**, *1270*, 28–40. [CrossRef] [PubMed]
11. Tolosa, J.; Font, G.; Mañes, J.; Ferrer, E. Nuts and Dried Fruits: Natural Occurrence of Emerging *Fusarium* Mycotoxins. *Food Control* **2013**, *33*, 215–220. [CrossRef]
12. Turner, N.W.; Bramhmbhatt, H.; Szabo-Vezse, M.; Poma, A.; Coker, R.; Piletsky, S.A. Analytical Methods for Determination of Mycotoxins: An Update (2009–2014). *Anal. Chim. Acta* **2015**, *901*, 12–33. [CrossRef] [PubMed]
13. Arroyo-Manzanares, N.; Huertas-Pérez, J.F.; Gámiz-Gracia, L.; García-Campaña, A.M. A New Approach in Sample Treatment Combined with UHPLC-MS/MS for the Determination of Multiclass Mycotoxins in Edible Nuts and Seeds. *Talanta* **2013**, *115*, 61–67. [CrossRef] [PubMed]
14. Lai, X.; Ruan, C.; Liu, R.; Liu, C. Application of Ionic Liquid-Based Dispersive Liquid-Liquid Microextraction for the Analysis of Ochratoxin A in Rice Wines. *Food Chem.* **2014**, *161*, 317–322. [CrossRef] [PubMed]
15. Amoli-Diva, M.; Taherimaslak, Z.; Allahyari, M.; Pourghazi, K.; Manafi, M.H. Application of Dispersive Liquid–Liquid Microextraction Coupled with Vortex-Assisted Hydrophobic Magnetic Nanoparticles Based Solid-Phase Extraction for Determination of Aflatoxin M_1 in Milk Samples by Sensitive Micelle Enhanced Spectrofluorimetry. *Talanta* **2015**, *134*, 98–104. [CrossRef] [PubMed]
16. Víctor-Ortega, M.D.; Lara, F.J.; García-Campaña, A.M.; del Olmo-Iruela, M. Evaluation of Dispersive Liquid–Liquid Microextraction for the Determination of Patulin in Apple Juices Using Micellar Electrokinetic Capillary Chromatography. *Food Control* **2013**, *31*, 353–358. [CrossRef]
17. Juan, C.; Chamari, K.; Oueslati, S.; Mañes, J. Rapid Quantification Method of Three *Alternaria* Mycotoxins in Strawberries. *Food Anal. Methods* **2016**, *9*, 1573–1579. [CrossRef]
18. Commission Decision No. 2002/657/EC of 12 August 2002 Implementing Council Directive 96/23/EC Concerning the Performance of Analytical Methods and the Interpretation of Results (Text with EEA Relevance). 2002. Available online: http://data.europa.eu/eli/dec/2002/657/oj (accessed on 5 September 2016).
19. Cruz-Vera, M.; Lucena, R.; Cárdenas, S.; Valcárcel, M. Sample Treatments Based on Dispersive (Micro)Extraction. *Anal. Method* **2011**, *3*, 1719–1728. [CrossRef]
20. López, P.; Venema, D.; de Rijk, T.; de Kok, A.; Scholten, J.M.; Mol, H.G.J.; de Nijs, M. Occurrence of *Alternaria* Toxins in Food Products in the Netherlands. *Food Control* **2016**, *60*, 196–204. [CrossRef]

21. Noser, J.; Schneider, P.; Rother, M.; Schmutz, H. Determination of Six *Alternaria* Toxins with UPLC-MS/MS and their Occurrence in Tomatoes and Tomato Products from the Swiss Market. *Mycotoxin Res.* **2011**, *27*, 265–271. [CrossRef] [PubMed]

22. Prelle, A.; Spadaro, D.; Garibaldi, A.; Gullino, M.L. A New Method for Detection of Five *Alternaria* Toxins in Food Matrices Based on LC–APCI-MS. *Food Chem.* **2013**, *140*, 161–167. [CrossRef] [PubMed]

23. Hickert, S.; Bergmann, M.; Ersen, S.; Cramer, B.; Humpf, H.U. Survey of *Alternaria* Toxin Contamination in Food from the German Market, Using a Rapid HPLC-MS/MS Approach. *Mycotoxin Res.* **2016**, *32*, 7–18. [CrossRef] [PubMed]

24. Zhao, K.; Shao, B.; Yang, D.; Li, F. Natural Occurrence of Four *Alternaria* Mycotoxins in Tomato- and Citrus-Based Foods in China. *J. Agric. Food Chem.* **2015**, *63*, 343–348. [CrossRef] [PubMed]

25. Terminiello, L.; Patriarca, A.; Pose, G.; Fernandez Pinto, V. Occurrence of Alternariol, Alternariol Monomethyl Ether and Tenuazonic Acid in Argentinean Tomato Puree. *Mycotoxin Res.* **2006**, *22*, 236–240. [CrossRef] [PubMed]

26. Van de Perre, E.; Jacxsens, L.; Liu, C.; Devlieghere, F.; De Meulenaer, B. Climate Impact on *Alternaria* Moulds and their Mycotoxins in Fresh Produce: The Case of the Tomato Chain. *Food Res. Int.* **2015**, *68*, 41–46. [CrossRef]

27. Zhang, Y.; Zhang, X.; Jiao, B. Determination of Ten Pyrethroids in Various Fruit Juices: Comparison of Dispersive Liquid-Liquid Microextraction Sample Preparation and QuEChERS Method Combined with Dispersive Liquid-Liquid Microextraction. *Food Chem.* **2014**, *159*, 367–373. [CrossRef] [PubMed]

© 2016 by the authors. Licensee MDPI, Basel, Switzerland. This article is an open access article distributed under the terms and conditions of the Creative Commons Attribution (CC BY) license (http://creativecommons.org/licenses/by/4.0/).

toxins

MDPI

Article

Natural Occurrence of *Alternaria* Toxins in the 2015 Wheat from Anhui Province, China

Wenjing Xu [1,2], Xiaomin Han [1], Fengqin Li [1,*] and Lishi Zhang [2,*]

[1]   Key Laboratory of Food Safety Risk Assessment, Ministry of Health, China National Center for Food Safety Risk Assessment, Beijing 100021, China; wenjingxu_vip@163.com (W.X.); hanxiaomin@cfsa.net.cn (X.H.)
[2]   Institute of Nutrition and Food Hygiene, West China College of Public Health, Sichuan University, Chengdu 610041, China
*   Correspondence: lifengqin@cfsa.net.cn (F.L.); lishizhang_56@163.com (L.Z.);
    Tel.: +86-10-6777-6356 (F.L.); +86-138-0807-1034 (L.Z.); Fax: +86-10-6777-6356 (F.L.)

Academic Editor: Aldo Laganà
Received: 31 August 2016; Accepted: 19 October 2016; Published: 26 October 2016

Abstract: The exposure to *Alternaria* toxins from grain and grain-based products has been reported to be related to human esophageal cancer in China. In this study, a total of 370 freshly harvested wheat kernel samples collected from Anhui province of China in 2015 were analyzed for the four *Alternaria* toxins tenuazonic acid (TeA), tentoxin (TEN), alternariol (AOH) and alternariol monomethyl ether (AME) by high performance liquid chromatography-tandem mass spectrometry method (HPLC-MS/MS). TeA was the predominant toxin detected followed by TEN, AOH and AME. The concentrations of the four *Alternaria* toxins varied geographically. The samples from Fuyang district showed higher TEN concentration levels than the other regions studied ($p < 0.05$). Furthermore, 95% (352/370) of the wheat samples were positive for more than one type of *Alternaria* toxins. Positive correlation was observed between concentration levels of TeA and TEN, AOH and AME, TeA and AOH, and the total dibenzopyrone derivatives (AOH + AME) and TeA. Results indicate that there is a need to set the tolerance limit for *Alternaria* toxins in China, and more data on the contamination of these toxins in agro-products is required.

Keywords: *Alternaria* toxins; wheat; China; HPLC-MS/MS

---

1. Introduction

Alternaria species are pathogenic, endophytic and saprophytic fungi that have been reported to cause extensive spoilage of crops such as grain, tomato, potato, citrus, apple and sunflower seed in the field or after harvest [1–8]. The invasion of agricultural commodities by *Alternaria* species are universal and result not only in considerable loss of crops due to decay but also contamination by *Alternaria* toxins. *Alternaria* species can produce more than 70 toxins, of which only a small proportion have been chemically characterized. These *Alternaria* toxins can be classified into three main categories: the dibenzo-α-pyrones, which include alternariol (AOH), alternariol monomethyl ether (AME), altenuisol (AS) as well as altenuene (ALT); tetramic acid derivatives including tenuazonic acid (TeA); and the perylene derivatives altertoxins I, II and III (ATX I, II and III) [9,10]. *Alternaria* toxins show cytotoxic activity among mammalian cells, and fetotoxicity and teratogenicity among mice and hamsters [11]. Some individual mycotoxins such as AOH and AME, though not acutely toxic, are mutagenic and genotoxic in various in vitro systems [12]. TeA is considered to be of the highest toxicity among the *Alternaria* toxins, and has been proven to be toxic to several animal species, e.g., mice, chicken and dogs [8].

The key contribution of dietary exposure to *Alternaria* toxins is from grain and grain-based products, especially wheat products [12]. Wheat is one of the most important foods for the Chinese population. More than half of Chinese people live on wheat as a stable food, particularly in the north part of China [13]. For example, Anhui province is a major wheat-producing region and a major outbreak of human intoxication attributed to the consumption of wheat-based foods contaminated with *Fusarium* toxins (mainly deoxynivalenol) occurred in 1991, with over 130,000 people being affected by gastrointestinal disorder [14]. Our recent research showed that *Alternaria* species were the predominant invading fungi in wheat collected from Anhui province in 2015 [15]. Further, the intake of grains contaminated by *Alternaria* toxins has been related to human esophageal cancer in some areas of China [16]. Therefore, the *Alternaria* toxins in wheat represent a potential hazard to the Chinese population.

Due to limited data available on natural occurrence of these toxins in foods and their toxicity to humans and animals, regulations for *Alternaria* toxins in food and feed have not been set nationally or internationally. The purpose of this study is to elucidate the natural occurrence of four major *Alternaria* toxins, namely TeA, tentoxin (TEN), AOH and AME, using the wheat samples from Anhui province collected in 2015. The results obtained will provide a scientific basis for assessing the impact of *Alternaria* toxins on Chinese public health resulting from consumption of wheat products.

2. Results

2.1. High Performance Liquid Chromatography-Tandem Mass Spectrometry (HPLC-MS/MS) Method Validation

The multiple reaction monitoring (MRM) chromatograms of the four *Alternaria* toxins standards and a naturally contaminated wheat sample are shown in Figure 1. The mean retention times (RTs) of the four toxins were 4.85 min for TeA, 6.12 min for TEN, 6.00 min for AOH and 6.37 min for AME, respectively. The RTs of these four toxins in wheat samples were set within $\pm 2.5\%$ difference with those of standards materials, in compliance with the requirements of European Union Decision 2002/657/EEC [17]. Excellent linearity of the standard curve was obtained for all four *Alternaria* toxins (coefficient $r^2 > 0.99$ for all curves). The limits of detection (LOD) and limits of quantification (LOQ) for each toxin in wheat samples were 0.6 and 1.9 µg/kg for TeA, 0.1 and 0.4 µg/kg for TEN, 1.3 and 4.2 µg/kg for AOH, and 0.04 and 0.1 µg/kg for AME, respectively (see Supplementary Table S1). The recoveries at five spiked concentrations ranged from 87.0% to 103.3% for all these four toxins in wheat samples, and the relative standard deviation (RSD) for all these four toxins were from 1.6% to 10.7%, which were both within the acceptable range recommended by the European Commission (Table 1) [18]. Regarding repeatability, the RSDr were 4.9% for TeA, 9.8% for TEN, 10.6% for AOH and 6.3% for AME, respectively. As for reproducibility, the RSD_R were 4.2% for TeA, 5.6% for TEN, 8.1% for AOH and 9.6% for AME, respectively. Based on the above results, it is concluded that the HPLC-MS/MS method for *Alternaria* toxins detection meets the requirements of our study.

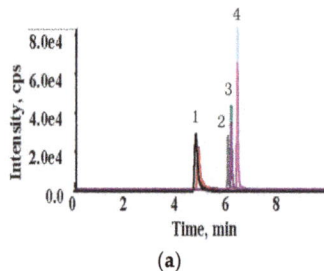

(a)

Figure 1. *Cont.*

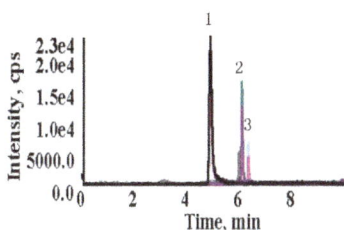

(b)

Figure 1. MRM chromatograms of (**a**) the four *Alternaria* toxins standards at concentrations of 250 μg/kg for TeA, 50 μg/kg for AOH, 50 μg/kg for TEN and 25 μg/kg for AME; (**b**) a naturally contaminated wheat sample with 112.9 μg/kg for TeA, 8.8 μg/kg for AOH, 3.3 μg/kg for TEN and negative for AME. 1 = TeA; 2 = AOH; 3 = TEN; 4 = AME.

Table 1. The recovery (%) of the four *Alternaria* toxins in spiked wheat samples (No. of repeat = 6).

Mycotoxin	Spiked Level (μg/kg)	Recovery (%, $\bar{x} \pm S$)	RSD, %
TeA	2	95.3 ± 8.1	8.5
	10	95.8 ± 1.7	1.8
	20	98.9 ± 4.3	4.3
	40	96.3 ± 2.6	2.7
	100	99.3 ± 1.6	1.6
TEN	5	101.2 ± 4.1	4.1
	25	97.1 ± 5.4	5.6
	50	95.2 ± 3.4	3.6
	100	95.0 ± 4.7	4.9
	250	95.0 ± 2.6	2.7
AOH	0.5	103.3 ± 8.4	8.1
	2.5	95.3 ± 6.2	6.5
	5	97.6 ± 7.1	7.3
	10	98.1 ± 3.4	3.5
	25	96.2 ± 4.9	5.1
AME	0.5	91.7 ± 5.0	5.5
	1	87.0 ± 9.3	10.7
	2	100.3 ± 9.6	9.6
	5	94.0 ± 4.0	4.3
	10	98.1 ± 2.4	2.4

2.2. Natural Occurrence of the Four Alternaria Toxins in Chinese Wheat

The contamination of the four *Alternaria* toxins in 370 freshly harvested wheat kernel samples are presented in Table 2. TeA is the predominant toxin in both frequency and concentration. It was found in all samples analyzed with an average level of 289.0 μg/kg (range: 6.0 μg/kg–3330.7 μg/kg, median = 150.0 μg/kg). Of all samples analyzed, 23 samples were contaminated with TeA at levels higher than 1000 μg/kg, 36 between 500 μg/kg and 1000 μg/kg, 184 between 100 μg/kg and 500 μg/kg, and 127 below 100 μg/kg. For TEN, 286 out of 370 (77%) samples were detectable, with levels ranging between 0.4 μg/kg and 258.6 μg/kg (mean = 43.8 μg/kg, median = 29.7 μg/kg). Among which, four samples were contaminated with TEN at a level higher than 200 μg/kg. One hundred and seventy three (47%) samples were positive for AOH, with a concentration range of 1.3 μg/kg–74.4 μg/kg (mean = 12.9 μg/kg, median = 7.9 μg/kg). There were three samples with AOH at a level higher than 50 μg/kg. AME was found in 55 (15%) samples at levels ranging from 0.3 μg/kg to 54.8 μg/kg (mean = 9.1 μg/kg, median = 4.2 μg/kg).

Table 2. Natural occurrence of the four *Alternaria* toxins in wheat kernel samples harvested in 2015 from Anhui province of China ($n = 370$).

Mycotoxin	Range (µg/kg)	Average (µg/kg)	Median (µg/kg)	Frequency, %
TeA	6.0–3330.7	289.0	150.0	100
TEN	0.4–258.6	43.8	29.7	77
AOH	1.3–74.4	12.9	7.9	47
AME	0.3–54.8	9.1	4.2	15

2.3. Geographical Distribution of the Four Alternaria Toxins in Anhui Province

Wheat samples analyzed were collected from eight regions that covered almost all wheat-producing areas in Anhui province (Figure 2). The concentrations of the four *Alternaria* toxins in samples from different regions are shown in Table 3. It was demonstrated that the contamination of wheat kernels by the four *Alternaria* toxins varied geographically. The higher concentrations of the four toxins in terms of either average or median were found in samples from Fuyang, an area where, following a large scale flood, a major outbreak of human red mold intoxication caused by moldy wheat occurred in 1991, with 130, 141 people affected [14]. Fuyang is also adjacent to Henan, a province with high incidence of human esophageal cancer and where the contamination of *Alternaria* toxins in wheat-based products is relatively high [19]. Particularly, all (93/93) wheat samples from Fuyang were positive for TeA with a mean level of 582.6 µg/kg, which was significantly higher than those from the other regions (Mann-Whitney U Test, $p < 0.05$) except Huainan. TEN was detected in 99% (92/93) of wheat samples from Fuyang with an average level of 77.2 µg/kg, significantly higher than those from the other seven regions ($p < 0.05$). AOH was detected in 67% (62/93) of wheat samples from Fuyang, with a mean level of 18.3 µg/kg, significantly higher than those from the other regions ($p < 0.05$) except Chuzhou. Besides, 38% (35/93) of wheat samples from Fuyang were positive for AME with an average level of 8.5 µg/kg, significantly higher than those from the other regions ($p < 0.05$) except Huainan and Lv'an. The geographical distribution of the mean concentration of the four toxins (as well as most of the median) from high to low were in the following order : Fuyang > Huainan > Others > Bozhou > Lv'an > Chuzhou > Bengbu > Suzhou for TeA; Fuyang > Bozhou > Others > Bengbu > Chuzhou > Lv'an > Suzhou > Huainan for TEN; Huainan > Fuyang > Lv'an > Chuzhou > Others > Bozhou > Bengbu > Suzhou for AOH; Lv'an > Suzhou > Bengbu > Huainan > Fuyang > Chuzhou > Bozhou > Others for AME, respectively. Notably, the germination, growth and toxin production of *Alternaria* species would be influenced by environmental conditions [10]. Although these eight regions were all situated in Anhui province, the differences in temperature and precipitation among these areas during wheat earring, flowering, maturing and harvest season may lead to the significant differences in *Alternaria* toxin production.

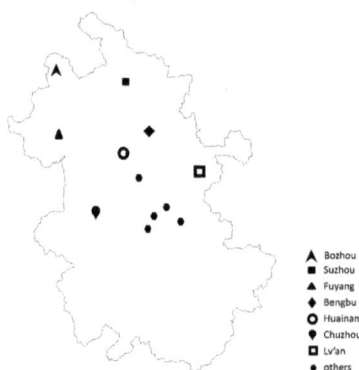

Figure 2. Geographical locations of eight sampling sites in Anhui province.

Table 3. Natural occurrence of the four *Alternaria* toxins in the 2015 wheat samples from different regions of Anhui province.

Region	Mycotoxin	Range (µg/kg)	Average (µg/kg)	Median (µg/kg)	Frequency, %
Chuzhou (n = 53)	TeA	6.0–3330.7	205.6	108.1	100 (53/53)
	TEN *	0.9–83.0	20.4	15.3	43 (23/53)
	AOH	1.3–57.1	12.1	7.8	87 (46/53)
	AME	2.1–17.6	6.2	3.9	9 (5/53)
Fuyang (n = 93)	TeA	53.7–2518.1	582.6	412.8	100 (93/93)
	TEN *	2.3–258.6	77.2	66.5	99 (92/93)
	AOH	1.3–74.4	18.3	16.0	67 (62/93)
	AME	0.3–46.1	8.5	4.2	38 (35/93)
Huainan (n = 13)	TeA	116.2–587.7	311.6	294.9	100 (13/13)
	TEN *	5.5–22.3	11.0	10.7	69 (9/13)
	AOH	4.3–47.3	26.8	26.6	46 (6/13)
	AME	1.4–16.4	8.8	10.6	38 (5/13)
Suzhou (n = 86)	TeA	13.2–832.3	125.0	86.5	100 (86/86)
	TEN *	0.4–76.7	14.2	8.7	58 (50/86)
	AOH	1.3–15.5	3.9	2.8	23 (20/86)
	AME	18.7	18.7	18.7	1 (1/86)
Bozhou (n = 35)	TeA	19.0–826.7	250.7	183.3	100 (35/35)
	TEN *	1.2–220.9	54.6	41.2	97 (34/35)
	AOH	1.6–30.3	6.5	3.2	46 (16/35)
	AME	0.9–9.8	4.2	3.1	11 (4/35)
Lv'an (n = 22)	TeA	21.1–1591.6	225.9	108.2	100 (22/22)
	TEN *	1.9–83.2	15.1	10.2	82 (18/22)
	AOH	1.3–33.9	12.5	9.6	50 (11/22)
	AME	0.9–54.8	20.7	13.4	18 (4/22)
Bengbu (n = 41)	TeA	21.2–616.2	131.1	93.4	100 (41/41)
	TEN *	0.5–101.5	21.6	14.1	80 (33/41)
	AOH	1.7–9.9	4.1	2.8	12 (5/41)
	AME	10.2	10.2	10.2	2 (1/41)
Others (n = 27)	TeA	47.3–1280.2	294.1	220.7	100 (27/27)
	TEN *	6.6–123.1	48.0	45.6	100 (27/27)
	AOH	2.3–12.8	7.3	7.5	26 (7/27)
	AME	0	0	0	0 (0/27)

* The concentration of TEN in samples from Fuyang was significantly higher than those from the other seven regions ($p < 0.05$).

2.4. Co-Occurrence of the Four Alternaria Toxins in Wheat Kernels

Co-occurrence of the four *Alternaria* toxins in wheat samples was frequent: 19% of samples (72/370) were contaminated by all four toxins, 30% (110/370) by three toxins, 46% (170/370) by two toxins and 5% (18/370) by only one toxin (TeA). Regarding the co-occurrence of the three toxins, 53 (48%, 53/110) samples were positive for TeA, TEN and AME, 20 (18%, 20/110) samples for TeA, TEN and AOH, 19 (17%, 19/110) samples for TeA, AOH and AME, and 18 (16%, 18/110) samples for TEN, AOH and AME, respectively. In terms of wheat co-contamination by the two *Alternaria* toxins, TeA and TEN was the most frequent combination detected in 60 (35%, 60/170) samples, followed by TeA and AME as well as TEN and AME, with the prevalence of 21% (35/170) for both combinations (Table 4). TeA and AOH occurred together in 14 (8%, 14/170) samples, whereas both AOH with TEN and AOH with AME in 13 (8%, 13/170) samples, respectively. Moreover, a significant linear correlation in concentrations was observed between TeA and TEN ($r = 0.675$, $p < 0.05$), AOH and AME ($r = 0.558$, $p < 0.05$), TeA and AOH ($r = 0.407$, $p < 0.05$), and also the total dibenzopyrone derivatives (AOH + AME) and TeA ($r = 0.431$, $p < 0.05$), respectively (Figure 3). These results are similar to those reported by Zhao et al. [19], of which a significant linear correlation in toxin concentration was found between AOH and AME ($r = 0.877$, $p < 0.01$), TeA and TEN ($r = 0.747$, $p < 0.01$), and the total dibenzopyrone derivatives (AOH + AME) and TeA ($r = 0.860$, $p < 0.01$).

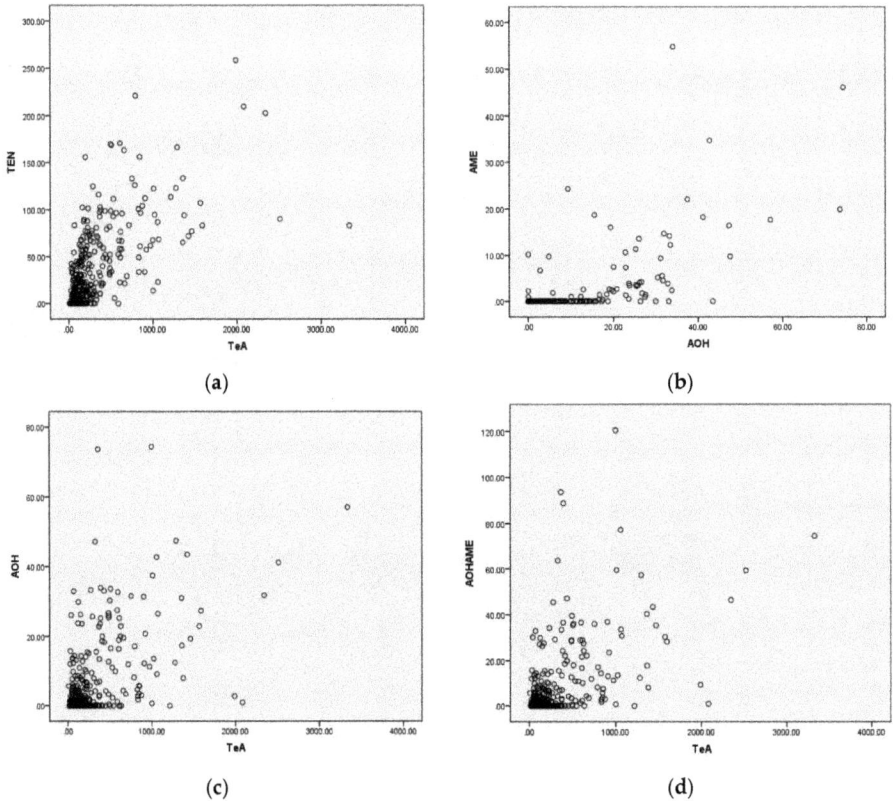

(a)

(b)

(c)

(d)

Figure 3. Correlation in concentrations of the four *Alternaria* toxins in wheat samples harvested in 2015 from Anhui province of China. (**a**) TeA vs. TEN, $r = 0.675$, $p < 0.05$, (**b**) AOH vs. AME, $r = 0.558$, $p < 0.05$, (**c**) TeA vs. AOH, $r = 0.407$, $p < 0.05$, (**d**) TeA vs. AOH + AME, $r = 0.431$, $p < 0.05$.

Table 4. Co-occurrence of the four *Alternaria* toxins in wheat samples harvested in 2015 from Anhui province of China.

Contamination by	Toxin Combinations	Frequency, %
Two mycotoxins (46%, 170/370)	TeA-TEN	35 (60/170)
	TeA-AME	21 (35/170)
	TEN-AME	21 (35/170)
	TeA-AOH	8 (14/170)
	AOH-TEN	8 (13/170)
	AOH-AME	8 (13/170)
Three mycotoxins (30%, 110/370)	TeA-TEN-AME	48 (53/110)
	TeA-TEN-AOH	18 (20/110)
	TeA-AOH-AME	17 (19/110)
	TEN-AOH-AME	16 (18/110)
Four mycotoxins (19%, 72/370)	TeA-TEN-AOH-AME	19 (72/370)

3. Discussion

Alternaria species are widely distributed in various habitats such as the surface of buildings, cellulose, paper, and textiles as well as in the soil as normal components of its microflora [20]. They are also known to frequently occur in agricultural commodities. As *Alternaria* mycotoxins can contaminate agro-products across the food chain, and little information on their toxicity and occurrence is available, it has become an emerging concern for public health.

The contamination trends of the four *Alternaria* toxins in Chinese wheat analyzed in our study is in line with those described by the European Food Safety Authority (EFSA), of which TeA, TEN, AOH and AME were generally found in grain and grain-based products and TeA was the most predominant. When compared with the results from a 10-year study of a total of 1064 wheat samples between 2001 and 2010 conducted by German scientists [21], the prevalence of TeA (100%), AOH (47%) and AME (15%) found in the present study was much higher than that in German wheat (30%, 8% and 3%, respectively). However, the maximum levels of TeA, AOH and AME obtained in our study (3330.7 μg/kg, 74.4 μg/kg and 54.8 μg/kg, respectively) were lower than those reported in Germany (4224 μg/kg, 832 μg/kg and 905 μg/kg, respectively). To our knowledge, the highest concentrations of TeA, AOH and AME were found in Argentinean wheat at 8814 μg/kg for TeA, 1388 μg/kg for AOH and 7451 μg/kg for AME [22]. Although the contamination rates of TeA, AOH and AME in Chinese wheat were higher than those of Argentinean wheat samples (19% for TeA, 6% for AOH and 23% for AME), the average and maximum concentrations were much lower in our study. Thus, the geographical location plays an important role in toxin production.

In contrast to the average levels of *Alternaria* toxins in 22 Chinese weather-spoiled wheat samples (2419 μg/kg for TeA, 335 μg/kg for AOH and 443 μg/kg for AME) reported by Li et al. [23], the average levels of TeA, AOH and AME in this study were much lower. Moreover, our findings demonstrated that levels of TeA and TEN in wheat kernels were higher than those in wheat flour samples being collected also from Anhui province in 2014 as reported by Zhao et al. [19]. The mean and maximum levels of TeA in this current study were 2.4-fold and 6.4-fold, respectively, higher than those in wheat flour samples; for TEN it was 1.2-fold and 2.7-fod higher, respectively. Contrarily, AOH mean and maximum concentrations in wheat flour were 2.9-fold and 1.3-fold, respectively, higher than those of wheat kernels. Whereas, the AME concentration was very similar in both wheat kernel and wheat flour samples. The industrial procedures may be a factor in the different levels of the four *Alternaria* toxins among wheat kernel and wheat flour samples.

It is known that *Alternaria* toxin production is associated with the *Alternaria* species that invade, and their competition for energy with other fungal types [12]. *Alternaria* species are widespread in both humid and semi-arid regions. Furthermore, different species of *Alternaria* might be of different toxigenic profile. For example, the species of *Alternaria alternata* isolated from Argentinean wheat kernels could produce TeA, AOH and AME, but the species of *Alternaria infectoria* only produce TeA [24]. Interactions between *Alternaria* and *Fusarium* species and consequently their impact on toxin production has been reported by Muller et al. [25]. It was found that the increased population of *Alternaria* on ripening ears of wheat coincided with reduced population of *Fusarium* species [25,26]. Therefore, the increase of *Alternaria* toxins in cereals was often accompanied by the decline of *Fusarium* toxins.

The type of crop and variety of wheat as substrates for fungal growth, the agricultural practice and the climate condition might be responsible for the different levels of *Alternaria* toxins contamination in different geographical areas. Study shows that the contamination level of *Alternaria* toxins in freshly harvested wheat kernels was likely to be dependent on preceding year crop type and tillage [21]. Additionally, environmental conditions may influence the germination, growth and toxin production of *Alternaria* species. According to Magan et al. [27], the optimum temperature for *Alternaria* species invasion and toxin production was 25 °C with a water activity (a_w) ranging between 0.88 and 0.99. The germination occurs within the temperature range of 5–35 °C, with the minimum a_w of 0.84–0.85 at 25 °C at pH 6.5. Therefore, a warm and humid climate during the period of wheat flowering, sprouting and heading stage (late April to June) in Anhui province is favorable to *Alternaria* fungal growth.

These factors contribute to the different type of co-occurrence and correlation among *Alternaria* toxins observed between our research and the German study [21].

The study demonstrates a frequent presence of *Alternaria* toxins and the co-occurrence of four *Alternaria* toxins on wheat grain samples from Anhui province in China. The finding raises concern given the high wheat consumption in many regions in China. There is a need for more occurrence data to estimate the exposure to *Alternaria* toxins from cereal and cereal-based products in Chinese populations. Further, an integrated risk assessment of these four toxins' contamination in Chinese wheat and the implications for public health needs to be carried out; strategies must be developed to reduce the risk from these contaminants, and specific regulations are expected to be established.

4. Materials and Methods

4.1. Chemicals and Reagents

Methanol and acetonitrile, both HPLC grade, were purchased from Fisher Scientific (Fair Lawn, NJ, USA). Ammonium bicarbonate and formic acid were of analytical grade (purity $\geq$ 95%) from Fluka (Steinheim, NRW, Germany). Pure water was obtained from a Millipore Milli-Q System (Millipore, Bedford, MA, USA). Standards of TeA (purity > 98%), TEN (purity > 98%), AOH (purity > 98%) and AME (purity > 98%) were supplied by Fermenteck Ltd. (Jerusalem, Israel). All experimental practice followed Environmental Health Safety Guidelines for the use of chemicals authorized by China National Center for Food Safety Risk Assessment. Samples extraction should be handled with care.

4.2. Samples Collection

A total of 370 wheat kernel samples were collected from local farmers who were randomly selected from main wheat production regions in Anhui province, China, namely Chuzhou, Fuyang, Huainan, Suzhou, Bozhou, Lv'an, Bengbu and a few "other" regions where a very small amount of samples were collected. All wheat samples were produced locally in 2015 and intended for human consumption. Fresh wheat kernels were sampled in the field during harvesting and threshing, one portion per household. Each portion was pooled from 4 subsamples (at least 600 grams each) that were collected from four areas of a field (north west, south west, north east and south east of the field). The sample were mixed thoroughly before a random quarter of the sample was selected, half of which was then finely ground to a powder of 20 meshes using a Blender 8010ES (Warning Commercial, Torrington, CT, USA). All ground samples were kept in the Ziploc plastic bags at 4 °C prior to analysis.

4.3. Toxin Analysis

All samples were analyzed for TeA, TEN, AOH and AME based on the methods reported previously by Zhao et al. [19] with modifications. Briefly, a finely ground sample (5 g) was mixed with 20 mL acetonitrile-water-methanol (45:45:10, $v/v/v$), adjusted pH between 3 and 4, and sonicated for 30 min before centrifugation. A 5 mL supernatant was loaded onto a HLB solid phase extraction (SPE) cartridge (Waters, Milford, MA, USA), eluted with 5 mL methanol followed by 5 mL acetonitrile. Both eluents were combined, dried and reconstituted in 1 mL methanol-water containing 2 mmol/L ammonium bicarbonate (10:90, v/v). Following centrifugation, 2 µL of the extract was analyzed for the four *Alternaria* toxins by HPLC-MS/MS method following the conditions described previously [28]. An *Alternaria* toxin-free wheat sample was used as a base for the matrix-matched calibration standards for quantification.

4.4. HPLC Conditions

HPLC-MS/MS system equipped with a Shimadzu 20A HPLC (Shimadzu, Kyoto, Japan), Triple Quard 3500 MS/MS system (AB Sciex, Foster City, CA, USA) was used to quantify the four *Alternaria* toxins simultaneously, and an Analyst Version 1.6.2 software (AB Sciex, Foster City, CA, USA) for data acquisition and analysis. Chromatographic separation was achieved using a C_{18} column

(2.1 mm × 100 mm, 1.7 μm bead diameter, Waters, Milford, MA, USA). The temperature of both column and autosampler were set at 24 °C. A binary gradient with mobile phase A (2 mmol/L ammonium bicarbonate) and B (methanol) was programmed, and the flow rate was 0.3 mL/min.

4.5. MS/MS Conditions

The MS was operated in the negative electrospray ionization (ESI⁻) mode, set as MRM with an optimized dwell time of 50 ms. The curtain gas was set to 35 psi, the collision gas to 7 psi, Gas1 to 60 psi and Gas2 to 50 psi. The source desolvation temperature was 600 °C. The ion spray voltage was −4500 V in negative mode. Parent and fragment ions (quantifier and qualifier) for each analyte were determined based on the optimal signal-to-noise ratios in a spiked sample. The parent ions (*m/z*) for TeA, TEN, AOH and AME are 196.0, 413.2, 256.9 and 270.9, respectively. The most intense product ion was employed for quantification, and the less intense signals were used for confirmation of toxin identity. The quantitative/qualitative daughter ions are 138.8/111.8 for TeA, 140.9/271.0 for TEN, 212.9/214.9 for AOH and 255.8/227.8 for AME, respectively (see Supplementary Table S2).

4.6. Method Validation

To evaluate the method proficiency, *Alternaria* toxin-free wheat samples were spiked with the four *Alternaria* toxins at five concentrations ranging from 2 to 100 μg/kg for TeA, from 5 to 250 μg/kg for TEN, from 0.5 to 25 μg/kg for AOH, and from 0.5 to 10 μg/kg for AME, respectively. The spiked samples (*N* = 6 repeat each level) were extracted and analyzed for recovery calculation. In terms of the matrix-matched calibration, standards of the four *Alternaria* toxins were added to *Alternaria* toxin-free wheat extracts to reach the final concentrations between 2.5 and 5000 μg/kg for TeA, 0.5 and 1000 μg/kg for both TEN and AOH, 0.25 and 500 μg/kg for AME, respectively. The LOD and LOQ were determined based on the signal-to-noise ratios of 3:1 and 10:1, respectively. An *Alternaria* toxins-positive wheat sample was extracted and analyzed 6 times within 1 day for interlaboratory RSDr calculation. With regard to reproducibility, a naturally contaminated wheat sample was analyzed for *Alternaria* toxins over 5 successive days according to the procedure described above. The data obtained were used for RSD_R calculation.

4.7. Data Analysis

Significant differences in concentrations of mycotoxins were tested using Kruskal-Wallis H test or Mann-Whitney U test. Relationships of concentrations between any of the four mycotoxins were tested with the Spearman correlation. All statistical analyses were computed using the SPSS statistical package (version 20.0, IBM, Armonk, NY, USA).

Supplementary Materials: The following are available online at www.mdpi.com/2072-6651/8/11/308/s1, Table S1: The parameters associated with the calibration curve and method sensitivity for the detection of the four *Alternaria* toxins, Table S2: The MRM parameters of MS/MS conditions.

Acknowledgments: This work was financially supported by the Special Foundation for State Basic Research Program, the Ministry of Science and Technology of the People's Republic of China (Grant Number: 2013FY113400). Grateful thanks also go to Professor Songxue Wang from Academy of State Administration of Grain for sample collection. We thank Michael N Routledge, from University of Leeds and Professor Séamus Fanning, from UCD-Center for Food Safety, School of Public Health, Physiotherapy & Population Science, University College Dublin (UCD), Ireland for their valuable suggestions.

Author Contributions: Fengqin Li and Lishi Zhang conceived and designed the experiments; Wenjing Xu performed the experiments; Wenjing Xu and Xiaomin Han analyzed the data; Wenjing Xu and Fengqin Li wrote the paper.

Conflicts of Interest: The authors declare no conflict of interest. The funding sponsors had no role in study design, sample collection, sample extraction and analyses, data interpretation, manuscript writing and decision making to publish the results.

References

1. Pavón Moreno, M.Á.; González Alonso, I.; Martín de Santos, Y.R.; García Lacarra, T. The importance of genus *Alternaria* in mycotoxins production and human diseases. *Nutr. Hosp.* **2012**, *27*, 1772–1781. [PubMed]
2. Kosiak, W.; Torp, M.; Skjerve, E.; Andersen, B. *Alternaria* and *Fusarium* in Norwegain grains of reduced quality—A matched pair sample study. *Int. J. Food Microbiol.* **2004**, *93*, 51–62. [CrossRef] [PubMed]
3. Hasan, H. *Alternaria* mycotoxins in black rot lesion of tomato fruit: Conditions and regulation of their production. *Mycopathologia* **1995**, *130*, 171–177. [CrossRef]
4. Van der Waals, J.E.; Korsten, L.; Slippers, B. Genetic diversity among *Alternaria solani* isolates from potatoes in South Africa. *Plant Dis.* **2004**, *88*, 959–964. [CrossRef]
5. Akimitsu, K.; Peerver, T.L.; Timmer, L.W. Molecular, ecologocal and evolutionary approaches to understanding *Alternaria* diseases of citrus. *Mol. Plant Pathol.* **2003**, *4*, 435–446. [CrossRef] [PubMed]
6. Rang, J.; Crous, P.W.; Mchau, G.R.A.; Serdani, M.; Song, S. Phylogenetic analysis of *Alternaria* spp. associated with apple core rot and citrus black rot in South Africa. *Mycol. Res.* **2002**, *106*, 1151–1162. [CrossRef]
7. Carson, M.L. Epidemiology and yield losses associated with *Alternaria* blight of sunflower. *Phytopathology* **1985**, *75*, 1151–1156. [CrossRef]
8. Ostry, V. *Alternaria* mycotoxins: An overview of chemical characterization, producer, toxincity, analysis and occurrence in foodstuffs. *World Mycotoxin J.* **2008**, *1*, 175–188. [CrossRef]
9. Bottalico, A.; Logrieco, A. *Alternaria* plant diseases in Mediterranean countries and associated mycotoxins. In *Alternaria: Biology, Plant Diseases and Metabolites*; Chełkowski, J., Visconti, A., Eds.; Elsevier Science: New York, NY, USA, 1992; pp. 209–232.
10. Barkai-Golan, R. *Alternaria* mycotoxins. In *Mycotoxins in Fruits and Vegetables*; Barkai-Golan, R., Nachman, P., Eds.; Academic Press: San Diego, CA, USA, 2008; pp. 185–203.
11. Visconti, A.; Sibilia, A. *Alternaria* toxins. In *Mycotoxins in Grain: Compounds Other Than Aflatoxin*; Miller, J.D., Trenholm, H.L., Eds.; Eagan Press: St. Paul, MN, USA, 1994; pp. 315–336.
12. European Food Safety Authority (EFSA). Scientific opinion on the risks for animal and public health related to the presence of *Alternaria* toxins in feed and food. *EFSA J.* **2011**, *9*, 1–97.
13. Wang, X.L.; Sun, J.M. Status and outlook of wheat consumption in China. *J. Triticeae Crops* **2015**, *35*, 655–661.
14. Li, F.Q.; Luo, X.Y.; Takumi, Y. Mycotoxins (trichothecenes, zearalenone and fumonisins) in cereals associated with human red-mold intoxications stored since 1989 and 1991 in China. *Nat. Toxins* **1999**, *7*, 93–97. [CrossRef]
15. Xu, W.J.; Han, X.M.; Zhang, J.; Pan, Z.; Li, F.Q.; Zhang, L.S. Survey on fungi contamination of wheat harvested in 2015 from Anhui province of China. *Health Sci.* **2016**, *54*, 92–96.
16. Liu, G.T.; Qian, Y.Z.; Zhang, P.; Dong, Z.M.; Shi, Z.Y.; Zhen, Y.Z.; Miao, J.; Xu, Y.M. Relationships between *Alternaria alternata* and oesophageal cancer. *IARC Sci. Publ.* **1991**, *105*, 258–262.
17. European Union Decision 2002/657/EC. Commission Decision of 12 August 2002 implementing Council Directive 96/23/EC concerning the performance of analytical methods and the interpretation of results. *Off. J. Eur. Common.* **2001**, *L221*, 8–36.
18. European Commission. SANTE/11945/2015. Guidance Document on Analytical Quality Control and Method Validation Procedures for Pesticides Residues Analysis in Food and Feed. Available online: http://ec.europa.eu/food/plant/docs/plant_pesticides_mrl_guidelines_wrkdoc_11945_en.pdf (accessed on 25 March 2016).
19. Zhao, K.; Shao, B.; Yang, D.J.; Li, F.Q.; Zhu, J.H. Natural occurrence of *Alternaria* toxins in wheat-based products and their dietary exposure in China. *PLoS ONE* **2015**, *10*, e0132019. [CrossRef] [PubMed]
20. Rotem, J. The genus *Alternaria*. In *Biology, Epidemiology and Pathogenicity*; APS Press: St. Paul, MN, USA, 1994; p. 326.
21. Muller, M.E.H.; Korn, U. *Alternaria* mycotoxins in wheat—A 10 years survey in the Northeast of Germany. *Food Control* **2013**, *34*, 191–197. [CrossRef]
22. Azcarate, M.P.; Patriarca, A.; Terminiello, L.; Fernandez Pinto, V. *Alternaria* toxins in wheat during the 2004 to 2005 Argentinean harvest. *J. Food Prot.* **2008**, *71*, 1262–1265. [PubMed]
23. Li, F.Q.; Yoshizawa, T. *Alternaria* mycotoxins in weathered wheat from China. *J. Agric. Food Chem.* **2000**, *48*, 2920–2924. [CrossRef] [PubMed]

24. Oviedo, M.S.; Sturm, M.E.; Reynoso, M.M.; Chulze, S.N.; Ramirez, M.L. Toxigenic and AFLP variability of *Alternaria alternata* and *Alternaria infectoria* occurring on wheat. *Braz. J. Microbiol.* **2013**, *44*, 447–455. [CrossRef] [PubMed]

25. Muller, M.E.; Urban, K.; Koppen, R.; Siegel, D.; Korn, U.; Koch, M. Mycotoxins as antagonistic or supporting agents in the interaction between phytopathogenic *Fusarium* and *Alternaria* fungi. *World Mycotoxin J.* **2015**, *8*, 311–321. [CrossRef]

26. Magan, N.; Lacey, J. The phylloplane microflora of ripening wheat and effect of late fungicide appications. *Ann. Appl. Biol.* **1986**, *109*, 117–128. [CrossRef]

27. Magan, N.; Lacey, J. Effect of water activity, temperature and substrate on interactions between field and storage fungi. *Trans. Br. Mycol. Soc.* **1984**, *82*, 83–93. [CrossRef]

28. Zhao, K.; Shao, B.; Yang, D.Y.; Li, F.Q. Natural occurrence of four *Alternaria* mycotoxins in tomato- and citrus-based foods in China. *J. Agric. Food Chem.* **2015**, *63*, 343–348. [CrossRef] [PubMed]

© 2016 by the authors. Licensee MDPI, Basel, Switzerland. This article is an open access article distributed under the terms and conditions of the Creative Commons Attribution (CC BY) license (http://creativecommons.org/licenses/by/4.0/).

MDPI AG

St. Alban-Anlage 66

4052 Basel, Switzerland

Tel. +41 61 683 77 34

Fax +41 61 302 89 18

http://www.mdpi.com

Toxins Editorial Office

E-mail: toxins@mdpi.com

http://www.mdpi.com/journal/toxins

www.ingramcontent.com/pod-product-compliance
Lightning Source LLC
Chambersburg PA
CBHW051853210326
41597CB00033B/5884